PROCESS DESIGN TOOLS FOR THE ENVIRONMENT

PROCESS DESIGN TOOLS FOR THE ENVIRONMENT

Subhas K. Sikdar, Ph.D.
Director, Sustainable Technology Division
National Risk Management Research Laboratory
US Environmental Protection Agency
Cincinnati, Ohio

Mahmoud M. El-Halwagi
Alumni Professor
Chemical Engineering Department
Auburn University

TAYLOR & FRANCIS
ALERE FLAMMAM
Founded 1798

Published in 2001 by
Taylor & Francis
29 West 35th Street
New York, NY 10001

Published in Great Britain by
Taylor & Francis
11 New Fetter Lane
London EC4P 4EE

Printed in the United States of America on acid-free paper.

Library of Congress Cataloging-in-Publication Data
Process design tools for the environment / edited by Subhas Sikdar & Mahmoud El-Halwagi.
p. cm.
ISBN 1-56032-824-X (alk. paper)
1. Environmental chemistry—Industrial applications.
2. Environmental management. I. Sikdar, Subhas K.
II. El-Halwagi, Mahmoud M., 1962–
TP155.2.E58 P76 2000
628—dc21 00-022007

To the memory of a little girl, Manjori Sikdar, who would have been 19 this year, and to her two younger siblings, Ronjan and Reena, who by their wonderfully funny and silly ways kept the parents amused and lightened much of the pain from the loss for which life prepares no one.
—S.K. Sikdar

To my parents
To my wife Amal
To my children Omar and Ali
For their unlimited inspiration, love, encouragement, and support.
—M. M. El-Halwagi

Contributors

Luke E. K. Achenie and Manish Sinha
Computer Aided Process and Product Design Laboratory, Department of Chemical Engineering, University of Connecticut, Storrs, CT 06269

A. Alva-Argáez, A. C. Kokossis, and R. Smith
Department of Process Integration, UMIST, Manchester M60 1QD, U.K.

Weiyang Cheong and Paul I. Barton
Department of Chemical Engineering and Energy Laboratory, Massachusetts Institute of Technology, Cambridge, MA 02139

Russell F. Dunn
SOLUTIA, Gonzalez, FL 32560

Yan Fu and Urmila Diwekar
Environmental Institute, Carnegie Mellon University, Pittsburgh, PA 15213

Nick Hallale
Department of Process Integration, UMIST, Manchester M60 1QD, U.K.

Peter M. Harper and Rafiqul Gani
Computer Aided Process Engineering Center (CAPEC), Department of Chemical Engineering, Technical University of Denmark, DK-2800 Lyngby, Denmark

Andreas A. Linninger and Aninda Chakraborty
Laboratory for Product and Process Design, Department of Chemical Engineering, University of Illinois, Chicago, Chicago, IL 60607

Subba R. Nishtala
Research Triangle Institute Center for Environmental Analysis, Research Triangle Park, NC 27709-2109

Derya B. Özyurt and Matthew J. Realff
School of Chemical Engineering, Georgia Institute of Technology, Atlanta, GA

Gautham Parthasarathy and Mahmoud M. El-Halwagi
Department of Chemical Engineering, Auburn University, Auburn, AL 36849

David W. Pennington
ORISE Research Fellow, Systems Analysis Branch, Sustainable Technology Division, National Risk Management Research Laboratory, U.S. Environmental Protection Agency, Cincinnati, OH 45268

David W. Pennington,* Gregory Norris,† Theresa Hoagland,‡ and Jane C. Bare‡
**ORISE Research Fellow, Systems Analysis Branch, Sustainable Technology Division, National Risk Management Research Laboratory, U.S. Environmental Protection Agency, Cincinnati, OH 45268*
†Sylvatica, Inc., North Berwick, ME 03906
‡Systems Analysis Branch, Sustainable Technology Division, National Risk Management Research Laboratory, U.S. Environmental Protection Agency, Cincinnati, OH 45268

Alan P. Rossiter
Process Improvement Consultant, 4421 Darsey Street, Bellaire, TX 77401

David R. Shonnard, Tony N. Rogers, Bruce A. Barna, Daniel A. Crowl, Eric J. Oman, Peter P. Radecki, Jon A. Herlevich, Jr., and Piyush B. Parikh
Department of Chemical Engineering and The Center for Clean Industrial and Treatment Technologies, Michigan Technological University, Houghton, MI 49931

George Stephanopoulos
Laboratory for Intelligent Systems in Process Engineering, Department of Chemical Engineering, Massachusetts Institute of Technology, Cambridge, MA 02139

Robert W. Sylvester
DuPont Company, Newark, DE 19898

C. G. Vassiliadis, S. K. Stefanis, and E. N. Pistikopoulos
Centre for Process Systems Engineering, Department of Chemical Engineering, Imperial College, London, U.K.

Douglas Young and Heriberto Cabezas
National Risk Management Research Laboratory, U.S. Environmental Protection Agency, Cincinnati, OH 45268

Renhong Zhao and Heriberto Cabezas
Systems Analysis Branch, Sustainable Technology Division, National Risk Management Research Laboratory, U.S. Environmental Protection Agency, Cincinnati, OH 45268

Contents

Foreword

Industrial civilization, which has given us higher living standards, comfort, and convenience than previous generations, has also resulted in widespread pollution of air, land, and waterways. The United States Environmental Protection Agency's (EPA's) toxic release inventory (TRI) data show that tremendous quantities of unhealthy organic, inorganic, and organometallic compounds are discharged to the environment everyday. Similar data are being published in other countries, raising serious public concerns everywhere. A major part of this pollution is attributed to industrial processes. Many industrial wastes (which end up contaminating air, water, land, and sediments) adversely affect human health and ecosystems.

In principle, one can devise end-of-pipe waste-treatment facilities such that all wastes are treated in a way that results in no harmful emissions or discharges from a defined plant boundary to the surrounding environment. Prohibitive cost penalties associated with such treatment options exclude the viability of this peripheral approach. Reliance on treatment technologies to satisfy environmental regulations has other problems. First, regulations change more frequently than processes can be modified. Second, unanticipated legal liabilities may appear in the future because of better understanding of environmental impacts of treated wastes that may be sequestered within the plant boundary. There is not a clear method of internalizing these externalities in the production cost structure. Third, competition in the marketplace becomes very difficult for those processes that are already burdened by the costs of wastes. Therefore, it is now widely recognized that the most fundamental and sustainable approach to minimizing industrial pollution is to address the root causes of pollution by redesigning processes, products, and operating practices to minimize negative impact on the environment.

Design changes can be made for new processes (grassroot) or for existing processes (retrofitting). In the past, much success was encountered when the primary motivator for redesigning processes and products was the pursuit of lower cost and higher quality. Incorporating environmental concerns (at the earliest stages of process design or for existing processes) is imperative in our times and will result in rapid increase in environmental gains. Most of the recent design endeavors targeting environmental friendliness were spotty and were not a result of systematic approaches. A critical need toward sustainable environmental friendliness of industrial processes is the development of systematic and generally applicable design procedures. We are still at an early and exciting stage of developing systematic process design tools for the environment. The challenges of developing environmentally preferable processes are great and would require concerted efforts in many disciplines. Some good work has already been done, as evidenced by pronouncements by several prominent companies that pollution prevention is preferable, even from a business viewpoint. Much further research is needed. This book attempts to capture significant results of recent research in the field.

Several approaches to cleaner process design are presented in this book. Some procedures are focused on relatively few unit operations and species in a flow sheet, whereas others tackle an entire process plant, and still others relate the process to the larger environmental surroundings. The common theme among these approaches is that they are both systematic and quantitative in nature. These procedures also attempt to reconcile metrics of cost, quality, safety, and environmental impact. This is the road to sustainable industrial processes.

We have been fortunate to have received a great deal of enthusiasm for the book from the chapter authors. We gratefully acknowledge their fullest cooperation. All the chapters in the book have received peer reviews and have been revised according to the suggestions of the referees. We are thankful to the reviewers who generously donated their time and showed great interest in the topic. Finally, this book would not have ben undertaken without the encouragement of the publisher's staff, who proposed the volume in the first place.

Subhas K. Sikdar
Director of National Risk Management Research Laboratory
United States Environmental Protection Agency
Cincinnati, Ohio 45268
USA

Mahmoud M. El-Halwagi
Alumni Professor of Chemical Engineering
Auburn University
Auburn, Alabama 36849
USA

Part One

INTRODUCTION

Process Design Tools for the Environment: Introduction

Subhas K. Sikdar
National Risk Management Research Laboratory, U.S. Environmental Protection Agency, Cincinnati, OH 45268

Mahmoud M. El-Halwagi
Chemical Engineering Department Auburn University, Auburn, AL 36849

MOTIVATION

Industrial operations have significant ecological impacts. Because of the characteristics of processed materials, it is inevitable that some wastes are discharged out of the processing facilities. Nonetheless, the type, magnitude, and toxicity of these wastes can be determined through process design and operation activities subject to techno-economic constraints. Once released to air, land, or waterways, industrial wastes enter into a biocomplex cycle that ultimately affects human health and the environment in some form. The very use of toxic and hazardous substances in manufacturing is the other concern, as they can also cause adverse impacts when the products containing these substances, through intended use, come in contact with humans and the ecosystems.

ENVIRONMENTAL PERFORMANCE OF THE PROCESS INDUSTRIES

Traditionally, industrial processes have been designed and operated to optimize profitability and product quality. Wastes were treated in an auxiliary facility to comply with environmental effluent guidelines or hazardous wastes regulations. Not so infrequently, wastes in the past were dumped in the environment without treatment, and attention was paid only when these improper discharges were discovered and health and safety issues became public concerns. With mounting public pressure, increasing stringency of environmental regulations, and a more environmentally responsible stewardship by the process industries, there has been a consistent trend to design and operate cleaner processes. Cleaner processes are those that incorporate concerns for human health and the environment in the process design stage, the outcome of which is the use of nontoxic materials or modification of processes to significantly reduce the generation of wastes or eliminate them altogether. This concept of proactively preventing pollution was the basis of the Pollution Prevention Act that the U.S. Congress enacted in 1990 (Habicht, 1992). Pollution prevention, according to the Act, means "source reduction" and other practices that reduce or eliminate the creation of pollutants through (i) increased efficiency in the use of raw materials, energy, water, or other sources, or (ii) protection of natural resources by conservation.

Perhaps the greatest incentive to pollution prevention came from the Toxics Release Inventory (TRI) provision of the Emergency Planning and Community Right-to-Know Act (also known as SARA title III). By this provision, certain industrial sectors are required to report annual emissions of toxic compounds from manufacturing facilities. Process industries that thrive on chemical changes tend to produce large quantities of wastes. Many of these wastes are useless byproducts and result from unwanted side reactions. Frequently, solvents (including water), cleaning agents, volatile reactants, and products are emitted from manufacturing plants and become hazardous wastes. In many industries producing highly valued products, such as semiconductors and pharmaceuticals, a majority of the chemicals used in reactions end up in waste streams. Achieving better atom economy in such cases will constitute examples of pollution prevention and cost savings. Table I.1 is a list of the major industrial polluters among those industries that are required to report TRI data. The list is adapted from TRI data published by the EPA in 1998 (Toxic Release Inventory, 1997).

The EPA publishes the release numbers to the public every year. Besides allowing a public discussion of the impact of industrial emissions on human health and ecosystems, TRI made industries aware, in a quantitative

TABLE I.1
Major industrial polluters

Industry	*1998 Projected emission, 10^6 lbs*	*Change (1996–1998), %*
Chemicals	10,711	6.8
Primary metals	4,157	−2.0
Petroleum	2,149	0.0
Paper	1,607	−0.5
Food	728	83.1

sense, what they were doing to the environment. This awareness triggered an aggressive industrial pursuit of pollution prevention. For instance, the Chemical Manufacturers Association (CMA) created the Responsible Care program of an industry-wide code of environmental ethics, emulating a Canadian initiative of the same name. CMA has recently decided to specify performance goals for companies under Responsible Care (see, for instance, *Chem. & Eng. News*, May 24, 1999, p. 18). Large chemical companies, such as DuPont, Monsanto, Solutia, General Electric, Dow, and 3M, created their own programs of internal audit for pollution avoidance. Because TRI reporting began in 1988, EPA reports a 43% reduction of TRI chemicals by 1997 (for example, refer to *Chem. & Eng. News*, May 17, 1999). The largest fraction of the TRI emissions is air pollutants, and they showed significant decline. The discharges to water and land, however, increased over the duration. It must be remembered, however, that the reduction (and increase) occurred in the face of significant increase in production capacities.

CONCEPTS AND TOOLS

Pollution prevention as a concept for creating environmentally preferable technologies evolved as a necessity for integrated multimedia solutions to waste emission and generation. A large number of problem-focused, single-medium regulations created complicated compliance problems for industry. Often several permitting requirements are necessary for an industrial operation, and the time and resources spent on meeting compliance goals are frequently daunting. Small- and medium-sized companies usually are ill equipped to tackle the environmental issues that confront them. Many large companies realized the dilemma. As a result they have been able to capitalize on the business opportunities that cleaner processes present. These companies have spent research and development resources to develop processes that incorporate pollution prevention concepts. A survey of large companies conducted by the Industrial Research Institute (IRI) indicated that, on average, 13% of their R&D resources go to solve environmental problems (Resetar, 1999). Although a significant fraction of this money is spent on pollution control, some definitely contribute to developing newer technologies that help these companies to go beyond compliance. Intel Corporation also determined that it saves significant time and money in developing technologies that exceed emission permit thresholds. Likewise, Xerox adopted the policy of total recycle of material to meet corporate environmental goals (Resetar, 1999). The EPA also realized that single media-focused regulations limit the industry's ability to innovate. In the recent past, the EPA has created many voluntary programs to encourage industry to develop cleaner technologies. Examples of this type of voluntary programs are Project XL, Green Chemistry, etc. In the Project XL program, for instance, the EPA allows certain regulatory relief to companies if they achieve overall reduction of emission of toxic compounds.

Recently, there have been remarkable advances in developing suites of tools for assessing and preventing pollution from industrial facilities. These tools can be roughly classified into three categories:

1. **Technical Tools.** Those methods that allow us to analyze processes to determine their environmental impacts and/or direct us in finding appropriate solutions in reducing these impacts belong to this category. This category is the primary focus of this book. In addition to this category, there are two categories of nontechnical tools that are also useful in preventing pollution. These are discussed below.
2. **Economic Tools.** Market-based economic instruments can sometimes be an effective way of inducing cleaner industrial outcomes. In the United States, sulfur oxide emissions from utilities have been successfully reduced by emission permit trading system. Carbon taxes have been discussed in this context for reducing the potential impact of global warming. Companies can buy and sell these permits to stay within permitted thresholds, with the cost of these permits, which are openly traded in the marketplace, being an incentive for finding innovative solutions. Though the ultimate solution is technical, market forces bring it about.
3. **Regulatory Tools.** So far, authorities have relied on the enforcement of regulation and legal actions for reducing pollution from fixed manufacturing and mobile sources. Mandating the elimination of lead in gasoline is a good example of pollution prevention in this context.

Although the economic and regulatory tools are not the focus of this book, they will be implicitly included. The ultimate denominator for successful cleaner technologies is cost. Hence, one important implied constraint of the analytical tools is that they lead to cost-effective solutions that reduce pollution to a level that meets or exceeds environmental regulations.

DESIGN TOOLS: CHALLENGES AND OVERVIEW

Design tools are needed to analyze the flowsheets of process industries for assessing wastes generated and systematically finding ways of redesigning existing processes or designing newer processes that are environmentally more preferable. There are numerous challenges in developing these design tools. To demonstrate the complexity and combinatorial nature of a typical design-for-the-environment activity, let us consider the following motivating example on the production of acrylonitrile (AN, C_3H_3N), which is manufactured via the vapor-phase ammoxidation of propylene:

$$C_3H_6 + NH_3 + 1.5O_2 \xrightarrow{catalyst} C_3H_3N + 3H_2O.$$

The reaction takes place in a fluidized-bed reactor in which propylene, ammonia, and oxygen are catalytically reacted at 450°C and 2 atm. The reaction is a single pass with almost complete conversion of propylene. The reaction products are cooled using an indirect-contact heat exchanger that

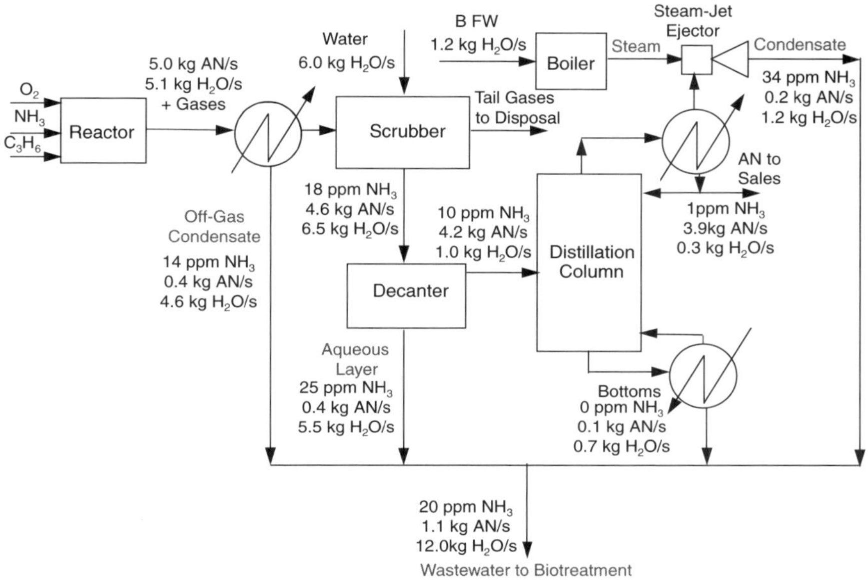

Figure I.1: A schematic flowsheet of the acrylonitrile process (El-Halwagi, 1997).

condenses a fraction of the reactor off-gas. The remaining off-gas is scrubbed with water and then decanted into an aqueous layer and an organic layer. The organic layer is fractionated in a distillation column under slight vacuum, which is induced by a steam-jet ejector. Figure I.1 shows a simplified process flowsheet along with pertinent material balance data.

In this case, the plant is seeking a significant reduction of the aqueous discharge in the most cost-effective manner.

Until recently, the traditional approach to solving this problem relied primarily on one of the following two methods:

1. **Brainstorming among experienced individuals.** A select few of the engineers and scientists most familiar with the process work together to suggest and synthesize several conceptual design scenarios (typically between three and five). Technical tools and experience are used to examine the feasibility of the suggested solutions. Finally, economics and other metrics are used to rank these options and choose a recommended solution. This recommended solution may be inaccurately referred to as the "optimum" solution. Actually, it is only optimum out of the few suggested alternatives. In reality it is not, and may not be close at all to, the true optimum solution to the original problem.

2. **Adopting/adapting earlier designs.** In this case, a related problem that has been solved earlier is identified. This problem may be at the same plant or at other plants. Then, its solution is either copied or adopted to suit the problem at hand.

Notwithstanding the usefulness of these traditional approaches in providing solutions that typically work, they have the following limitations when used in isolation of the more systematic techniques (which will be discussed later in this chapter):

- While working solutions are acceptable from a technical perspective, they may be significantly more expensive than optimum solutions. Except for limited cases of simple problems (or extreme luck!), brainstorming and adaptation of earlier solutions are not guaranteed to provide solutions that are in the proximity of optimal designs.
- Inability to enumerate the numerous alternative solutions (as will be shown later, in many cases there is an infinite number).
- Additional search to include more alternatives simply imply more time and money.
- In adopting solutions from other scenarios or other plants, it is often forgotten that there is not a single solution that fits all plants. Instead, there can be a single solution methodology that fits various plants. Differences (sometimes even small differences) in the characteristics of the problem can lead to significantly different solutions. By simply starting from another scenario or another plant without shedding light on the global insights and key characteristics of the process, any approach will be severely limited in terms of generating groundbreaking and novel ideas. These limitations can be eliminated when the brainstorming and adaptation approaches are integrated with systematic design techniques.

To further illustrate these issues, let us revisit the motivating AN case study and brainstorm for some alternative solutions. Because there is a need to reduce the wastewater discharge, it may be noted that some form of recycle/reuse is warranted. Furthermore, an overall water balance around the plant would suggest that in order to reduce effluent discharge, a reduction in the influent fresh water and/or a decrease in the generation of water by chemical reaction can be useful. Based on these observations, let us construct some solution scenarios. In all these scenarios, we will focus on the combinatorial nature of the various configurations and will not attempt to readjust the flows and compositions as a result of changing the process structure. In Figure I.2a, we show a poor recycle of wastewater. Because it does not contribute to reducing fresh usage or the water generation by reaction, at steady state it will find its way out of the plant. Let us assume that the production rate of AN and the reaction stoichiometry will remain unchanged. In this case, let us now focus on recycle alternatives that reduce fresh water consumption in the scrubber (Fig. I.2b) or the boiler (Fig. I.2c).

Now, what about eliminating fresh water from both the scrubber and the boiler (Fig. I.2d). What if the quality of recycled water is unacceptable for the boiler? Perhaps we can mix with fresh water to meet the quality constraints of the boiler (Fig. I.2e) or we can use a separation process to clean up the recycled stream and prevent the buildup of impurities (Fig. I.2f). But why

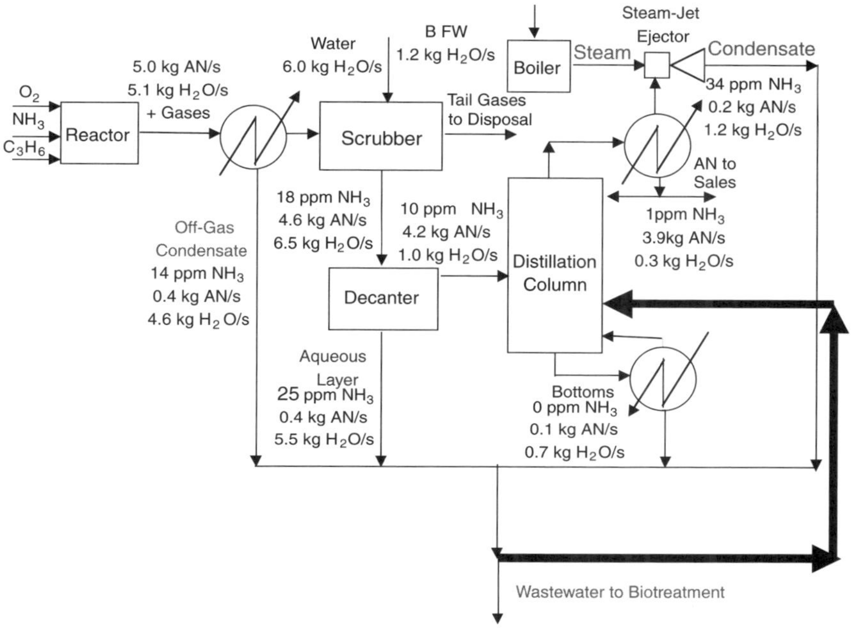

Figure I.2a: A poor recycle.

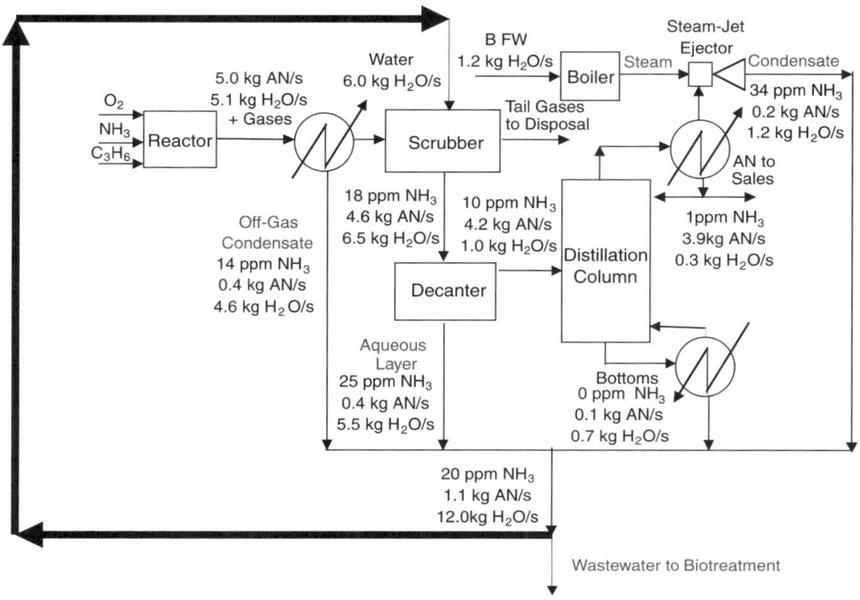

Figure I.2b: Recycle to eliminate fresh water usage in the scrubber.

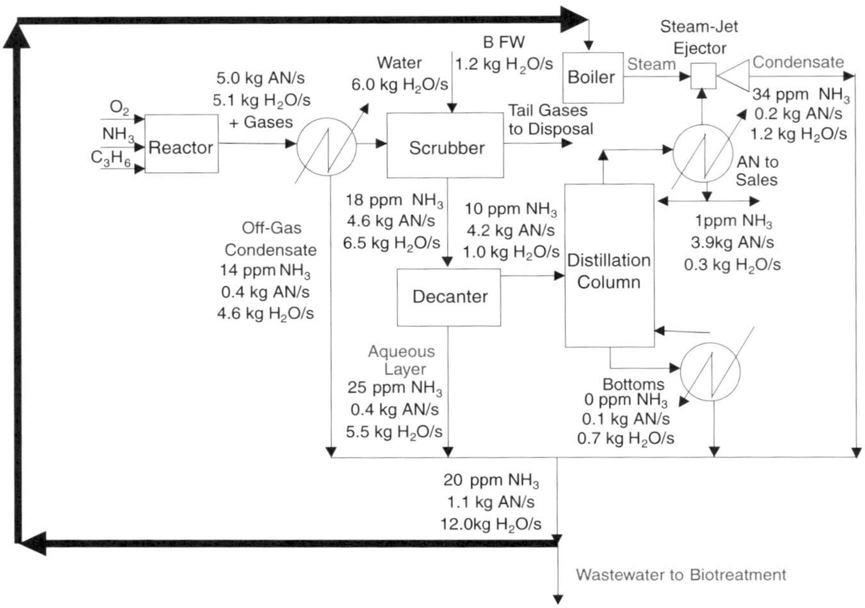

Figure I.2c: Recycle to eliminate fresh water usage in the boiler.

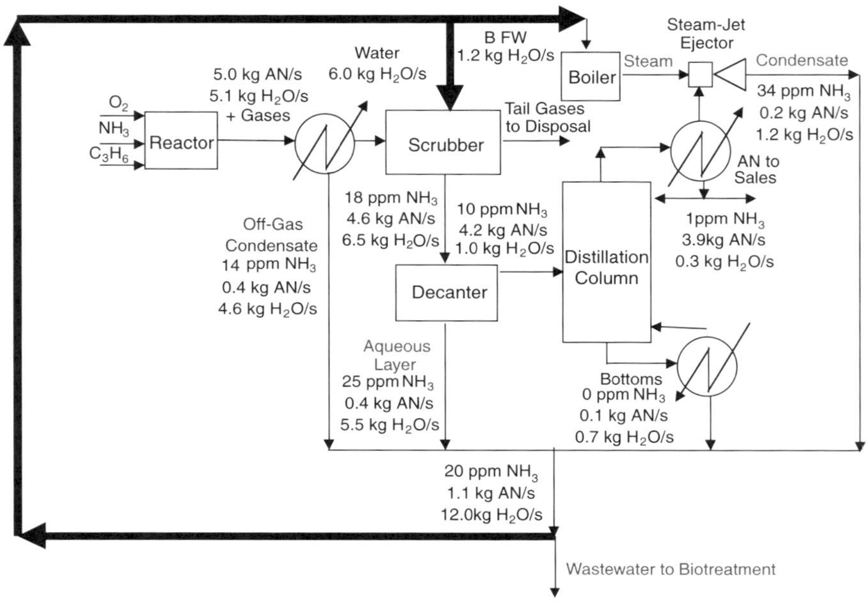

Figure I.2d: Recycle to eliminate fresh water usage in the scrubber and the boiler.

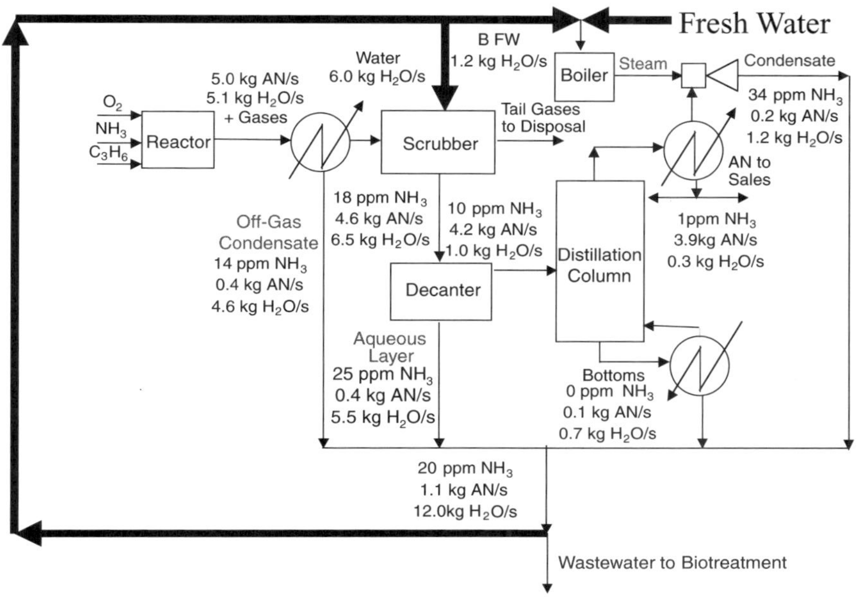

Figure I.2e: Recycle to eliminate fresh water usage in the scrubber and reduce it in the boiler.

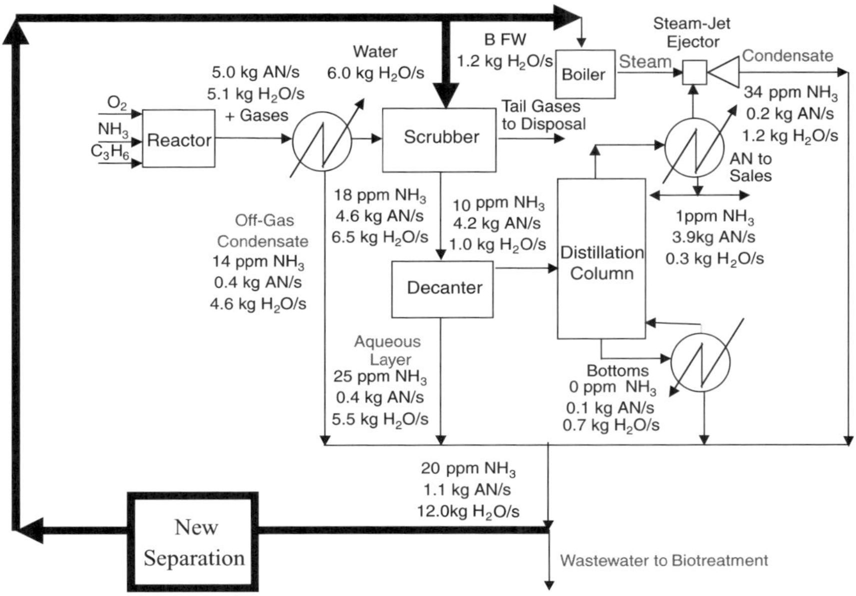

Figure I.2f: Recycle with separation of recycled stream.

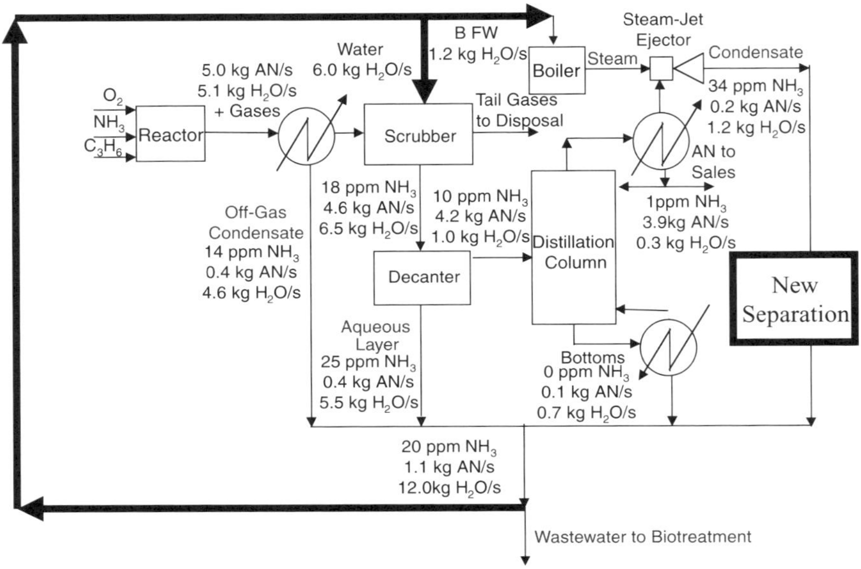

Figure I.2g: Recycle with separation of condensate.

does the separation process have to operate on the recycled stream? Should it be used to separate pollutants from another stream such as the condensate (Fig. I.2g)? Or, perhaps we can use more than one separation unit applied to several streams, such as the recycle, the off-gas condnesate, and the jet-ejector condensate (Fig. I.2h), but what are those separation technologies? Other plants have separately used extraction, stripping, and ion exchange. Perhaps, we can place them as shown in Figures I.2i, I.2j, I.2k, and I.2l. What about other separation technologies? These are just a few configurations. One can suggest far more!

Another level of complexity is the use of candidate materials. For instance, which mass separating agents (MSAs) (e.g., solvents, adsorbents, and stripping agents) should be used in the selected separation columns? Typically there are numerous MSAs that can be employed to remove certain impurities. How do we screen them?

In addition to the numerous configurations and species, there also are numerous continuous degrees of freedom. For instance, in Figure I.2l, how should the separation load of the pollutants in the jet-ejector condensate be distributed between ion exchange (e.g., 99%–1%, 98%–2%, . . . , 1%–99%)? Also, how should the separation load be distributed among the various separated streams, such as the recycle stream, the condensate, etc. Because the distribution of loads is a continuous variable, we can have infinite possibilities. In light of the intractable number of alternatives, it is impractical to attempt to enumerate them.

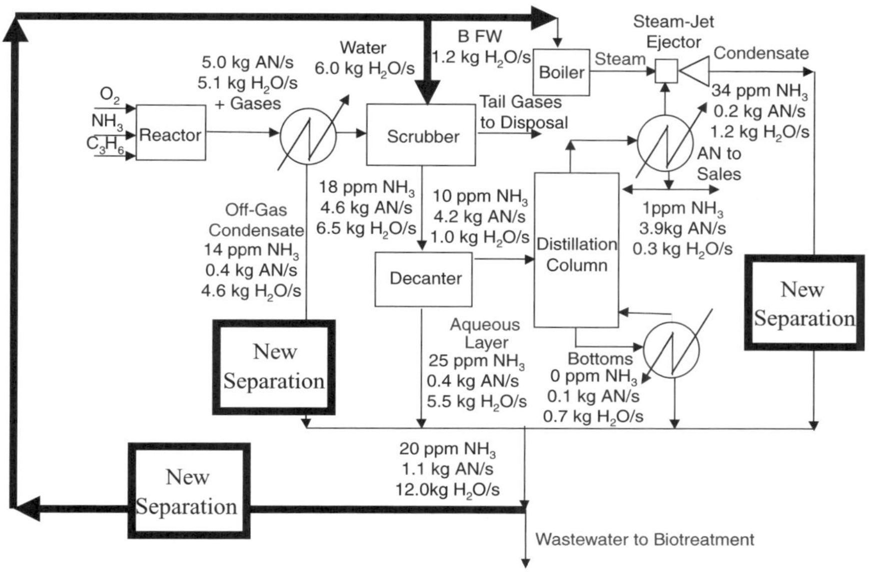

Figure I.2h: Recycle with multiple separations.

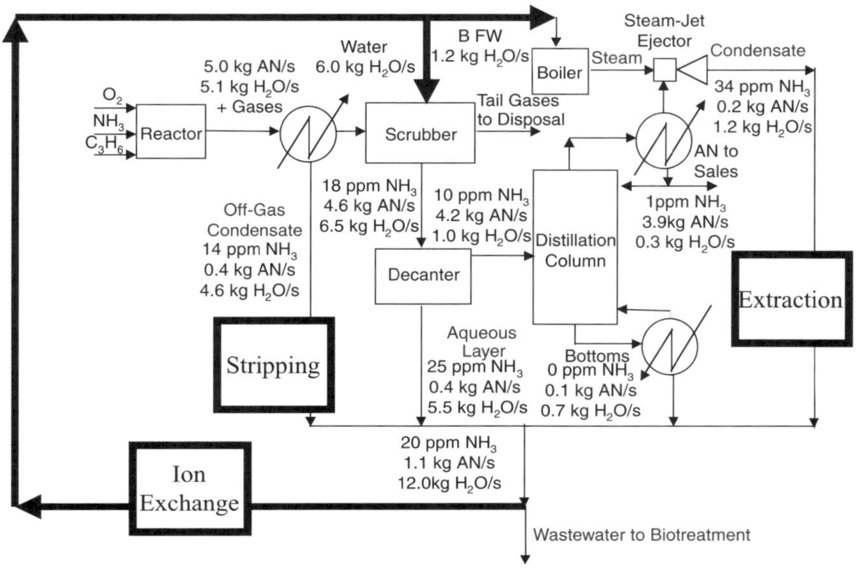

Figure I.2i: Recycle with selected separation technologies.

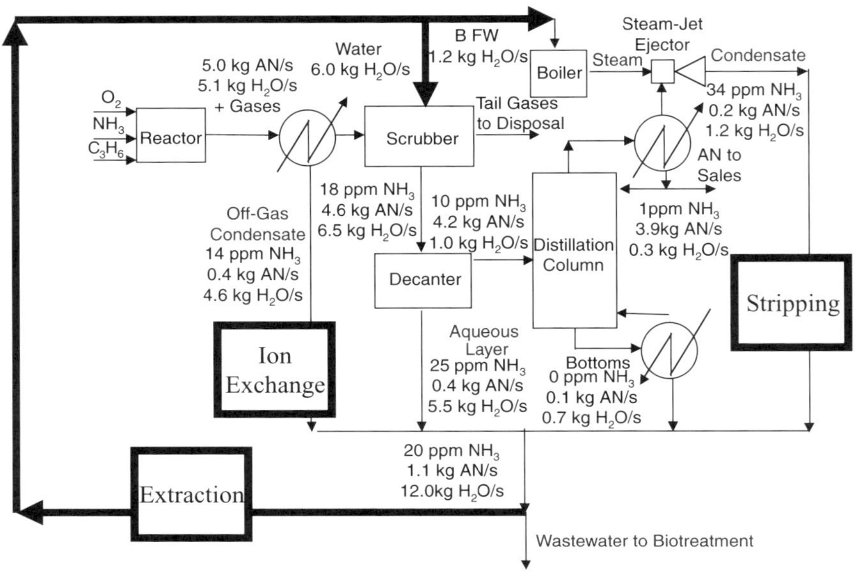

Figure I.2j: Recycle with an alternative arrangement of separation technologies.

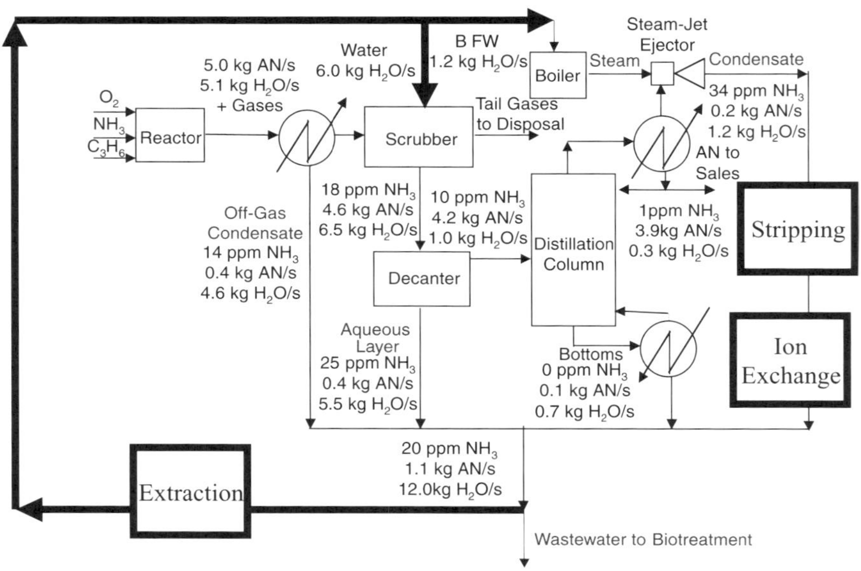

Figure I.2k: Recycle with an alternative arrangement of separation technologies (including hybrid units for the condensate).

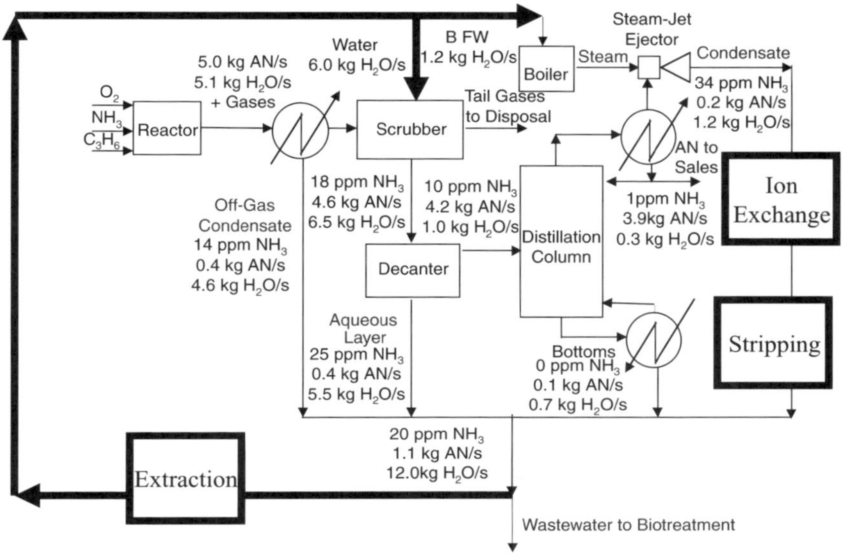

Figure I.21: Recycle with an alternative arrangement of separation technologies (including reversing the order of the hybrid units for the condensate).

It is interesting to consider the optimal solution to this case study, which is shown in Figure I.3. It is not an intuitively obvious solution that can be readily guessed. The challenges associated with the huge number of alternatives for process configurations, new units, material utilities, and species underscore the critical need for systematic design tools for the environment. Therefore, it is necessary to be able to develop systematic design tools that can overcome the large dimensionality of the problem and render it in a tractable domain.

The foregoing discussion illustrates that an effective design-for-the-environment tool should encompass the following aspects:

1. **Task Identifications.** The design tool should identify process tasks to be undertaken, including answering the following questions: Which streams should be rerouted? To which units? What transformations are needed (e.g., separation, biotreatment, etc.)? To remove what? How much should be removed? From which streams? Should we change operating conditions of some units? Which units and which operating conditions? To what extent should they be changed? These questions are important in effectively guiding the design activities.

2. **Unit Selection.** What type of technologies should be selected? What type of units and internals? What type of material and energy utilities? For instance, in the AN case study, should we use extraction, stripping, ion

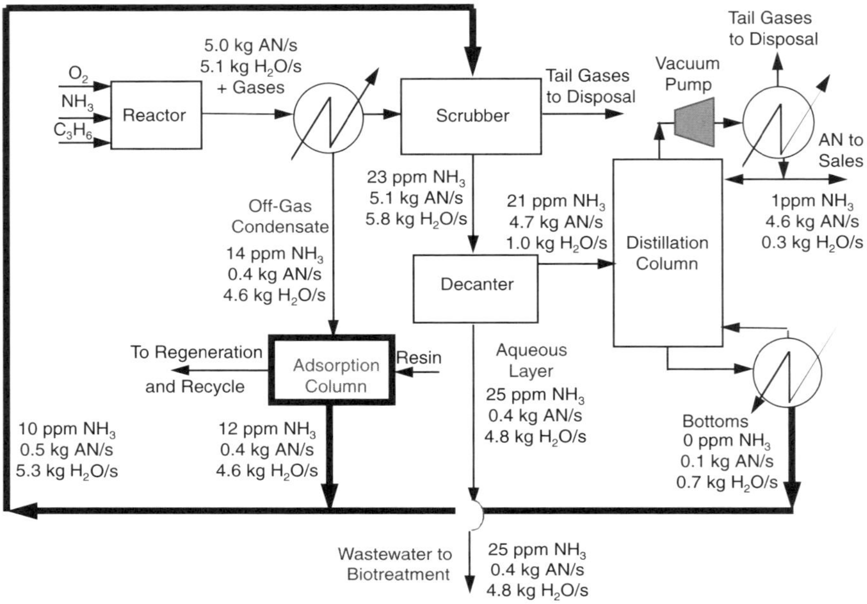

Figure I.3: Optimal solution to the AN case study (El-Halwagi, 1997).

exchange, or a combination? What type of columns? Which MSAs should be utilized?

3. **Generation and Interconnection of Alternatives.** The identified tasks and candidate units must be synthesized to generate an interconnected process flowsheet that addresses the problem. The numerous alternatives must be incorporated in this framework in an embedded or explicit manner.

4. **Screening.** The synthesized alternatives must be screened based on certain objectives (e.g., economics, environmental impact, etc.).

These aspects can be systematically included in design tools as described in the following section.

CRITICAL ELEMENTS IN DESIGN TOOLS

Over the past decade, significant contributions have been made in developing systematic design tools in general (Seider et al., 1999; Turton et al., 1998; Biegler et al., 1997; Smith, 1995; Douglas, 1988), and for pollution prevention in particular (El-Halwagi, 1997). These design tools provide systematic, fundamental, and generally applicable techniques that can be learned and

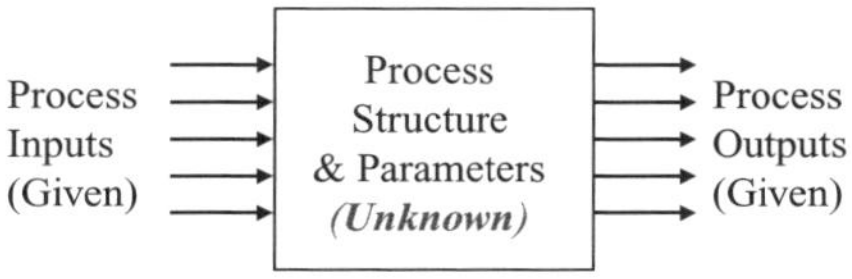

Figure I.4: Process synthesis problems.

applied to identify optimal designs for improving process objectives, including the environmental performance. In this regard, four elements are useful in a comprehensive design tool: process synthesis, simulation, integration, and optimization. The following is a brief discussion of these elements.

PROCESS SYNTHESIS

Process synthesis is concerned with the activities in which the various process elements are combined and the flowsheet of the system is generated so as to meet certain objectives. Hence, in process synthesis we know process inputs and outputs, and we are required to revise the structure and parameters of the flowsheet (for retrofitting design of an existing plant) or create a new flowsheet (for grass-root design of a new plant). This is shown schematically in Figure I.4. Detailed discussions on process-synthesis techniques are found in the literature (e.g., Seider et al., 1999; Turton et al., 1998; Biegler et al., 1997; El-Halwagi, 1997; Smith, 1995; Douglas, 1988).

PROCESS ANALYSIS

Process analysis is aimed at predicting how the synthesized process will perform. It involves the decomposition of the process into its constituent elements (e.g., units) for individual study of performance. Hence, once a process is synthesized, its detailed characteristics (e.g., flowrates, compositions, temperature, pressure, etc.) are predicted using analysis techniques. Therefore, in process analysis, we are given the process inputs and flowsheet and are required to predict process outputs (Fig. I.5). Analysis techniques include computer-aided process simulation tools (e.g., ASPEN Plus, HYSIM, PRO/II). In addition, process analysis may involve tailored mathematical models, empirical correlations, and even the use of experimental means to predict and validate performance.

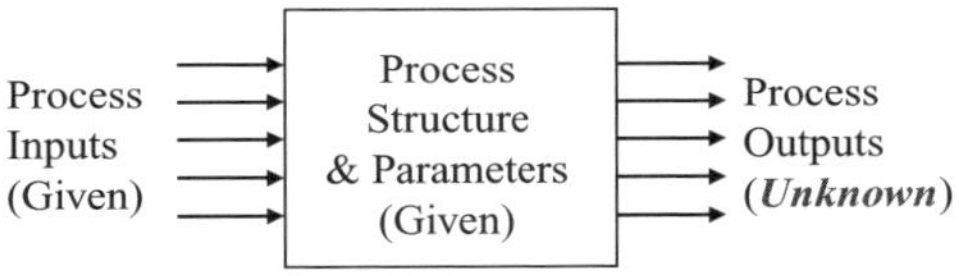

Figure I.5: Process analysis problems.

PROCESS INTEGRATION

Process integration is a holistic approach to process design, retrofitting, and operation that emphasizes the unity of the process. It can be broadly classified into mass integration and energy integration. *Mass integration* is a systematic methodology that provides a fundamental understanding of the global flow of mass within the process and employs this understanding in identifying performance targets and optimizing the generation and routing of species throughout the process. Because of the mass-related issues associated with pollution, mass integration is an instrumental tool in addressing environmental problems. *Energy integration* is a systematic methodology that provides a fundamental understanding of energy utilization within the process and employs this understanding in identifying energy targets and optimizing heat-recovery and energy-utility systems along with their gaseous emissions. Additional information on process integration techniques and tools are found in the literature (e.g., El-Halwagi and Spriggs, 1998; El-Halwagi, 1997; Smith, 1995; Linnhoff, 1993).

OPTIMIZATION

Optimization techniques have emerged recently as critical elements in design-for-the-environment tools. This trend has been accelerated as a result of the tremendous growth in computing and the recent advances in developing large-scale optimization algorithms. Optimization involves the task of identifying the value of an n-dimensional vector, x, that minimizes (or maximizes) a certain quantity called the objective function, $f(x)$, subject to a number of equality and inequality constraints. For instance, an optimization program may aim at minimizing the discharged wastes out of a plant or maximizing the economic potential of the process. Examples of equality constraints include material and energy balances, process modeling equations, and thermodynamic requirements. On the other hand, the nature of inequality constraints may be environmental (e.g., concentration of certain pollutants should be below specific levels), technical (e.g., pressure, temperature, or flowrate should not exceed some given values), and thermodynamic (e.g., driving force for mass, heat, or momentum transfer should be positive).

An optimization problem in which the objective function as well as all the constraints are linear is called a linear program (LP); otherwise it is referred to as a nonlinear program (NLP). The nature of optimization variables also contributes to the classification of optimization programs. An optimization formulation that contains continuous (real) variables (e.g., pressure, temperature, flowrate) as well as integer variables (e.g., 0, 1, 2, . . . , etc.) is called a mixed-integer program (MIP). Depending on the linearity characteristics of the problem, MIPs can be further classified into mixed-integer linear programs (MILPs) and mixed-integer nonlinear programs (MINLPs). A particularly useful class of integer variables is the zero/one. These binary-integer variables can be used to model logical events and decisions. For instance, a 0/1 variable may be designated to assume the value of one when an event occurs (e.g., a certain pollution prevention technology or unit is used), and zero when that event does not occur. The principles of optimization theory and algorithms are covered by various books. A recent coverage is provided by Floudas (1999).

STRUCTURE OF THE BOOK

The act of preventing pollution by industries of all sizes is facilitated by the availability of generic, widely applicable organizing principles and tools. The kinds of systematic and generic tools that are useful in developing environmentally preferable processes are many. Most of these are in various stages of development. For convenience, we have grouped the contributed subjects of this book into five categories:

I. **Environmental-Based Assessment and Review of Design.** This category includes top-level and management frameworks for assessing the environmental performance of a process as well as incorporating environmental issues into design. These frameworks can be based on heuristics, knowledge base, and mathematical approaches.

II. **Process Integration for the Environment**. The emphasis in this category is on holistic approaches which emphasize the unity of the process. Mass and energy integration methodologies are presented to develop optimal waste reduction strategies.

III. **Process Synthesis and Screening of Technologies for Pollution Prevention.** In this class, several methodologies and tools are presented to synthesize environmentally acceptable flowsheets. The scope covers retrofitting (for existing plants) as well as grassroot (for new plants) scenarios. Furthermore, this category provides systematic techniques for screening a multitude of waste-reduction technologies based on their technical, environmental, and economic performance.

IV. **Macroscopic Design and Environmental Impact Assessment.** In this case, the boundaries of the targeted system extend beyond the plant. The applied tools track the effect of industrial wastes on the environment. Examples include

   a. Life cycle assessment, which is a cradle-to-grave accounting of environmental impact, energy use, and cost.

   b. Environmental impact assessment, which accounts for the impact of emissions and discharges on human health, ecosystems, global warming, ozone depletion, or other effects.

   c. Interplant waste exchange, where wastes from certain plants are passed to other plants as feedstocks and additives.

V. **Molecular Design and Selection.** Here the focus is on the species more than on the process units. Activities in this area involve the design of nontoxic intermediates or products having acceptably similar properties to materials to be replaced. Examples include

a. Design of environmentally acceptable solvents, products, refrigerants, and nontoxic media for conducting reactions and separations.

b. Green synthesis and catalysis, where environmentally benign reactions are synthesized and harmful byproducts are minimized or eliminated.

The details of these tools, along with their applications, are covered in the ensuing chapters. The contributions provide state-of-the-art approaches and applications. The chapters are written by leading experts in the academic and industrial worlds. It is hoped that these chapters provide the readers with a wide spectrum of established and developing design tools for the environment as well as critical needs and on-going endeavors in the area.

REFERENCES

Biegler, L. T., Grossmann, I. E., and Westerberg, A. W. (1997). *Systematic Methods of Chemical Process Design*. Prentice-Hall, Englewood Cliffs, NJ.

Douglas, J. M. (1988). *Conceptual Design of Processes*. McGraw-Hill, New York.

El-Halwagi, M. M. (1997). *Pollution Prevention through Process Integration: Systematic Design Tools*. Academic Press, San Diego.

El-Halwagi, M. M., and Spriggs, H. D. (1988). Solve design puzzles with mass integration. *Chem. Eng. Prog.* 94:22–44.

Floudas, C. A. (1999). Deterministic global optimization: Theory and methods. *Nonconvex Optimization and Its Applications*, Vol. 37. Kluwer Academic Publishers, Dordrecht, The Netherlands.

Habicht, H.F., II (1992). U.S. Environmental Protection Agency, U.S. EPA Memorandum: EPA Definition of Pollution.

Linnhoff, B. (1993). Pinch analysis: A state of the art review. *Trans. IChemE* 71(A): 503.

Resetar, S. (1999). *Technology Forces at Work*. Rand Corporation.

Seider, W. D., Seader, J. D., and Lewin, D. R. (1999). *Process Design Principles*. Wiley, New York.

Smith, R. (1995). *Chemical Process Design*. McGraw-Hill, New York.

Toxic Release Inventory (1997). Public Data Release, U.S. EPA, Washington, D.C.

Toxic Release Inventory (1996). Public Data Release, EPA 745-R-96-002, U.S. EPA, Washington, D.C.

Turton, R., Bailie, R. C., Whiting, W. B., and Shaeiwitz, J. A. (1998). *Analysis, Synthesis, and Design of Chemical Processes*. Prentice-Hall, Englewood Cliffs, NJ.

Part Two

ENVIRONMENTAL BASED ASSESSMENT AND REVIEW OF DESIGN

CHAPTER 1

Environmental Reviews in the Design Process

Robert W. Sylvester

DuPont Company, Newark, DE 19898

Abstract

An environmental design review is a structured review of a chemical process with emphasis on waste generation and management. Like the pollution prevention review, it includes substantial consideration of elimination of waste generation by process change. It differs by including identification and evaluation of treatment and disposal options. The design environmental review should be expected to identify opportunities for pollution prevention, improve process economics, and provide assurance of regulatory compliance.

Keywords: pollution prevention, regulatory compliance, environmental design

1 WHY DO AN ENVIRONMENTAL REVIEW?

The addition of an environmental review to a project may seem like an unnecessary administrative burden to a project team charged with ambitious objectives, a tight budget, and scheduling difficulties. Why would a research group which is seeing success in the development of a new product accept the distraction of a review of wastes? Why would a process group divert attention from expansion and utility improvement plans to review emissions? Potential reasons are presented in Table 1.1, Purposes of an environmental review. The reasons are arranged in order from primary economic through the attainment of voluntary environmental goals, to achieving compliance with regulatory requirements (Chemical Manufacturers Association, 1993).

The necessity of obtaining environmental permits and complying with regulations is obvious, but will an environmental review really identify opportunities to save investment and operating costs? The DuPont Chambers Works facility conducted a published environmental review of 15 processes (U.S. EPA, 1992). The DuPont assessment was that improvements identified in this review could save $14,900,000 per year. It is informative to look at

TABLE 1.1
Purposes of an environmental review

Identify opportunities to reduce process wastes and improve profitability	
Identify ways to reduce the costs of treating waste	
Identify opportunities to recycle materials which would otherwise become wastes	
Identify opportunities to recover wastes as salable products	↑ Increasing economic drivers
Meet voluntary business commitments to waste minimization	Increasing legal/ public acceptance drivers
Meet Responsible Care® and other trade association commitments	↓
Identify opportunities to simplify permitting and regulatory compliance	
Ensure communication within the project team of permitting requirements	
Meet local and regional requirements for pollution prevention reviews	
Ensure compliance with emissions permits	

the sources of these savings:

- \$1,645,000 from variable cost waste disposal savings;
- \$2,185,000 from improved product recovery;
- \$11,070,000 from process improvements identified during the design review, including increased availability, increased capacity, and improved product quality.

These savings suggest that in at least some cases, the major justification for environmental review is economic. The Chambers Works example illustrates the potential for reexamination of the inefficiencies of a mature process to identify not only waste generation, but to understand why the waste is generated.

A more common form of environmental review generally takes place during the process design stage of a project. The project team typically brings in environmental specialists who define allowable emissions limits and permitting requirements for releases to air and water and requirements for disposal of solid and hazardous wastes. These specialists may also guide the selection of emissions treatment. Such a review is a necessary step for the permitting processes and should be done very early in design to ensure that permitting timetables can be met. Often, the process design engineers do not understand detailed permitting requirements, and the environmental specialists do not understand design options that will reduce emissions and simplify permitting. Interaction via a thorough design environmental review may identify the opportunity to eliminate routine venting from storage tanks or the potential to recover solvents to eliminate a priority wastewater pollutant or hazardous wastes, design changes which may result in substantial savings in treatment system costs and also reduce materials losses.

Project reviews often result in the conclusion that more benefits could have been achieved if a review had been held during process development. The project team may not have the authority to substitute solvents to reduce environmental impact or to simplify permitting, but the process researcher and developers do. Researchers are often not aware of waste treatment costs which may profoundly affect the value of their work. Simple reviews throughout the development process allow feedback on environmental considerations and ensure that permitting schedules do not become the limiting step in commercialization.

2 WHAT IS AN ENVIRONMENTAL DESIGN REVIEW?

An environmental design review is a structured review of a chemical process with emphasis on waste generation and management. Although various formats may be possible, this chapter is built on the assumption that a process flowsheet is available which provides identification of emission and discharge points, an inventory of emissions, and sufficient process information to allow understanding of the source or cause of wastes. Additional understanding of wastes from maintenance and fugitive emissions may be required. The structure begins with identification of treatment requirements which are sufficient to allow discharge or disposal of the wastes and environmental permitting of the facility.

Environmental specialists contribute an understanding of permitting requirements, knowledge of the treatment facilities to meet these requirements, and understanding of how the economics of treatment respond to changes in the emissions inventory. Process specialists provide understanding of the emissions inventory and causes of waste. The review leader ensures understanding of the purpose of the review, documentation, and appropriate follow-up to the review. A meeting facilitator is often utilized to support the structure of the review. When the identification of process changes to accomplish pollution prevention is an objective of the review, creative process specialists from outside the design team are encouraged to challenge assumptions inherent in the flowsheet and to suggest alternatives which might accomplish pollution prevention.

The review should generate a better understanding of the nature of wastes from the process, mutual awareness of waste treatment processes necessary for permitting, and identification of options which would reduce waste and improve process economics. The most promising options will be identified at the end of the review and thoroughly evaluated for feasibility, environmental benefit, and economic benefit before incorporation into the process design.

The structure of the review, proceeding from emissions limits to waste treatment to waste generation to process changes which reduce waste, may seem backwards at first. Why not start with consideration of wastes at the point of generation? There are several reasons to start at the point of release (El-Halwagi, 1997; Douglas, 1988):

- Released emissions and those sent away for disposal represent the environmental impact of the process on the world.
- The process will not be permitted to operate unless the released emissions meet legal limits.
- The project or development team needs an understanding of permitting schedules.
- Awareness of the investment and operating costs associated with waste treatment increases the willingness of the project team to consider design changes which reduce costs.
- Changes made to waste treatment will not change the product. Changes such as solvent recovery and product isolation may result in product change, but this risk can often be managed by close attention to specifications of recovered materials. Changes made in the product formation process, particularly changes to chemical reactor design, are seen as carrying greater risk of unexpected quality, yield, or operability problems, and thus require more time to evaluate and will not be implemented late in the design process.

The design environmental review is very similar in structure to a pollution prevention review. Like the pollution prevention review, it includes substantial consideration of elimination of waste generation by process change. It differs by including identification and evaluation of treatment and disposal options. A pollution prevention review is not sufficient to ensure adequate waste treatment to meet permitting requirements. The design environmental review includes consideration of permitting requirements and treatment options and should provide assurance of regulatory compliance.

3 BASIC STEPS FOR A SUCCESSFUL ENVIRONMENTAL REVIEW

Common steps in a design environmental review or a pollution prevention review are as follows:

3.1 CHARTERING

Higher organization levels confirm a commitment to waste minimization and to compliance with environmental regulations. Expectations concerning commitments beyond regulatory requirements, schedule limitations, openness to process change, and economics should be clearly communicated.

It is a rare environmental design review which will propose or attempt to act on opportunities outside of this charter. Indeed, the chartering phase of an environmental review might well take less time than any other step and yet have the most profound impact on the ultimate success of the effort.

A hint for better environmental reviews: Invite a business leader with overall financial responsibility to give the charter. Ask for the charter to include clear boundaries defining the freedom to make process changes and a challenge to reduce the expected investment and cost of treatment facilities by 50% with no more wastes while meeting permitting requirements. This charter empowers the design team to consider nonroutine options to treatment.

Engineers and scientists are trained to think in terms of decades, and this thinking spills over into environmental regulations. In a scale of decades, there is no zero. Thus, to a great extent the technology community is not able to envision zero emissions and thinks of emissions reduction by some percentage or to some deminimus concentration. This thinking is also displayed in U.S. environmental requirements. For example, the MACT standards of the Clean Air Act may require 98% control of potential emissions of hazardous air pollutants; however, pollution prevention can often identify opportunities to achieve zero emissions. The chartering phase can encourage environmental reviews which target emissions reduction, or those which target emissions elimination.

Is there a perception in parts of industry that reduced environmental impact only comes at a cost? Additional treatment of wastes typically requires more investment and operating costs. Indeed, treatment beyond a certain level may result in greater global waste generation associated with remote energy generation and treatment chemical consumption (Mulholland and Dyer, 1999), but what about pollution prevention options? Why not propose a charter which asks for two options: traditional pollution prevention followed by treatment to meet regulations and economical proposals for near zero wastes, with minimal treatment? Such a charter may well inspire more creative approaches to waste minimization.

Changes in a process are necessary to accomplish pollution prevention but may also change the nature of the product. There is a risk associated with all changes. This risk must be articulated and fully understood before process changes to accomplish waste minimization are incorporated in a process. These concerns must be addressed in the chartering phase to allow the broadest consideration of pollution prevention opportunities.

3.2 PROCESS DESCRIPTION

Pollution prevention cannot be part of the environmental review unless participants have a good understanding of how the products are made. The ideal process description will include a broad process overview of interest to all participants. This broad review should provide an understanding of how wastes are generated. Do the wastes originate as unwanted components in the process feeds, like sulfur in coal? Are they unavoidable byproducts of reaction, like water in condensation polymerization? Are they undesired byproducts, like tars from a coupling reaction? Understanding of the mechanism of generation is necessary for pollution prevention.

A second, more detailed tier of process description should be provided to the process specialists. This more detailed description should include details such as optimization studies of reactions and general operating procedures and control strategies for major process steps. Many sophisticated pollution prevention options are not implemented because of inadequate identification and communication of research results and operating practices.

3.3 WASTE STREAM SELECTION

This step is inherent in the creation of an emissions inventory. Priority will be given to the streams which have the largest amount of wastes, which have the greatest environmental impact, and which will cost the most to treat, but in the review of a new plant design, every permitting requirements of every emission point must be considered. The environmental specialist must guide the review to an understanding of key emissions and the environmental impact and treatment characteristics which make them key. For example, atmospheric emissions may be regulated by an allowable annual mass emission of hazardous air pollutants, but the cost of emissions treatment is often dominated by the volume of contaminated air and is often nearly independent of the pollutant mass. The environmental review of such an emission should consider ways to reduce pollutant mass, to reduce air volume, and to substitute materials with less environmental impact.

A hint for better environmental reviews: Distribute information about the process, the major waste streams, and the planned control technology at least a week before the assessment. Provide an approximate estimate of the investment and operating costs associated with treatment, and the sensitivity of these costs to the dominant variables. Give everyone time to become familiar with the charter and process.

3.4 ASSESSMENT

Options for minimizing wastes are generated, prioritized, evaluated, and selected for implementation. Idea generation is done in a structured brainstorming format. The structure is attained in part by analyzing the entire process in a sequence. Boundaries should be defined before brainstorming begins, not after. An example boundary is that "no new utilities will be created at the site, so that natural gas is not available." Also, criteria for evaluating ideas should be explained before the brainstorming begins so that the participants have an idea of what constitutes a "good" idea. Examples are that "ideas must reduce emissions, not create new safety risks, and reduce the net present costs of waste treatment and the overall project. Treatment technologies much be commercially available and must not delay startup."

A good way to begin is to review the process flowsheet, from raw materials receiving to product shipping and waste treatment systems. The waste treatment systems should be based on technology and designs which are certain to achieve environmental performance requirements. A very useful approach is to define the sensitivity of net present cost (or other economic parameters, if net present cost is not the appropriate economic metric) to key waste characteristics. This frequently will encourage participants to consider a wider range of options. Having ensured that everyone understands the origin of the wastes, brainstorming is begun, starting at the wastewater treatment systems (or product shipping if no wastewater treatment is present), reversing the sequence presented in the flowsheet review.

A hint for better environmental reviews: Encourage questions to improve understanding of the process. Split the meeting, with a review beginning at noon of one day and brainstorming beginning the next morning.

Another hint for better environmental reviews: Negative comments in a brainstorming session, even an appeal to remember the basis of evaluation, will inhibit participation. If the leader sees this as a problem, start the meeting by giving everyone five $1 bills. Anyone who makes a negative comment is "fined" one dollar per comment. Anyone who has not made a suggestion in two hours is fined a dollar. Use the collected fines to pay for lunch or the break. Each person keeps what they have left at the end of the section.

The suggestions presented in the brainstorming session will be limited by the experience of the reviewers. Seek diversity among reviewers by including mechanics, operators, researchers, wildcards, etc. Avoid filling the meeting with people known for reluctance to present ideas. A meeting size of 8 to 16 seems to allow diversity while ensuring enough "airtime" for everyone to participate. Questions during the brainstorming session should be asked to ensure common understanding of ideas, not to suggest that an idea will not work. Ideas should be presented one at a time and recorded. Participants should seek to build on previous ideas in addition to generating new ideas.

A typical brainstorming session reviewing a chemical process with three steps, storage, and treatment might generate 100 to 200 ideas over 6 to 10 hours. These ideas should range from the simple, such as "provide vapor return from storage tanks to the truck spots to reduce vent losses," to the complex, such as "conduct a mass integration assessment to reduce the water content of hazardous wastes."

Occasionally, design engineers become concerned about the chaos associated with brainstorming. Mutually opposed ideas will sometimes be recorded right next to each other. How can we possibly act on these ideas without delaying the design schedule? Screening evaluations are the answer to this concern.

Screening evaluations can take several forms. All should begin with a reminder of the evaluation criteria. Perhaps the simplest evaluation is to give each participant 10 sticky-colored dots. Each person then puts a dot by ideas which best meet the criteria. Ideas are then listed in order of most to least dots. Ideas with no dots need no further discussion, as no one in the evaluation process sees these ideas as among the 10 best.

The supported ideas can then be individually evaluated against the criteria.

A second, still relatively simple approach is to create a ranking system. For example:

Four = a good idea which meets the evaluation criteria and which can be acted on without extensive study. Do this now.

Three = possibly a good idea but formal study is needed to confirm feasibility or to confirm economics; an idea which might not be feasible but would have a major impact if it is. Study this.

Two = possibly a good idea, but probably not feasible within time constraints, the effort to develop the idea is not justified by the benefits. Postpone consideration.

One = the potential benefits do not justify the efforts to evaluate or prove feasibility. Apparently impractical.

Ideas are briefly discussed to ensure that everyone understands the idea, its expected benefits, and its feasibility. Each participant votes on each idea, with the average score rounded to two digits. A normal conclusion is to adopt all ideas with a score of 3.7 or higher. Most industrial firms have a formal method of evaluating the economics, safety, and environmental impact of new ideas. This formal evaluation should be applied to ideas ranked above 2.5. The 1 to 4 ranking system can also be used as a second evaluation following the colored dot technique.

Formal evaluation can be the most difficult and time consuming step of the assessment but is an essential step to the consideration and implementation of new ideas with major environmental benefit. These formal evaluations should be done by an experienced evaluator who is not the main proponent of the idea. Major inputs to the evaluation are

- Expected environmental benefit.
- Operating costs and investment to implement the proposal.
- Cost of data acquisition to permit design.
- Avoided investment and costs of alternative treatment facilities.
- Additional benefits of capacity, reliability, or resource recovery.
- Intangible benefits, including goodwill and avoided liability.
- Future savings if the design can be reused at other facilities.
- Savings from avoided permitting costs.
- The value of capital.

The DuPont Chambers Works Waste Minimization Project (U.S. EPA, 1993) illustrates how to do formal evaluations. It suggests several common mistakes:

- Failure to consider non-waste minimization improvement options.
- Overestimating sales, savings, or waste minimization.
- Underestimating additional operating costs.
- Underestimating required capital investment.
- Taking credit for cost savings that merely shift costs to other areas of the plant.
- A tendency of manufacturing teams to favor familiar solutions which provide incremental benefit over innovative strategies which have greater potential benefit and uncertainty.

Formal evaluations are often an iterative process, first based on relatively optimistic assumptions which are refined with additional data until uncertainties are eliminated or the idea is clearly not justified.

The EPA Pollution Prevention Guide (U.S. EPA, 1993) offers several tools for evaluation. This includes weighted-sum rating methods which allow simultaneous consideration of multiple criteria.

3.5 IMPLEMENTATION

Implementation is the process of acting on the assessment. Good implementation actually begins with the assessment. Ideas should be recorded and evaluations documented during the assessment. At the end, responsibility for follow-up or evaluation should be assigned by the meeting leader. All of this information is provided with the meeting minutes.

Some changes to the process or design may be without major expenditure or risk. These may be accomplished quickly. Other implementations require substantial expenditures, laboratory testing, demonstration, and capital projects. The success of the review will be defined by the economic and environmental benefits. Unless the benefits are well understood, resources will not be committed to pursue the most innovate ideas. Periodic progress reports should be made summarizing the status of remaining active ideas, economic benefits of accepted ideas, and environmental impact.

4 AN ILLUSTRATIVE EXAMPLE

A simple problem involving the control of emissions from storage tanks illustrates the project environmental review process (U.S. EPA, 1993).

Methanol, toluene, and hexane are required in large quantities by a chemical complex and are stored in a number of 50 cu meter tanks. These tanks are fixed roof and have a pressure rating of plus or minus 0.03 bar.

New regulations require 95% control of vent losses; 90% of the vent losses are the result of filling; the remainder result from tank breathing due to temperature and atmospheric pressure changes. The maximum rate of filling is 500 gpm (2,000 l/m), resulting in the displacement of less than 100 cfm (160 cu m/h) of vapor-laden air. How can the tanks be brought into compliance with these requirements? The complex manager charters a group to find the best answer to this problem. Boundaries identified in the charter are that preferred solutions will not reduce the safety of the plant, will require minimum testing to prove compliance with the rules, and will have the least possible net present cost. Preference is to be given to pollution prevention suggestions.

An environmental review is organized, including design engineers, a mechanic, an operator, a permitting specialist, and a research chemist. This team generates the following options:

1. Connect all of the tanks to a common vent pipe and take this to a flare.
2. Eliminate solvent storage. Pump solvents into the process directly from rail cars.
3. Replace these volatile, highly regulated solvents with less volatile or less regulated solvents. For example, replace both the toluene and hexane with heptane, which is not a hazardous air pollutant. Replace the methanol with acetone, which is not a hazardous air pollutant or a volatile organic compound.
4. Connect all of the tanks to a common vent pipe and take this to a water scrubber.
5. Connect all of the tanks to a common vent pipe and take this to a refrigerated condenser.
6. Connect all of the tanks to a common vent pipe and take this to a carbon adsorber.
7. Replace the fixed roof tanks with floating roof tanks which do not vent.
8. Replace the low pressure rating tanks with three tanks, one for each solvent, each rated for operation to 3 bar positive pressure and one-half bar of vacuum. Operate the tank levels to stay between these pressures without venting.
9. Confirm the maximum allowable tank pressure. If necessary, stiffen the tanks to increase the allowable maximum pressure to 0.06 bar positive pressure. Install vapor return lines and vent the tanks back to rail cars during filling.
10. Paint the tanks white to reduce the daily vapor space temperature change.
11. Insulate the top of the tanks to reduce the daily vapor space temperature change.
12. Chill the liquid in the tanks to reduce the vapor pressure.

13. Buy solvent from the adjacent chemical plant. Have it delivered by pipeline.
14. Maintain a fixed inventory of solvents in the storage tanks to minimize filling losses. Concentrate inventory in a minimum number of tanks which are nearly full to minimize breathing losses. Pump solvent from the rail cars directly to the process. Use the solvent in the tanks only when no rail car is available.
15. Connect all storage tank vents to a common impervious bladder placed inside an old tank so that all breathing losses and most filling losses are contained. The bladder will fill as the combined liquid level rises and then empty as the levels drop. Periodically drain condensed solvents from the bladder and burn this condensate for the fuel value.
16. Float ping pong balls in the tanks to reduce the wetted surface which saturates the vapor space.
17. Place bladders within each tank so that no evaporation occurs.
18. Use only one tank for each solvent. Vent these tanks to the bottom of a fourth tank. Rely on the fact that the solvent vapors are all heavier than air so that they will stay in the tank.

While some of these ideas may not be feasible or capable of achieving 95% control, interrupting the review to make these points would merely have shut down the flow of ideas. Other ideas which would be attractive for a new plant may not be particularly cheap for a retrofit. Application of the 1 to 4 rating criteria might lead to the following conclusions:

3.5 or higher: none

2.5–3.5: 1, 3, 5, 6, 9, 14.

1.5–2.5: 2, 7, 8, 10, 11, 12, 13, 15.

Less than 1.5: 4, 16, 17, 18.

Idea 3, solvent substitution, may be possible without investment but requires acceptance of major changes to the manufacturing process. This idea offers the opportunity to eliminate all hazardous air emissions and to greatly simplify permitting. Idea 9, vapor return, may be relatively inexpensive if tank stiffening is not actually necessary. Detailed analysis may be required to verify this conclusion, to determine investment, and to verify whether additional controls are needed to meet the performance objective. Both of these ideas are examples of pollution prevention. Ideas 1, 5, and 6, highly rated in the initial review, require some new investment, and probably increase the risk of a fire spreading through the tank farm (Clark, 1996). Idea 14 may not require any investment but requires a willingness to change operating procedures and may result in substantially increased demurrage charges. Thus, we see the need for a detailed evaluation to identify the best idea from several good ideas.

5 ECONOMICS OF POLLUTION CONTROL DEVICES

Environmental reviews will generate suggestions to install end-of-pipe control devices. These devices require additional investment, add to operating costs, and may introduce new safety hazards. Several control devices (e.g., flares, thermal oxidizers, and catalytic oxidizers) may be functionally capable of meeting performance objectives but will vary in operating costs and initial investment. Increasingly, vendors are proposing to lease and operate large new treatment systems for both gaseous and aqueous emissions. How can the optimum among several treatment proposals be identified and compared with pollution prevention options?

The evaluation of ideas may be simplified by developing guidelines which identify the most cost effective control devices that meet environmental objectives. Dyer and colleagues have published such guidelines (Dyer and Mulholland, 1994a; Dyer and Mulholland, 1994b; Dyer and Taylor, 1994; Keller and Dyer, 1998) for the control of VOCs (volatile organic compounds), halogenated VOCs, particulate control, and certain wastewater emissions. These guidelines use net present value to simultaneously consider investment and future operating costs in a single financial metric. Mukhopadhyay (Mukhopadhyay and Moretti, 1993) provides additional economic information.

Dyer and his colleagues evaluated 15 technologies for the control of VOC-containing gas streams. The investment in a minimum net present cost control system is \$40 to \$1000 per standard cubic foot per minute (scfm) of gas flow treated (\$25–\$600 per normal cubic meter per hour), and annual operating costs are 10% to 30% of this value. Investment and operating costs per unit of gas flow for the minimum net present cost technology decrease with increasing flow. The minimum absolute investment using their estimating methods is on the order of \$100,000 for flow rates of 100 cfm and less. Net present cost, including both investment and operating costs, is much higher if particulates, aerosols, or halogenated compounds are present in the gas stream.

Net operating costs for wastewater treatment are more variable than costs for gas treatment. Important waste characteristics include biological oxygen demand, chemical oxygen demand, suspended solids, priority pollutants, and possibly dissolved ions. A limiting case is treatment of traditional dilute aqueous wastewater with aerobic activated sludge technology. Incremental economics for *new* treatment capacity are at least \$6000 investment and \$2000 per year annual operating costs per pound per hour of biological oxygen demand (\$13,000 investment and \$4500 annual cost per kg/h of BOD).

Solids disposal costs also vary substantially with composition, and for some wastes, with region. Nonhazardous solid (office wastes, clean packaging, and clean construction debris) disposal costs may be obtained from local waste haulers and may be as low as \$10/ton. Hazardous waste which is toxic, flammable, corrosive, or otherwise an RCRA hazardous material will incur costs from \$200/ton for cleanly burning solvents to \$2000 or more per ton for highly halogenated wastes ($220–$2200 per ton). Pollution prevention suggestions will often reduce the consumption of raw materials and intermediates per unit of product, and may even increase the production capacity. Economic evaluation of pollution prevention is often based on the purchase cost of raw materials and cost of manufacture of finished goods. In some cases, such as when raw materials availability is limited or a process is sold out, the proper basis of evaluation is product selling price less additional operating costs associated with pollution prevention.

6 FREQUENTLY EMPLOYED STRATEGIES FOR POLLUTION PREVENTION

Certain problems, such as the storage of volatile liquids, recur in many environmental reviews. Ideas and sometimes portions of evaluations may be simplified by using the results of other reviews. The U.S. EPA Risk Reduction Laboratory has published pollution prevention studies for a number of industries, including

The Pesticide Formulating Industry	EPA/625/7-90/004
The Paint Manufacturing Industry	EPA/625/7-90/005
The Fabricated Metal Products Industry	EPA/625/7-90/006
Printed Circuit Board Manufacturing	EPA/625/7-90/007
The Commercial Printing Industry	EPA/625/7-90/008
Selected Hospital Waste Streams	EPA/625/7-90/009
Research and Educational Institutions	EPA/625/7-90/010
The Photoprocessing Industry	EPA/625/7-90/012
The Automotive Repair Industry	EPA/625/7-91/013
Fiberglass Reinforced and Composite Plastics	EPA/625/7-91/014
The Marine Maintenance and Repair Industry	EPA/625/7-91/015
The Automotive Refinishing Industry	EPA/625/7-91/016
The Pharmaceutical Industry	EPA/625/7-91/017
Mechanical Equipment Repair	EPA/625/R-92/008
The Metal Casting/Heat Treating Industry	EPA/625/R-92/009
The Metal Finishing Industry	EPA/625/R-92/011

Materials substitution may be utilized to eliminate hazardous air pollutants identified by the Clean Air Act Amendments and by other regulations to eliminate priority pollutants from wastewater streams, to eliminate RCRA designations of certain hazardous solids waste streams, and, in general, to reduce the environmental impact of processes (Joback, 1994; Gessner, 1998). Materials substitution may bring the additional benefits of reducing flammability hazards and worker exposure to toxic substances, thus increasing the inherent safety of processes. Frequently considered solvent substitutions include

- Water in place of hydrocarbons (e.g., water plus surfactants in latex paints).
- Carbon dioxide or acetone to eliminate VOCs and HAPs from processes and products (acetone has a very low potential to form smog and is not regulated in the United States as a VOC).
- Heptane or less volatile alkanes in place of hexane or other hydrocarbon solvents.
- Xylene in place of toluene or benzene.
- Hydrocarbons in place of chlorocarbons (e.g., as a vapor degreaser).
- Methylene chloride or ortho-dichlorobenzene is used if a chlorocarbon is absolutely necessary (e.g., as a solvent for halogenation reactions).
- Carbon dioxide as a weak acid acidulant in place of acetic acid.
- Sodium hydroxide, sodium carbonate, and sodium bicarbonate as bases in place of ammonia.

Dyer and Mulholland, (1999) present a review of pollution prevention opportunities organized in part by unit operation. These include

- For batch operations, options associated with vessel charging, operation, discharging, and cleaning.
- For equipment and parts cleaning, options including increased employee awareness, several cleaning strategies, and ways to reduce cleaning frequency.
- For reactors, raw material quality, recycle of byproducts in equilibrium reactors, improved control, and better mixing and temperature control for fast reactions.
- Solvent substitution.
- Ideas for water conservation.
- Control of pH as a pollution prevention tool.
- Reduction of emissions from vacuum processes.
- Reduction of emissions from the ventilation of manufacturing areas.
- Reduction of emissions from volatile organic liquid storage.
- Reduction of fugitive emissions.
- Selection of separations technologies to minimize emissions.

7 CONCLUSION

An environmental design review is a structured review of a chemical process with emphasis on both reduction of waste generation and on waste management. It can provide a framework for input from manufacturing staff, R&D, and environmental engineers. It can be used to charter more quantitative studies, including heat and materials integration. A frequently unanticipated benefit of environmental design reviews for existing process is identification of capacity and quality improvements. The design environmental review should be expected to identify opportunities for pollution prevention, improve process economics, and provide assurance of regulatory compliance.

REFERENCES

Chemical Manufacturers Association. May 1993. *Designing Pollution Prevention into the Process: Research, Development and Engineering*. Chemical Manufacturers Association.

Clark, D. G., and Sylvester, R. W. (1996). Ensure Process Vent Collection System Safety. *Chemical Engineering Progress* 92:65–77.

Douglas, J. M. (1988). *Conceptual Design of Chemical Processes*. McGraw-Hill, New York.

Dyer, J. A., and Mulholland, K. L. (1994). Toxic Air Emissions: What is the Full Cost to Your Business? *Environmental Engineering: A Special Supplement to February 1994 Chemical Engineering* 101(suppl.):4–8.

Dyer, J. A., and Taylor, W. C. (1994). Waste Management: A Balanced Approach. *Proceedings of the Meeting and Exhibition of the Air and Waste Management Association*, Cincinnati, OH.

Dyer, J. A., and Mulholland, K. L. (1994). Follow This Path to Pollution Prevention. *Chemical Engineering Progress* 94:34–42.

El-Halwagi, M. M. (1997). *Pollution Prevention Through Process Integration*. Academic Press, San Diego.

Joback, K. B. (1994). Solvent Substitution for Pollution Prevention: Pollution Prevention via Process and Product Modifications. *AIChE Symposium Series* (M. M. El-Halwagi and D. P. Petrides) 90:98–103.

Gessner, A. W. (1998). Switch Materials to Prevent Pollution. *Chemical Engineering Progress* 94:59–64.

Keller, R. A., and Dyer, J. A. (1998). Abating Halogenated VOCs. *Chemical Engineering* 105:100–105.

Mukhopadhyay, N., and Moretti, E. C. (1993). *Current and Potential Future Industrial Practices for Reduction and Controlling Volatile Organic Compounds*. American Institute of Chemical Engineers, New York.

Mulholland, K. L., and Dyer, J. A. (1999). *Pollution Prevention: Methodology, Technologies and Practices*. American Institute of Chemical Engineers, New York.

U.S. Environmental Protection Agency (1992). *Facility Pollution Prevention Guide*. EPA/600/R-92/088. Washington, DC, U.S. EPA, Office of Research and Development.

U.S. Environmental Protection Agency (1992). *DuPont Chambers Works Waste Minimization Project*. EPA/600/R-93/203. Washington, DC, U.S. EPA, Office of Research and Development.

U.S. EPA Emissions Standards Division (1993). *EPA Control Techniques Guideline Series—Control of Volatile Organic Compound Emissions from Volatile Liquid Storage in Floating and Fixed Roof Tanks*. Research Triangle Park, NC, U.S. EPA, Office of Air and Radiation, OAQPS.

CHAPTER 2

Integrated Assessment Methodologies and Software Tools for Process Design: Economic, Environmental, Safety, and Decision Analyses

David R. Shonnard, Tony N. Rogers, Bruce A. Barna, Daniel A. Crowl, Eric J. Oman, Peter P. Radecki, Jon A. Herlevich, Jr., and Piyush B. Parikh

Department of Chemical Engineering and The Center for Clean Industrial and Treatment Technologies, Michigan Technological University, Houghton, MI 49931

Abstract

The chemical industry has led all others in developing process unit simulation and plant-wide material and energy accounting capabilities; however, existing simulators have critical omissions that prevent their effective use in economic, environmental, and safety calculations. Process design assessment and evaluation software tools and methodologies have been developed by the authors to enhance the capabilities of commercial simulator packages. The key features of these design tools are presented and they are applied in concert for the conceptual design of a number of energy and cogeneration design alternatives to satisfy facility heat and power requirements. The Base Case is the existing facility (natural gas boilers for heat and purchase power), whereas Case B (convert boilers to burn wood and purchase power) and Case D (natural gas-powered engines for heat and power) are alternatives. One of the software tools predicts five economic measures; another tool calculates up to nine environmental indexes, and a third tool provides two safety metrics. In addition, in this study two more design criteria were included—processability and societal aspects—with multiple assessment metrics in each category. A final software tool provides Analytical Hierarchy Processing (AHP) weightings to each assessment metric and calculates a final ranking of process alternatives. The Base Case option was deemed most desirable among the competing design alternatives based on the AHP weightings of the individual process design metrics.

Keywords: multicriteria process design assessment, software tools

1 INTRODUCTION TO PROCESS DESIGN AND ASSESSMENTS

Process design involves multiple steps, beginning with process creation and development of a base case design, progressing through detailed process synthesis and detailed design, and, finally, to plant-wide controllability assessment (Seider et al., 1999). Process assessment and evaluation tools are necessary to enhance and make more efficient the process design activity. In this chapter we describe assessment methodologies and software tools for economic, environmental, safety, and decision analysis that can be used in both preliminary conceptual design and more detailed design using commercial process simulation. In the first two sections we present details of the assessment methodologies and their potential integration with commercial process simulators. In the third section we present results from the application of these tools to conceptual design of a facility utility generation system.

1.1 BACKGROUND ON PROCESS SIMULATOR CAPABILITIES

The chemical industry has led all others in developing process unit simulation and plant-wide material and energy accounting capabilities. Current steady-state process flowsheet simulation is highly deterministic. One typically devises the basic plant configuration and uses the simulator to size unit operations and estimate the process energy requirements, product yields, and chemical separation profiles. Existing simulators are excellent at material and energy balance bookkeeping for certain well-characterized plant layouts, but they have critical omissions that prevent their effective use in economic, environmental, and safety calculations. First, a number of key waste separation and treatment technologies are not a part of most commercial simulators. Second, environmental data and parameters generally are not tabulated. Third, estimates of reaction byproduct formation and releases of other forms of pollution are lacking. Fourth, the incremental economic effects of pollution prevention strategies are difficult to assess, and typically there is no accounting of the noncost consequences of emitting pollution (e.g., global warming, ozone depletion, etc.). Fifth, commercial simulators do not provide access to experience-based information regarding the use of pollution prevention technologies. Finally, information on process safety is not available to the design engineer while synthesizing the flowsheet. The process design enhancement tools described in this article address these deficiencies.

1.1.1 Economic Assessment

The long-term success of individuals, companies, institutions, and even countries can be closely linked to their selection of sound investment strategies that ensure the best use of scarce human and capital resources. In economic assessment we are primarily concerned with the allocation of the capital resources.

In both retrofit and "green-field" projects, standard industrial practice is to prepare preinvestment feasibility studies to predict the future performance of potential investments. These economic feasibility studies use estimates of investment, working capital, sales revenue, and operating expenses to assess the long-term impact of the project on the investor's treasury. Frequently the results are translated into a variety of performance yardsticks that can then be used to sort, eliminate, or accept investment opportunities.

Some examples of such yardsticks include

- Simple return on investment
- Discounted cash flow evaluation
  - Net present value
  - Cash flow rate of return
  - Benefit/cost
  - Equivalent uniform annual worth
- Payback
- Breakeven

The widespread availability of cheap computer power and spreadsheets has led to the near universal adoption of cash flow, payback, and breakeven analyses in the chemical process industries. Because each of these gives some additional insight into the project performance, it is common to compute them all as part of the analysis. The common starting point is the preparation of an annual cash flow table constructed from the estimates of investment, working capital, etc.

Standard practice in the evaluation of mutually exclusive alternatives or in optimizations is to perform an incremental analysis. The incremental cash flows resulting from an incremental investment are then judged against some minimum acceptable standard of performance, typically a discounted cash flow yardstick. Usually the assumption is made that sufficient capital is available to implement all of the acceptable increments.

For "green-field" designs (new plants on new sites), uncertainties in market information or other cost projections can become a primary concern. Some form of risk analysis is frequently added to the preinvestment economic evaluation to help judge the risk. Common approaches include

- Best case/worst case
- Single parameter sensitivity

- Monte Carlo simulation
- Probability analysis

These approaches are shown in the order of simplest to most complex, and the latter two usually are not performed unless the uncertainties in the project are large.

One problem that occurs in comparing numerous alternatives with differing economic, environmental, and safety characteristics is the availability of rapid and accurate cost estimating tools. Detailed designs or vendor quotes usually are not an option because of time and cost constraints at this stage of design. Companies frequently rely upon the experience and expertise of their design staff to ensure that sufficient alternatives are considered and that a suitable screening process can reduce the number of detailed designs necessary. Process simulators are not yet to the point where they are of much use in this screening phase. Improved tools in this area would likely lead to better environmental and safety performance while still maximizing economic performance.

1.1.2 Environmental Assessment

Interest in the development of methodologies for assessing environmental and human health impacts of chemical process designs has grown in recent years. This interest is an outcome of current trends for managing environmental impacts of industrial processes and products by reducing pollution generation and release within manufacturing facilities, rather than relying solely on end-of-pipe pollution control. Efforts to develop environmental assessment methodologies of chemical process designs have lagged behind economic and safety assessments because of this traditional reliance on pollution control strategies. In addition, environmental assessments of processes have lagged behind assessments of products. Life Cycle Assessment methodologies have been applied to the manufacture of products (SETAC, 1993), but fewer applications of impact assessment have been developed for chemical processes and their designs.

Nonetheless, the recent engineering literature has seen an increased frequency of papers dealing with various aspects of environmental assessment of chemical processes. A central issue in environmental assessment is the inclusion of impact indicators and answering such questions as, What impacts are included in the assessment? How are the indexes derived? Prominent among these methods for environmental assessment of chemical processes are (a) the method of Minimizing Environmental Impact (MEI) (Pistikopoulos, 1994; Stephanis et al., 1995; Romero-Hernandez, 1998), (b) the WAste Reduction (WAR) algorithm, developed by the U.S. Environmental Protection Agency (Hilaly and Sikdar, 1994; Mallick, 1996; Cabezas et al., 1997), and (c) the Environmental Fate and Risk Assessment Tool (EFRAT) (Shonnard and Hiew, in press).

There are many similarities between the three environmental assessment methods listed. All include one or more indexes for global and regional scale environmental impacts, such as global warming, stratospheric ozone depletion, photochemical smog formation, and acid rain. The WAR algorithm and EFRAT include both human and ecotoxicity indexes (Cabezas et al., 1997; Shonnard and Hiew, in press); however, there also are many differences in these methodologies. Only the method of MEI includes life cycle stages other than the chemical processing stage in the impact assessment, whereas EFRAT and WAR have only been applied thus far to the chemical processing stage. EFRAT incorporates air emission estimation methods in the methodology, whereas WAR and MEI do not provide information regarding emissions estimation. EFRAT calculates multimedia environmental fate of emitted chemicals, whereas WAR assumes that the pollutants emitted into a particular environmental compartment (air, water, soil) exert the impact in only that compartment (Cabezas et al., 1997).

1.1.3 Process Safety Assessment

A chemical manufacturing facility is **inherently safer** if it "reduces or eliminates the hazards associated with materials and operations used in the process, and this reduction or elimination is permanent or inseparable" (AICHE, 1996). With inherently safer design, the hazards are low and constant, independent of any layers of protection around the hazard. Four methods are used to implement inherently safer design: minimize, substitute, moderate, and simplify. To minimize means to reduce the quantities of hazardous materials or energy in a plant; substitution means to replace hazardous chemicals with less hazardous chemicals; moderate means to reduce operating temperatures and pressures; simplify means to reduce the complexity of the process. Inherently safer design is easiest to apply during the conceptual design of a process, although it can be applied at any point during the life of a plant.

A hazard assessment study is used specifically to determine the hazards in the design and operation of a chemical plant that could result in a fire, explosion, or toxic release of chemicals. Table 2.1 shows some of the common procedures adopted for hazard identification and evaluation (AICHE, 1992). Descriptions of each of these standard techniques are found in standard textbooks (Crowl and Louvar, 1990). The first four techniques (i.e., Safety Review, Checklist Analysis, Preliminary Hazard Analysis, What-if Analysis) are commonly used during the preliminary hazard evaluation study to give a general overview of the presence of hazards within a plant. The next three techniques [i.e., What-if/Checklist Analysis, HAZard and OPerability (HAZOP) Analysis, Failure Mode and Event Analysis (FMEA)] are used to perform a more detailed analysis to identify potentially hazardous situations. The next four techniques [Fault Tree Analysis (FTA), Event Tree Analysis (ETA), Cause Consequence Analysis (CCA), Human Reliability Analysis (HRA)] are performed at a high level of detail and are part of Quantitative Risk Analyses (QRAs). QRAs usually are conducted by highly trained individuals and require a greater amount of time and effort. The Relative Ranking method falls in the middle of Table 2.1 and is

TABLE 2.1
Summary of hazard evaluation techniques

Technique	*Description*
Safety reveiw	Qualitative hazard reveiw
Checklist	
Preliminary hazard analysis	
What-If	
Relative Ranking	**Dow Indices**
What-If/Checklist	
HAZard and OPerability study	Using guide words
Failure mode and event analysis	Consequences of equipment failure
Fault tree analysis (FTA)	Top event—using and/or boolean
Event tree analysis (ETA)	Initial event—using and boolean
Cause consequence analysis	Combination of FTA and ETA
Human reliability analysis	Consequence of human-related errors

Increasing level of detail ↓

deemed appropriate for the information required, the level of effort, and the value of the results. This approach will be the method of choice for this work.

The Substance Hazard Index (SHI), Material Hazard Index (MHI), Threshold Planning Quantity (TPQ), and the Dow Fire and Explosion (F&EI) and Chemical Exposure Indices (CEI) are commonly used relative ranking indices. Although these indices are generally customized to fit company needs, the F&EI and the CEI, developed at Dow Chemical Company, are considered industry standards. The F&EI and the CEI are discussed in depth here and are the basis for the software developed. The F&EI is a system to quantify the expected property damage and business interruption in the event of an accident (Dow Chemical, 1994a). The CEI guide accounts for the risk due to an accidental chemical release. It is a relatively new development of the Dow Chemical Company, developed in 1986 (Dow Chemical, 1994b).

1.2 PROCESS DESIGN DECISION ANALYSIS METHODS

As additional performance indices in the safety and environmental arenas are considered, the need also arises for more sophisticated decision assistance tools. Such tools may help the designer to rapidly assess the economic trade-offs that may be associated with increased performance in these other areas.

Normally, cost and profit are used as measures of desirability in plant design, but there are many decisions where additional measures of process performance are needed. Design engineers and decision-makers often must take into account economic, pollution prevention, safety, and environmental attributes. The difficulty with taking multiple criteria into account in process design is the need to reach a decision among competing design alternatives. For over 20 years, decision-makers and engineers have researched and developed advanced decision-making techniques. Several techniques have become accepted (MacCrimmon, 1973), such as weighting methods, sequential elimination methods, mathematical programming methods, spatial proximity methods, and the Analytical Hierarchy Process (AHP) Saaty (1982). The advantages and disadvantages of these different methods are summarized in Table 2.2.

The Analytical Hierarchy Process (AHP) was first published by Saaty (1982) as a technique for multicriteria decision-making. AHP is a methodology for breaking down a complex, unstructured situation into its component parts, then arraying these parts into a hierarchical order and assigning numerical values to subjectively judge the relative importance of each variable. AHP has the ability to manage several alternatives with multiple attributes, with the possibility of interaction between the attributes. AHP allows the users to make simple comparisons of the attributes in pairs to determine their relative weights based on the objective, knowledge, and background of the decision-makers. The user can also execute contemplative "what if" scenarios with AHP to examine the effects possible changes in the situation will have on the decision (Min, 1994). There have been at least two conferences and eight special issues of journals devoted to AHP indicating that it is widely popular in various fields (Finnie et al., 1993); however, little has been published regarding the application of AHP for multicriteria decision-making in chemical process plant design, and it is likely, therefore, that its application in this field has been limited.

Some previous applications of AHP to process and product design include the analysis of a new metal-shaping process that reduced the generation of mixed waste from the process (Reid and Christianson,1994). AHP has also been applied to a multiobjective optimization model for waste management of the petrochemical industry (Alidi, 1995). These two previous studies illustrated the benefits of using AHP to solve multicriteria decision analysis problems; however, the goal of both studies was constrained to pollution prevention without consideration of competing criteria, such as profitability or safety. One study using AHP to evaluate multiple competing criteria was performed for the analysis of cogeneration of steam and electricity (Herlevich, 1997).

In the next section, an AHP method will be presented for use within the integrated suite of software tools. This suite of tools is intended to couple easily with commercial process simulators such that the material and energy balance information and list of equipment are available as input. The output from the enhancement tools is a set of scientifically based process indexes (or attributes) for subsequent decision analysis.

TABLE 2.2
Advantages and disadvantages of multicriteria decision methods (MacCrimmon, 1973; Saaty, 1982)

Methods	*Summary*	*Advantages*	*Disadvantages*
Spatial proximity	Graphical method to judge proximity of alternatives to ideal configurations	Can be applied to different combinations of attribute values	Not amenable to programming
Mathematical programming	An algorithm to converge an objective function onto an optimum configuration	Lets users specify desired levels on single or multiple attribute values	Difficult to develop utility functions
Sequential elimination	A process for sequentially comparing alternatives on the basis of attribute values so that alternatives can be either eliminated or retained	Discards minor alternatives and leads users to the right choice in a small number of options	Cannot be applied to complex problems
Weighting methods	A process of obtaining a numerical scaling for attributes, numerical weightings across attributes, objective function definition, and rules for choosing alternatives	Simple to apply Has ability to manage several alternatives	Has the possibility of loss of information among interacting attributes
Analytical hierarchy process (AHP)	Similar to weightings methods except that attributes are segregated into a set of tiers in a hierarchy	Helps users to combine various measured performances into a single ranking Easily adaptable to a variety of situations Can manage several alternatives with the possibility of interactions between attributes	The inconsistencies increase when the number of factors increases

2 DESCRIPTION OF THE CLEAN PROCESS ADVISORY SYSTEM METHODOLOGIES AND TOOLS

2.1 DESIGN OPTIONS RANKING TOOL METHODOLOGY

Currently, the evaluation of a chemical process is a multistaged procedure. First, the equipment is sized using commercial process simulators. Cost data for the required equipment are often gathered from diverse and inconsistent sources. The total capital investment, annual operating expenses, and income are then estimated from the equipment data. Finally, all the information calculated is entered into a discounted cash flow spreadsheet and the process economics are determined. At each step in the process there is the opportunity to lose track of various cost items and introduce errors into the evaluation. These errors may become even more pronounced when comparing design alternatives, as there is no guarantee that the calculations will be performed exactly the same from one alternative to another.

The Design Option Ranking Tool (DORT) is an integrated economic evaluation software with three main parts. First, the software contains a comprehensive database of equipment costs and has the ability to manipulate these data and derive cost correlations. Next, the software is an equipment design and evaluation tool using the stored cost correlations to calculate the contribution of each piece of equipment to total capital investment. Finally, it is a project economic evaluation and comparison tool performing discounted cash flow analysis on single projects and incremental analysis on competing processes. This integration ensures that no information is lost from one step to another. The software's factored estimation system is user-tunable with many default parameters that can be modified and then saved for later use. This approach allows a company to set the program variables to preferred values and then let any user within the company use this "default" set to analyze different designs. This provides a consistent platform for the evaluation of alternative process designs. Once these parameters are set, the program "remembers" to include cost items that might otherwise be overlooked using conventional design evaluation procedures. Figure 2.1 shows the interaction of the DORT software with commercial process simulators and with the process decision analysis software (DEAR).

2.2 ENVIRONMENTAL FATE AND RISK ASSESSMENT TOOL METHODOLOGY

A brief overview of the Environmental Fate and Risk Assessment Tool (EFRAT) methodology will be provided here. Figure 2.2 shows the information flows included in the methodology and software. In a typical application, a commercial process simulator would provide mass and energy balances, a list of chemicals, and a list of unit operations. From databases resident to the software, various properties for each chemical and unit operation would be retrieved. Three calculation routines make up the heart of the software: the Air Emissions Calculator, the Partition Calculator, and the Relative Risk Index

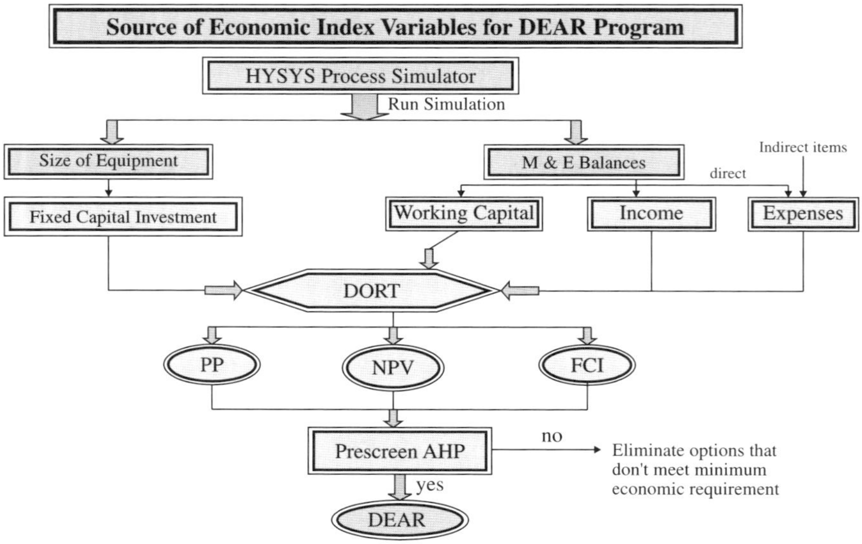

Figure 2.1: Features of the DORT software and relation to process simulators and decision analysis. PP—payback period; NPV—net present value; FCI—fixed capital investment.

Calculator. The Air Emissions Calculator uses emission factors, correlations (U.S. EPA, 1998; Hiew, 1998; EF, 1992), and process simulator stream information to provide an estimate of the unit-specific and fugitive air emission rates for each chemical within the process.

The Partition Calculator provides the dimensionless mole fraction of the emitted chemical that partitions into the air, water, soil, or sediment compartments of the environment at equilibrium using a modified "Level I" multimedia compartment model (Mackay, 1991; Mackay et al., 1992). For example, the mole fraction of the chemical that partitions into the air compartment at equilibrium (D_A) is given by the following equation:

$$D_A = \frac{\theta_a(\theta_{aa}K_H + \theta_{aw} + \theta_{as}\rho_{as}K_{ds})\,(\text{air})}{\begin{gathered}\theta_a(\theta_{aa}K_H + \theta_{aw} + \theta_{as}\rho_{as}K_{ds})\,(\text{air}) \\ + \theta_w(\theta_{ww} + \theta_{ws}\rho_{ws}K_{ds} + \theta_{ww}\rho_F BCF_F)\,(\text{water}) \\ + \theta_s\left(\rho_{ss}K_{ds} + \theta_{sw} + \theta_{sa}K_H + A_s\rho_p\frac{K_{sp}}{F_{fps}}K_{ds}\right)(\text{soil}) \\ + \theta_b(\theta_{bw} + \theta_{bs}\rho_{bs}K_{ds} + \rho_b BCF_b)\,(\text{sediment})\end{gathered}}. \quad (1)$$

Several parameters are needed for each chemical in the prediction (Mackay, 1992; Verschueren, 1996): Henry's constant (K_H), octanol-water partition coefficient (K_{OW}), bioconcentration factor (BCF), and several distribution coefficients, such as the water/solids coefficient (K_{ds}), and the soil/plant

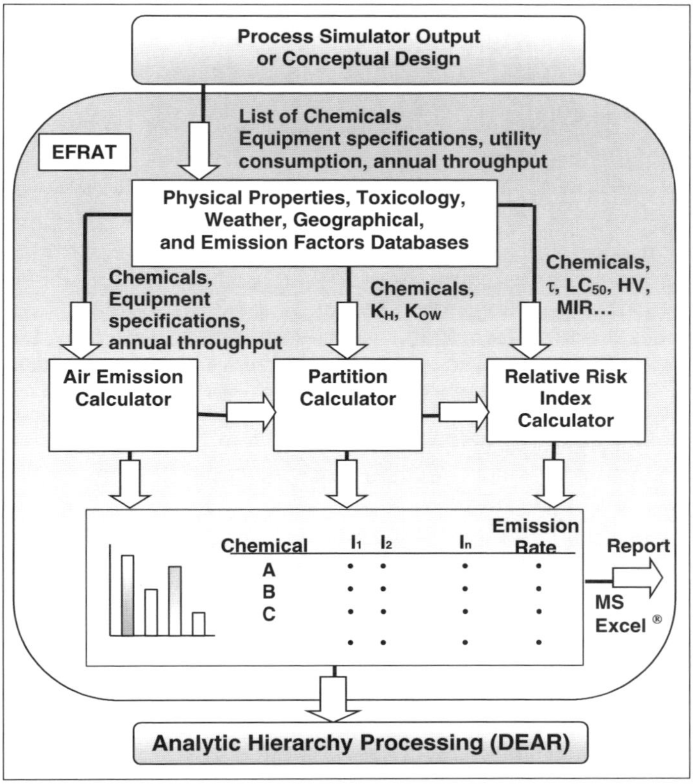

Figure 2.2: Information flow diagram for the EFRAT software and relation to process simulator and decision analysis.

coefficient (K_{sp}), however, because BCF is equal to 0.05 K_{OW} (Mackay, 1992), the distribution coefficients are related to the organic carbon partition coefficient and fraction organic carbon ($K_d = K_{oc}\ f_{oc}$), and because soil/plant coefficient can also be related to K_{OW} ($K_{sp} = 7.7\ K_{OW}^{-0.578}$) (McKone, 1986), only two parameters are needed in the prediction (K_H and K_{OW}). Other landscape parameters in the model, such as volume fraction (θ), fraction organic carbon, plant areal density (ρ_p) [kg dry plants/m^2 area], plant surface area coverage (A_S) [m^2/m^3 soil], and mass density (ρ) in each compartment, are specific to different geographical locations. Similar equations for D in the water, soil, and sediment compartments are easily obtained from equation (1).

The Relative Risk Index Calculator generates nine environmental and health impact indexes for the entire process. These indexes include impacts for (1) global warming, (2) ozone depletion, (3) smog formation, (4) acid rain, (5) and (6) human inhalation and human ingestion route noncarcinogenic toxicities, (7) and (8) human inhalation and human ingestion route carcinogenic toxicities, and (9) ecotoxicity (fish mortality).

All of the indexes are constructed using the following general formula:

$$I_i^* = \frac{[(D)(\tau)(IIP)]_i}{[(D)(\tau)(IIP)]_{benchmark}} \quad \text{(dimensionless)}, \tag{2}$$

where D is the mole fraction partitioned to the air, water, soil, or sediment compartment (eqn. 1 unitless), τ is the reaction residence time of the chemical in the impacted compartment (hours), and *IIP* is the inherent impact parameter for the index. Some examples of *IIPs* are Global Warming Potential (GWP) for the global warming index, Lethal Dose 50 % mortality (LD_{50}) for the human ingestion noncarcinogenic toxicity index, and Weight of Evidence classification schemes for the two carcinogenic toxicity indexes. The logic for constructing these indexes in this way is as follows. Relative to the benchmark compound, if a chemical does not distribute significantly (low D), has a short reaction residence time (small τ), and has a low impact potential in the compartment where the impact occurs (small *IIP*), the relative risk is low for this emitted chemical.

The entire process can be represented with the same nine relative risk indexes using equation (2):

$$I = \sum_{i=1}^{n} (I_i^* \times E_i). \tag{3}$$

E_i is the cumulative emission rate of chemical i from the entire process and n is the number of chemicals emitted from the process. Thus, the process index (I) has units of emission rate (mass/unit time) and represents the equivalent emission of the benchmark compound having the same impact as the actual process emissions.

The global warming index for compounds with tabulate GWPs (IPCC, 1996) is

$$I_{GW}^* = GWP. \tag{4}$$

For compounds that do not have GWPs but mineralize readily to CO_2, the global warming index is

$$I_{GW}^* = N_c \times \frac{Mw_{CO_2}}{Mw_i}, \tag{5}$$

where N_c is the number of carbons in the compound and *Mw* are the molecular weights. GWP takes the place of both τ and *IIP* in the general impact index equation listed above.

The ozone depletion index is given by

$$I_{OD}^* = ODP, \tag{6}$$

where *ODP* is the tabulated ozone depletion potential of the chemical (WMO, 1992).

The smog formation index equation is

$$I_{SF}^{*} = \frac{MIR_i}{MIR_{Formaldehyde}}, \tag{7}$$

where *MIR* is the tabulated maximum incremental reactivity of the compound (Carter, 1994).

The equation for acid rain index is

$$I_{AR}^{*} = \frac{ARP_i}{ARP_{SO_2}}, \tag{8}$$

where *ARP* is the tabulated acid rain potential of the chemical (Heijungs, 1992).

The inhalation and ingestion noncarcinogenic human toxicity indexes are

$$I_{INH}^{*} = \frac{LC_{50,Toluene}}{LC_{50,i}} \times \frac{\tau_{i,A}}{\tau_{Toluene,A}} \times \frac{D_{i,A}}{D_{Toluene,A}}, \tag{9}$$

$$I_{ING}^{*} = \frac{LD_{50,Toluene}}{LD_{50,i}} \times \frac{\tau_{i,W}}{\tau_{Toluene,W}} \times \frac{D_{i,W}}{D_{Toluene,W}}, \tag{10}$$

where LC_{50} and LD_{50} are the lethal concentration and lethal dose values for 50 % mortality of animals (rats) in an acute exposure study (NTP, 1997) and τ is the reaction half-life in air (subscript *A*) and water (subscript *W*) (Mackay, 1992).

Similar equations are used for human carcinogenic toxicity indexes:

$$I_{CINH}^{*} = \frac{HV_i}{HV_{Benzene}} \times \frac{\tau_{i,A}}{\tau_{Benzene,A}} \times \frac{D_{i,A}}{D_{Benzene,A}}, \tag{11}$$

$$I_{CING}^{*} = \frac{HV_i}{HV_{Benzene}} \times \frac{\tau_{i,W}}{\tau_{Benzene,W}} \times \frac{D_{i,W}}{D_{Benzene,W}}, \tag{12}$$

where *HV* are the hazard values (Davis et al.,1994) obtained from the Weight of Evidence classification scheme of the U.S. EPA and of the International Agency for Research on Cancer (U.S. EPA, 1997; OSHA, 1997).

Finally, an ecotoxicity index, represented by fish mortality, is given by

$$I_{FT}^{*} = \frac{LC_{50f,PCP}}{LC_{50f,i}} \times \frac{\tau_{i,W}}{\tau_{PCP,W}} \times \frac{D_{i,W}}{D_{PCP,W}}, \tag{13}$$

where the subscript PCP represents pentachlorophenol, the benchmark compound.

This same methodology is easily applied to other waste streams in a chemical process other than air emission streams, as long as rates of release or waste generation can be estimated. Presently there are no generally accepted methods for estimating pollution release rates from chemical processes involving wastewater, liquids, or solids.

2.3 DOW SAFETY ASSESSMENT TOOL METHODOLOGY

The software tool, Dow Indices, has been developed, and it uses the Dow F&EI and CEI relative ranking methods to provide the engineer with tools to include safety considerations during the design of a process (Parikh, 1996; Parikh and Crowl, 1998). The Dow F&EI is a system to quantify the expected property damage and business interruption in the event of an accident. It is based on process safety heuristics and helps identify equipment that has a high hazard potential. It aids in selecting techniques to control loss and is useful in determining plant layout based on the calculated unit hazard distances. The CEI accounts for the risk due to an accidental chemical release. It generates a risk index and also determines potential hazard distances based on several consequence modeling parameter assumptions.

The procedure to calculate the F&EI and its related parameters is shown in Figure 2.3. The calculation begins by selecting the pertinent process units for which an F&EI is to be calculated. This step is followed by entering the materials present in each unit and selecting the types of hazards present from the general and special hazard lists. Next, one must select the Loss Control Credit Factors (LCCF) based on the safety control features available for that unit. Credit is applied for the presence of these features to the unit's economic evaluation—property damage and business interruption—based on the unit's economic data entered. The F&EI can be used to determine the relative degree of hazard, and values above a certain standard (127 for Dow) generally determine the need for further review of the process unit. The F&EI procedure also provides important unit risk analysis information through the radius of exposure and the Business Interruption (BI) loss parameters. The value of equipment present within the area of exposure is the base Maximum Probable Property Damage (MPPD). The product of the LCCF and base MPPD results in the actual MPPD. This, coupled with the value of production per month, gives the Maximum Probable Days Outage (MPDO), which in turn yields the BI of the process.

Figure 2.3 summarize the procedures to perform a CEI calculation. Similar to the F&EI calculation, a list of possible incidents is first determined, followed by selecting the compound present in each scenario. The Airborne Quantity (AQ) for each release incident is then calculated. A chemical release may either be in the gas or liquid form. The AQ for a gas release is based on the sonic gas flow equation from a hole (Crowl and Louvar, 1990). The AQ for a liquid release is based on the flash fraction and rate of evaporation from the pool of liquid formed. The CEI and hazard distances are calculated using the highest AQ based on a wind speed of 5 m/s and neutral weather conditions. A CEI value of 200 or higher would require further review at the Dow facilities. The hazard distances are calculated for each of the three Emergency Response Planning Guide (ERPG) values.

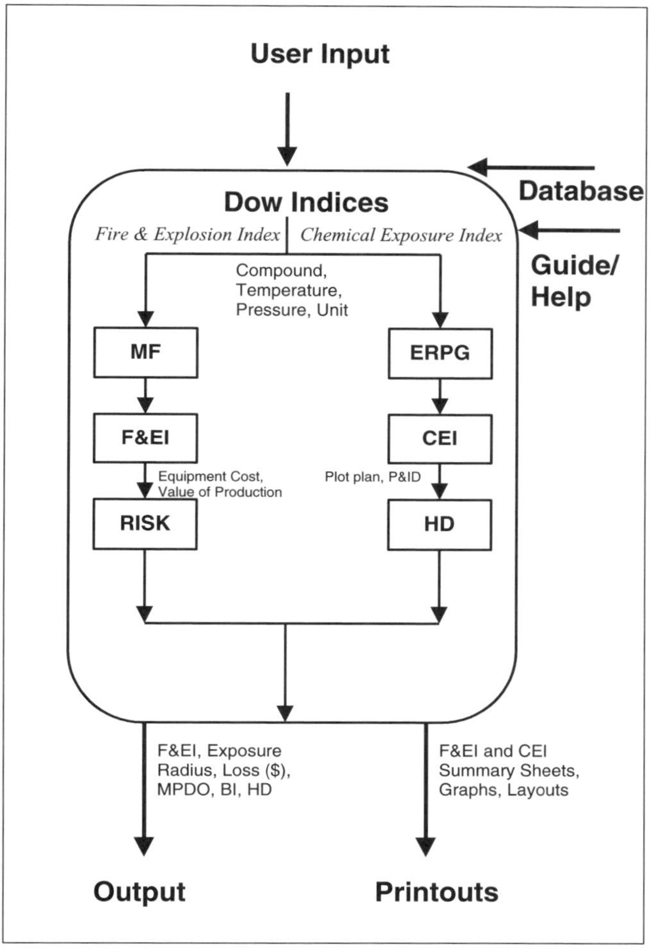

Figure 2.3: Flow of data through the Dow Indices software tool. MF—material factor; ERPG—emergency response planning guide; HD—hazard distance; MPDO—maximum probable days outage; MI—business interruption loss.

The Dow Indices tool attempts to serve three main purposes:

1. Ease the application of the F&EI and CEI by providing a user-friendly interface and an on-line help system
2. Enable the user to save the parameters used to perform the calculation and generate printouts similar to the summary forms in the index document

3. Enable the user to visually determine the largest contributor and the extent of hazard and to easily modify the impact values.

The software interface guides the user through the steps of the F&EI and CEI method (Parikh, 1996). The Dow Indices tool includes the material property data from a database, presence of general and special hazards, economic data, and loss control features for each process unit. The user can also obtain printouts of summary forms that resemble the forms found in the AIChE guides (AIChE, 1992).

2.4 AHP METHODOLOGY: THE DEAR SOFTWARE

Hierarchy components are defined as "tiers" or "levels," and as applied in this approach, consist of both qualitative (subjective) and quantitative (objective) weightings. At the bottom of the hierarchy is a set of quantitative metrics that is calculated from process conditions. In our study, those metrics are generated by DORT, EFRAT, and the DOW software tools. There are seven steps to the AHP analysis:

Step 1: Classify and construct a "tiered hierarchy" with respect to all decision attributes

Step 2: Survey to gather qualitative weightings for management and engineering preferences

Step 3: Do pair-wise comparisons of all attribute types

Step 4: Compare the process alternatives (quantitative comparison) with respect to their attributes

Step 5: Test consistency of the weighting factors

Step 6: Determine vectors of weighting priorities

Step 7: Calculate final process ranking

In Step 3, the paired attributes (A and B) are compared on a scale of 9 (absolute dominance of A over B) to 1 (equivalence of A with B), and further to 1/9 (absolute dominance of B to A).

The extended development of AHP in this study has led to several enhancements from the published methodology (Dechapanya, 1998). The main enhancements and discoveries are (a) bottom-tier automatic scoring of pair-wise comparisons, and (b) Monte Carlo simulation capability being added to produce statistical information about the uncertainty in overall composite ranking. The relationship among the integrated process design enhancement tools is shown in Figure 2.4.

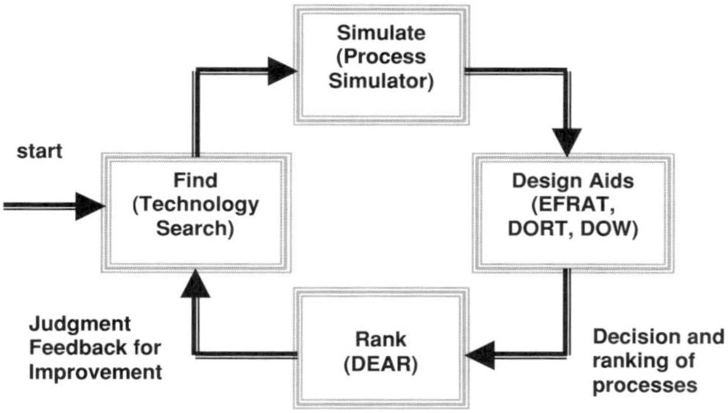

Figure 2.4: Generalized chemical process design, simulation, ranking, and improvement methodology using the integrated suite of software tools.

3 EXAMPLE APPLICATIONS OF INTEGRATED CPAS TOOLS

3.1 COGENERATION OF HEAT AND POWER

Cogeneration is the simultaneous production of heat and power from a single integrated process. Whereas conventional power production loses over 60% of the total energy contained in fuel to the environment as waste heat, process-matched cogeneration loses only between 10% and 20% of the total (Herlevich, 1997). This improved efficiency is accomplished by the recovery of low-grade heat in the form of steam for process heating, cooling, or machine driving. Successful industrial applications of cogeneration have been demonstrated in the chemical, pulp and paper, and petroleum refining industries. If the entire industrial energy demand (about 45% of U.S. demand) for heat and power were met via a typical cogeneration process, annual CO_2 emissions would be reduced by about 7% and industry would be able to sell 7 quadrillion BTU of excess power to the residential and utility sectors (U.S. DOE, 1998). In addition to the global warming reduction potential (CO_2 reductions), cogeneration would allow proportional decreases in precursors to acid rain and acid deposition, such as NO_x and SO_x, and to smog formation (NO_x).

For this initial application of the integrated process simulator enhancement tools, we will investigate the feasibility of replacing the current facility that generates heat for the campus at Michigan Technological University (Base Case) with either of two modifications, termed Case B or Case D. The Base Case consists of four direct fired boilers that burn natural gas and No. 2 fuel oil (as a backup) for heat demand and purchases power from a local utility. In Case B, three of the four boilers would be retrofitted for combustion of locally generated sawdust/wood chips with additional storage, handling, and combustor feed system. Power would be purchased from the local utility as in the Base Case. Case D replaces the four natural gas-fired boilers with five 1,000-kW diesel engines burning natural gas and retains two of the original natural gas boilers for backup and peak load. Power is generated from alternators attached to the engines, and it is enough to satisfy the entire power demand of the campus. Excess power is sold to the local utility. Exhaust heat from the engines is recovered to make steam for campus heating. Other cases were considered in the original study, but only the three cases listed above survived initial economic and technology screening (Herlevich, 1997). The choice of cogeneration on a university campus as the first application of the integrated suite of design tools was prompted by the easy availability of the data needed to conduct the analyses. We have applied these process assessment tools for the recovery of volatile organic compounds from gaseous waste streams (Shonnard and Hiew, in press; Dechapanya, 1998).

3.1.1 Economic Assessment of Cogeneration Designs

Table 2.3 shows three of the most important economic indexes (based on the AHP survey) used in this study for the Base Case, Case B, and Case D. For investment, the Base Case is the most favorable design; for expenses, Case D is the best; and for Net Present Value (NPV), Case D is also ranked the highest. This initial survey of economic performance might suggest that Case D would be the best choice, if all of the economic metrics were to be weighted equally, as it is superior for two out of the three economic indexes (ranks 3,1,1 for the indexes). Relative ranking of the other two cases using

TABLE 2.3
Economic analysis of three cogeneration processes

	*Process scenario evaluated**		
Economic indicators	*Base case*	*Case B*	*Case D*
Investment (MM$)	0	2.65	5.11
Expenses (MM$)	74.3	72.6	60.4
NPV† (MM$)	−42.6	−44.5	−36.2

* Base Case = Four natural gas-fired boilers supplying heat. Power purchased from a local utility;
Case B = Wood combustion, handling, and storage equipment addition to facilities described in Base Case;
Case D = Five 1000-kW natural gas-fired diesel engine generators feeding two heat recovery boilers. Two of the fired boilers described in the base case are also used for emergency and peak loads. Excess power is sold to the local utility at their avoided cost.

† NPV assumes no inflation, no taxes, a university project life of 20 years, and a minimum acceptable rate of return (MAR) of 6% (here the MAR represents the cost of borrowing and does not include a safety margin for technological risk).

economic indexes is more difficult, as the Base Case and Case B are 1,3,2 and 2,2,3 respectively; however, the Base Case is slightly favored, compared to Case B, based on these economic indexes.

3.1.2 Environmental Assessment of Cogeneration Designs

In this assessment of environmental impacts, we have restricted the analysis to uncontrolled releases for combustion of the various fuel types to satisfy both on-campus heating requirements and power demand, whether those releases were generated on site or not. In other words, if power was purchased from the local utility, combustion emissions from the off-site power generation facility were included in the impact assessment. Coal was assumed to be the fuel type for utility power releases because it is the major fossil fuel source used in power generation (U.S. DOE, 1998). Releases of volatile organic compounds from boiler units or fugitive sources are negligible compared to the byproducts of fuel combustion, and so these emission sources were neglected. Efficiencies of 35%, 70%, and 85% were assumed for utility power generation, boiler heating, and cogeneration, respectively. Fuel combustion emission factors were obtained from government sources (U.S. EPA, 1998; U.S. DOE, 1998), and are resident to the software tool (EFRAT).

Figure 2.5 shows the relative risk indexes for global warming, smog formation, acid rain, inhalation noncarcinogenic toxicity, and particulate matter

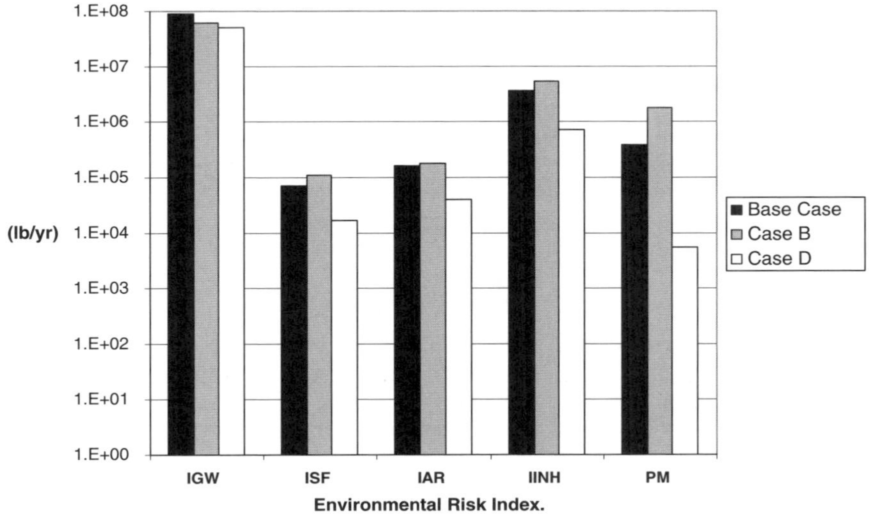

Figure 2.5: Environmental risk indexes for the Base Case, Case B, and Case D. Each index represents the equivalent emission rate of the benchmark compound for the entire process. I_{GW}—global warming index; I_{SF}—smog formation index; I_{AR}—acid rain index; I_{INH}—inhalation noncarcinogenic human toxicity index; PM—particulate matter emission rate.

TABLE 2.4
Process safety analysis of three cogeneration processes

	Process scenario evaluated		
Safety indicators	***Base case****	***Base case B****	***Base case D****
Backup fuel	Fuel oil	Wood	Fuel oil
Quantity† (MM lb/month)	1.16	4.76	1.79
F&EI (severe = 198)	36	99	38
Radius of exposure (ft)	30	84	32

* Primary fuel for Base Case and Case D are natural gas; for Case B is wood.
† Fuel capacity based upon on-site storage (one month capacity).

for each of the cases. Case D has lowest values for each index and for particulate matter and is clearly superior to the Base Case and Case B for environmental impacts. It is interesting to note that Case D is superior to Case B (wood combustion) for global warming, even though the CO_2 emissions were not counted for wood combustion (Case B), as this fuel source is renewable. This outcome is attributable to the high-energy efficiency of cogeneration compared to traditional technologies.

3.1.3 Safety Analysis

Table 2.4 lists the F&EI for each case. Case B, with an F&EI of 99, has a moderate degree of hazard, whereas both the Base Case and Case D have relatively low potential risks for fire and explosion. The moderate degree of hazard for Case B is due to the large inventory of wood chips required. Because none of the F&EI indexes is in excess of 127, one might interpret the Dow F&EI results for this case study to be of limited significance in terms of decision making; however, the fact that a significant fire and explosion hazard does not exist is of enormous value to the process designer.

3.1.4 Multicriteria Decision Analysis of Cogeneration

In addition to the economic, environmental, and safety indicators listed above, the cogeneration process decision was also affected by two additional indicators: societal and process-related. Figure 2.6 shows a complete list of performance indicators and three levels of decision criteria for these indicators. To provide weightings to individual indicators, an internal survey of 12 "experts" was conducted to assess the relative importance of indicators at level 3 (investment versus operating costs, etc.) and the relative importance of classes of indicators at level 2 (economic versus environmental, etc.). Among the survey respondents, three were chemical engineering professors having separate expertise in process design, engineering economics, and environmental risk assessment; six were graduate students participating

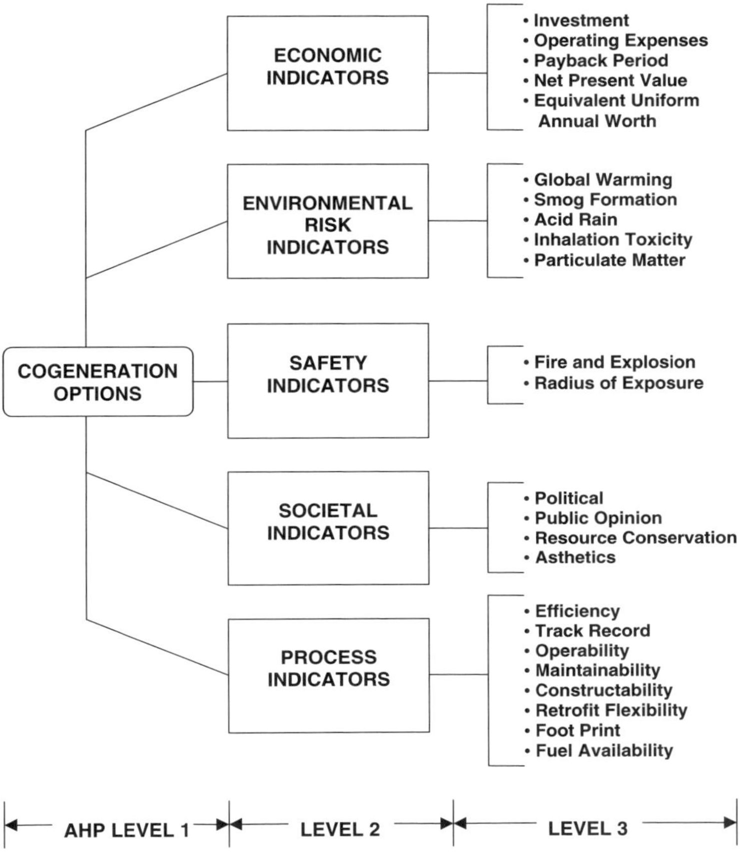

Figure 2.6: Cogeneration process evaluation decision hierarchy.

in the project; and two were nonaffiliated graduate students. One was a staff research engineer holding a doctorate degree in chemical engineering who was knowledgeable in process design.

Figure 2.7 shows the final multicriteria AHP ranking of the three process designs for generation of heat and power on the campus of Michigan Technological University. The data in Figure 2.7 show the average AHP ranking based on the 12 survey responses and the process economic, environmental, and safety indices. The error bars represent the 95% confidence intervals. A higher AHP value means a more desirable process. The Base Case scored the highest (0.44 ± 0.036), Case D (cogeneration) scored intermediate (0.42 ± 0.038), and Case B scored the lowest (0.14 ± 0.02). The error bars for the Base Case and Case D overlap, suggesting that there is no statistical difference between the AHP rankings for these cases. Therefore, neither case

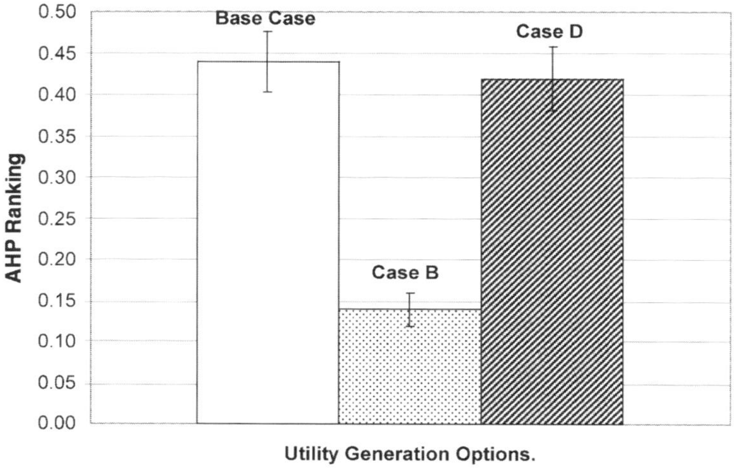

Figure 2.7: AHP results for utility process design for the Base Case, Case B, and Case D. Error bars represent the 95% confidence interval on the AHP results.

should be discarded in favor of higher-ranked options. Case B, on the other hand, is ranked much lower in comparison and would not be considered for subsequent detailed design calculations.

3.1.5 Discussion of Multicriteria Decision for Cogeneration

It is illustrative to consider the factors driving the AHP decision process. Table 2.5 lists the percentage of contributions of each class of indicators (at level 2) on the AHP decision, based on the survey responses. These percentages of contributions were calculated by considering the constant AHP weightings and the case-dependent values of the indexes for each attribute.

TABLE 2.5
Percent contributions of the level 2 components in the AHP analysis

	Percentage of contribution		
Level 2 attributes	*Base case*	*Case B*	*Case D*
Economics	42.78	45.09	45.24
Safety	22.97	17.32	29.28
Process	15.13	14.01	10.02
Environmental	14.39	14.15	9.31
Societal	4.74	9.42	6.15

Economics is the dominant factor in the multicriteria decision analysis, accounting for between 43 % and 45 % of the decision contribution, depending upon design alternative. Economics is followed in decreasing order with safety, process, environmental, and societal. The economic indicators for each heat and power case are assigned the highest weightings as a group, compared to the other categories of indicators; however, the individual economic indexes are not weighed equally. The five economic indicators were assigned weightings according to the AHP survey in the following order of descending importance: payback period > investment > annual operating expenses > net present value > equivalent uniform annual worth.

From Table 2.3, Case D has the lowest expenses and the highest NPV, but the Base Case has the lowest investment. Because of the higher relative weighting for investment, the final AHP decision was influenced by this index more than by NPV and EUAW, which favored Case D more than the Base Case. Referring to the safety indexes on Table 2.4, the Base Case and Case D are comparable and moderately preferred to Case B. Because the safety category of indexes was ranked second only to economics, the favorable safety rating of the Base Case contributed to its positive overall ranking. The environmental indexes favored the Case D (cogeneration of heat and power) over the Base Case and Case B. The societal category of indexes favored the Base Case and Case D over Case B (data not shown) (Herlevich, 1997). Finally, the process category of indexes favored the Base Case over both Case B and D.

If each of the five decision criteria (i.e., categories) and each index within these categories were weighted equally, Case D would be the most promising option; however, when AHP weightings were applied to these metrics, the Base Case scored highest among the three competing design options. This "switching" of the apparent outcome demonstrates the power of the AHP method to influence decisions by including the attitudes, values, and expertise of the decision-makers in the design process.

4 CONCLUSIONS

In this chapter we have outlined a framework for multicriteria assessment of chemical process designs based on economic, environmental, and process safety attributes. We have included Analytical Hierarchy Processing (AHP) as a decision analysis tool for comparing process design options. We have demonstrated the practical use of the framework with a case study for reductions in energy consumption and waste generation involving competing heat and power-producing technologies. Comparisons were made among three competing designs based on separate consideration of economic, environmental, and safety metrics.

The integrated suite of design enhancement software tools is able to significantly extend the capabilities of commercial chemical process simulators by bringing a wealth of useful design information to the process design activity. The information is easily accessible to the design engineer and the tools have been developed with the intent to quickly provide calculated performance indexes. In current projects we are extending the capabilities of the suite of tools by coupling them automatically with a commercial process simulator in such a way that simulator output is automatically entered into the process assessment tools.

ACKNOWLEDGMENTS

This research was supported by the U.S. EPA, National Risk Management Research Laboratory, the Center for Clean Industrial and Treatment Technologies (CenCITT), and by the Department of Energy (DOE), Office of Industrial Technologies, by grant no. CR824506-01. This research has not been subjected to U.S. EPA's required peer and policy review and therefore does not necessarily reflect the views of U.S. EPA, DOE or CenCITT, and no official endorsement should be inferred.

REFERENCES

AIChE (1992). *Guidelines for Hazard Evaluation Procedures*, ed. 2. American Institute of Chemical Engineers, New York.

AICHE (1996). *Inherently Safer Chemical Processes: A Life Cycle Approach*. American Institute of Chemical Engineers, New York.

Alidi, A. S. (1995). A multiobjective optimization model for the waste management of the petrochemical industry. *Journal of Applied Math Modeling* 20:925–933.

Cabezas, H., Bare, J. C., and Mallick, S. K. (1997). Pollution prevention with chemical process simulators: The generalized waste reduction (WAR) algorithm. *Computer & Chemical Engineering* 21:5305–5310.

Carter, W. P. L. (1994). Development of ozone reactivity scales for volatile organic compounds, *Journal of the Air and Waste Management Association* 44:881–899.

Crowl, D. A., and Louvar, J. F. (1990). *Chemical Process Safety: Fundamentals and Applications*. Prentice-Hall, Englewood Cliffs, NJ.

Dechapanya, W. (1998). Modifying Analytic Hierarchy Processing for Multi-Criteria Chemical Process Analysis of Economic and Environmental Performance, Master of Science Thesis, Michigan Technological University, Houghton, MI.

Dechapanya, W., Rogers, T. N., Baker, J. R., and Radecki, P. P. (1999). Tools and Concepts for Integrating Environmental Considerations Into Product and Process Design Decisions. *Tools and Methods for Pollution Prevention* (Subhas Sikdar). Kluwer Academic Publishers, Dordrecht, The Netherlands.

Dow Chemical Company (1994a). *Dow's Fire and Explosion Index Hazard Classification Guide*, ed. 7. American Institute of Chemical Engineers, New York.

Dow Chemical Company (1994b). *Dow's Chemical Exposure Index Guide*, ed. 1. American Institute of Chemical Engineers, New York.

EF, Energy Facts (1992). Energy Information Administration, U.S. Department of Energy, DOE/EIA-0469(92).

Finnie, G. R., Wittig, G. E. and Petkov, D. I. (1993). Prioritizing software development productivity factors using the analytic hierarchy process. *The Journal of Systems and Software* 22:129–139.

Heijungs, R. (1992). Environmental Life Cycle Assessment of Products: Backgrounds—October 1992. Netherlands Agency for Energy and the Environment (NOVEM) National Institute of Public Health and Environmental Protection (RIVM).

Herlevich, J. A., Jr. (1997). Conceptual Design Stage Chemical Process Evaluation to Optimize Economic, Safety, and Pollution Prevention Attributes: A Comprehensive Assessment of Heat and Power Cogeneration Processes. Master of Science Thesis, Michigan Technological University, Houghton, MI.

Hiew, D. S. (1998). Development of Environmental Fate and Risk Assessment Tool (EFRAT©) and Application to Solvent Recovery from a Gaseous Waste Stream. Master Thesis, Michigan Technological University, Houghton, MI.

Hillaly, K. H., and Sikdar, S. K. (1994). Pollution balance: A new methodology for minimizing waste production in manufacturing processes. *Journal of Air & Waste Management Association* 44:1303–1308.

IPCC (1996). *Radiative Forcing of Climate Change*. Climate Change 1995—The Science of Climate Change (WMO), Contribution of Working Group1 to the Second Assessment Report of the Intergovernmental Panel on Climate Change, Cambridge University Press, pp. 65–131.

MacCrimmon, K. R. (1973). An Overview of Multiple Objective Decision Making *Multiple Criteria Decision Making* (J. L. Cochrane and Zeleny), University of South Carolina Press, Columbia, SC, pp. 18–44.

Mackay, D. (1991). *Multimedia Environmental Models, The Fugitive Approach*, p. 257. Lewis Publishers, Inc., Chelsea, MI.

Mackay, D., Shiu, W., and Ma, K. (1992). *Illustrated Handbook of Physical-Chemical Properties and Environmental Fate for Organic Chemicals*, ed. 1, vols. 1–4. Lewis Publishers, Inc., Chelsea, MI.

Mallick, S. K., Cabezas, H., Bare, J. C., and Sikdar, S. K. (1996). A Pollution Reduction Methodology for Chemical Process Simulators. *Industrial & Engineering Chemistry Research* 35(11):4128.

McKone, T. E., and Layton, D. W. (1986). Screening the potential risk of toxic substances using a multimedia compartment model: Estimation of human exposure. *Regulatory Toxicology and Pharmacology*, pp. 359–380.

Min, H. (1994). Location and planning of airport facilities using the analytic hierarchy process. *Logistic and Transportation Review* 30:79–94.

NTP (National Toxicology Program) (1997). NTP Chemical Health & Safety Data, http://ntp-db.niehs.nih.gov/Main_pages/Chem-HS.HTML, National Toxicology Program.

OSHA (Occupational Safety & Health Administration) (1997). http://www.osha-slc.gov/ChemSamp_toc/ChemSamp_toc_by_chn.html, Chemical Sampling Information, Table of Contents by Chemical Name, U.S. Department of Labor, 7/15/97.

Parikh, P. B. (1996). Implementation and Application of the Dow Hazard Evaluation Indices in a Computer-Based Environment. Thesis for Master of Science degree, Department of Chemical Engineering, Michigan Technological University, Houghton, MI.

Parikh, P. B., and Crowl, D. A. (1998). Implementation and Application of the Dow Hazard Evaluation Indices in a Computer-Based Environment. International Conference and Workshop on Reliability and Risk Management, San Antonio, TX.

Pistikopoulos, E. N., Stefanis, S. K., and Livingston, A. G. (1994). A methodology for minimum environmental impact analysis, pollution prevention via process and product modifications. *AIChE Symposium Series* 90(303):139–150.

Reid, R. A., and Christianson, D. C. (1994). Evaluate decision criteria systematically. *Chemical Engineering Progress*, pp. 44–49.

Romero-Hernandez, O., Pistikopoulos, E. N., and Livingston, A. G. (1998). Waste treatment and optimal degree of pollution abatement. *Environmental Progress* 17(4):270–275.

Saaty, T. L. (1982). *Decision Making for Leaders: The Analytical Hierarchy Process for Decisions in a Complex World*. Lifetime Learning Publications, Belmont, CA.

Seider, W. D., Seader, J. D., and Lewin, D. R. (1999). *Process Design Principles: Synthesis, Analysis, and Evaluation*. John Wiley & Sons, New York.

SETAC (1993). Guidelines for Life-Cycle Assessment: A Code of Practice. *Society for Environmental Toxicology and Chemistry Workshop*, Sesimbra, Portugal.

Shonnard, D. R., and Hiew, D. S. Environmental assessment and optimization of chemical process designs I: Assessment framework. *Environmental Science and Technology* (in press).

Stephanis, S. K., Livingston, A. G., and Pistikopoulos, E. N. (1995). Minimizing the environmental impact of process plants. *Computers and Chemical Engineering* 19(S):39–44.

U.S. DOE (1998). Annual Energy Review 1997, Energy Information Administration, Office of Energy Markets and End Use, U.S. Department of Energy, Washington DC, DOE/EIA-0384(97).

U.S. EPA, U.S. Environmental Protection Agency, Integrated Risk Information System (IRIS), List of Substance on IRIS, http://www.epa.gov/ngispgm3/iris/Substance_List.html, 7/15/97.

U.S. EPA (1998). Compilation of Air Pollutant Emission Factors, Volume I, Fifth Edition, AP-42, Air CHIEF CD-ROM, Clearinghouse for Inventories and Emission Factors, Air CHIEF CD-ROM (EFIG/EMAD/OAQPS/EPA), Version 6.0, U.S. Environmental Protection Agency, Research Triangle Park, NC, EPA-454/F-98-007.

Verschueren, K. (1996). *Handbook of Environmental Data on Organic Chemicals*, ed. 3. Van Nostrand Reinhold Company, Cincinnati.

WMO (1992). *Ozone Depletion and Chlorine Loading Potential: Scientific Assessment of Ozone Depletion: 1991*. World Meteorological Organization, Report Number 25, Vol. 1.

CHAPTER 3

An Introduction to Metrics for the Environmental Comparison of Process and Product Alternatives

David W. Pennington,* Gregory Norris,† Theresa Hoagland,‡ and Jane C. Bare‡

**ORISE Research Fellow, Systems Analysis Branch, Sustainable Technology Division, National Risk Management Research Laboratory, U.S. Environmental Protection Agency, Cincinnati, OH 45268*

†Sylvatica, Inc., North Berwick, ME 03906

‡Systems Analysis Branch, Sustainable Technology Division, National Risk Management Research Laboratory, U.S. Environmental Protection Agency, Cincinnati, OH 45268

Management standards, government initiatives, nongovernment organizations, and corporate guidelines emphasize the strong need for the development of methodologies to facilitate the environmental comparison of process design and product alternatives. This chapter helps address the need to support these "burden-based" assessments by presenting a discussion of metrics commonly used for the relative comparison of alternatives in 11 impact categories, such as ozone depletion, smog formation, and acidification. Each section of the chapter provides a historical introduction, an overview of the state of the art, identifies key limitations, and illustrates one common approach. The illustrations are based on two alternatives for the production of 1,4-butanediol (BDO), considered both from a site-specific processing and a product life cycle perspective.

INTRODUCTION

COMPARISON METRICS

Impacts associated with process or product changes can be considered in the context of natural resources, human health, ecological health, social welfare, and economics. All of these factors are taken into consideration under the umbrella of sustainable development. In this chapter we focus on the first three and describe environmental comparison metrics (potentials, potency factors, equivalency factors, or characterization factors) for 11 associated categories: global warming, stratospheric ozone depletion, tropospheric ozone (smog) creation, eutrophication, acidification, human carcinogenicity, toxicological impacts to humans, toxicological impacts to ecosystems, land use, water use, and the depletion of fossil fuel resources. Although these 11 categories are commonly addressed in available environmental comparison case studies, this list is not comprehensive.

Emissions or resource consumption data are typically multiplied by environmental comparison metrics to help represent their relative importance within a given category. A significant number of methodologies are available that can be used to derive environmental comparison metrics in application domains like product life cycle assessment (LCA), process design, and for the regulatory prioritization of chemicals (Swanson and Socha, 1997; Gameson, 1998). These methodologies range in complexity (data intensity, knowledge requirements), comprehensiveness (breadth or scope of representation), sophistication (relevance to and depth of representation of the environmental mechanisms), and accuracy (uncertainty inherent to the model, scenario selected and associated with input data). As a result, the selection of a methodology often remains subjective and strongly influenced by resource availability (in-house knowledge, input data availability).

The metrics illustrated in this chapter are commonly used in relative comparison applications or as target (objective) functions for design optimization. No endorsement of the approaches or suggestion that they are the best available practice is intended. The results do not indicate that an actual impact will occur but often reflect differences in implicit concern, as in the precautionary principle. In some cases the metrics may reflect intrinsic differences at a common midpoint in a cause-effect chain (environmental mechanism), as illustrated in Figure 3.1, for acidification.

There is a tendency to define indicators at common midpoints to help minimize (parameter or data) uncertainty. For example, this midpoint modeling preference would imply that there is no need to forecast specific effects associated with global warming. Comparison can be made in terms of radiative forcing and half-life differences (inherent chemical properties); however, comparison at midpoints may not always account for all factors in a cause-effect chain (increasing model uncertainty), as is the case of acidification metrics in Figure 3.1, and can result in increased complications associated with the resolution of trade-offs, if any, across impact categories like acidification and global warming.

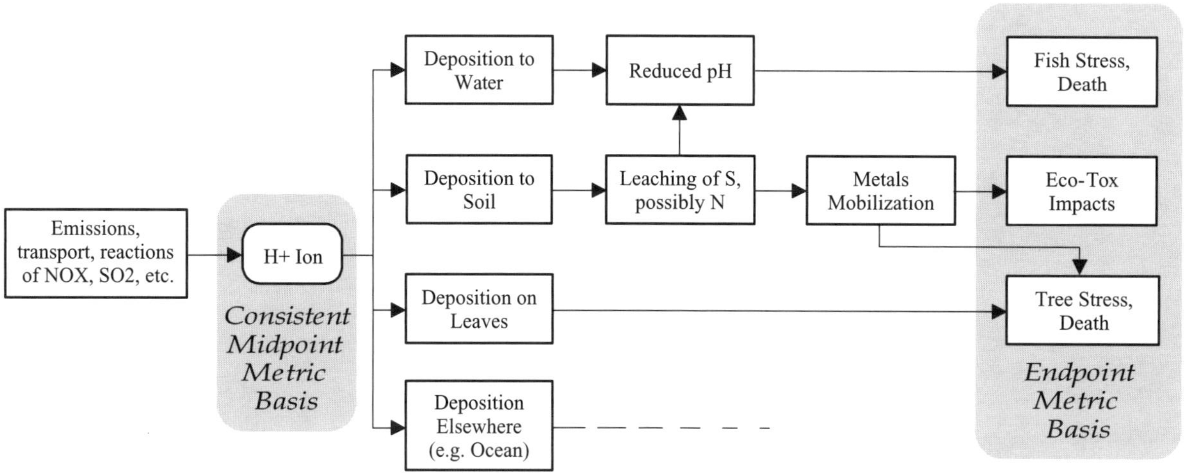

Figure 3.1: Midpoints and endpoints in the simplified cause-effect chain for acidification. (Chemicals can behave differently at each point in the chain.)

Comparison using endpoint metrics may increase the number of categories requiring consideration (e.g., forecasting different types of impacts to human health associated with global warming) but may provide a set of metrics that is more readily comparable across categories using available "valuation" tools due to their observed nature. For example, human health effect endpoints are often associated with different levels of perceived severity. Although considered a developing approach and requiring additional data, the results for different human health endpoints can be combined into a single human health measure using approaches like QALYs (quality-adjusted life years) and DALYs (disability-adjusted life years) (Hofstetter, 1998). Commonly, however, differences in severity in the context of toxicological impacts to human and ecological health are not taken into consideration (effects are inherently considered equal).

HOLISTIC ASSESSMENT

Life cycle assessment (LCA) is one framework for holistically evaluating the inputs and emissions associated with all the stages in a product's life cycle from cradle to grave (raw material acquisition, manufacturing, use, and disposal). Similar to site- or process-specific comparison of process designs, an LCA consists of iterative steps (U.S. EPA, 1995; SETAC, 1996; ISO, 1997). Having established the product system boundaries and aims of the assessment (goal and scope definition), associated resource consumption and emissions must be identified and tabulated in a data sheet (inventory analysis). These inventory data are then considered according to the substance's potential contribution within an impact category, such as global warming and ozone depletion. Associated elements include

- Selection of impact categories, category indicators, and characterization models
- Assignment of inventory analysis results to impact categories (classification)
- Calculation of category indicators (characterization) using associated characterization factors (comparison metrics) derived using characterization models

Calculating the magnitude of category indicator results relative to reference information like geographic- or industry-specific emissions (normalization), sorting or ranking impact categories (grouping), and aggregating indicator results across impact categories (weighting, valuation or multiobjective decision making) remains a controversial topic that cannot be addressed in sufficient detail in this chapter.

It should be noted that the inability to compare design alternatives in terms of actual impacts can be compounded in LCA by additional factors, including

- The inventory data reflecting only a portion of the total inputs/emissions at given sites
- The unknown and changing location of many sites in a life cycle

CASE STUDY

The U.S. EPA (1997) commissioned a life cycle case study to help explore the implications of BDO (1,4-butanediol) production from bioderived feed stocks compared to natural gas. Two system boundaries are addressed in this chapter:

- The processing stage (the gate-to-gate or site-specific perspective, including emissions associated with energy consumption)
- The cradle-to-gate life cycle (from extraction of the raw materials to the final deliverable product)

The product life cycle stages addressed for the BDO alternatives are summarized in Figures 3.2 and 3.3 (U.S. EPA, 1997). As the final product is functionally identical in both life cycles, a cradle-to-gate approach was selected. The functional unit, the performance or service basis used to help consistently compare alternatives, was defined as 1 lb of BDO produced and ready for shipment. This functional unit and the associated inventory data can be adjusted to reflect annual production rates for the U.S., thus being more relevant from a national emissions and resource consumption perspective; however, this multiplication will not influence the results in a relative comparison context and will still not provide an absolute indicator of actual impacts.

It should be noted that the BDO life cycle analysis presented represents a "snapshot in time," and that all the stages are subject to change, particularly the developing bio-based life cycle. It remains open to debate whether such a static comparison is always appropriate, although practical alternatives are currently limited in availability and it is difficult to forecast future developments (improvements in available technologies, novel processes, alternatives in reaction chemistry, solvent substitutions) that may alter the outcome. The merit of such studies lies in the fact that areas for improvement are identified.

UNCERTAINTY AND VARIATION

Before discussing the methodologies for assessing individual impact categories (acidification, eutrophication, photochemical oxidant formation, toxicological impacts to human health and the environment, ozone depletion, global warming, and resource depletion), it is important to address uncertainty and variation. To represent plausible variation (e.g., natural variation in body weights in a population) and uncertainty (e.g., ignorance about the actual value or range for a given parameter, like the body weight of an undefined individual), practitioners have increasingly been moving away

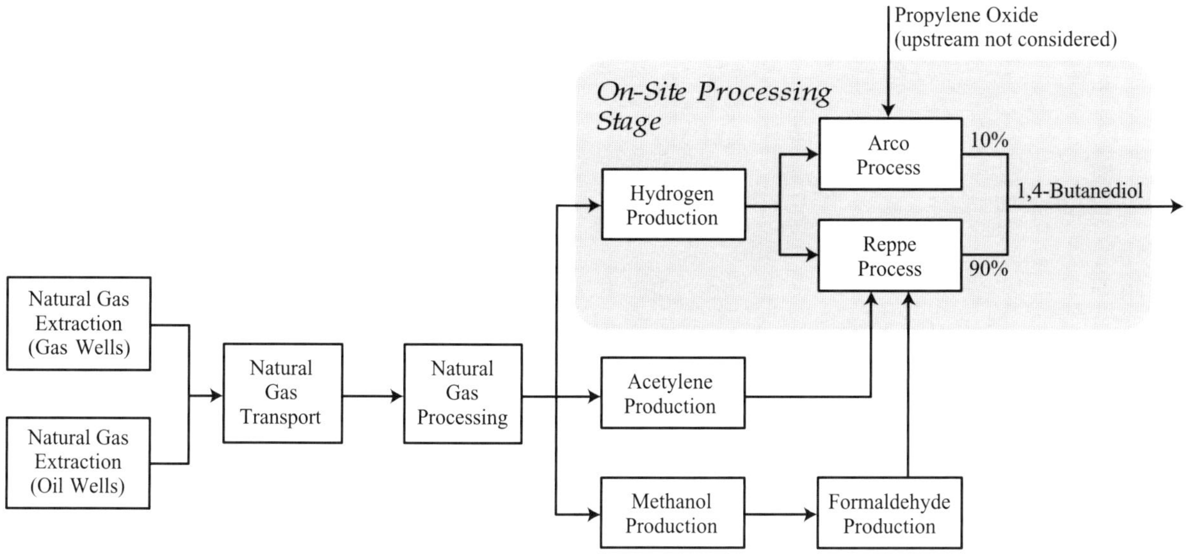

Figure 3.2: Simplified life cycle flow diagram for BDO derived from natural gas (energy consumption and associated processes not shown). Adapted from U.S. EPA (1997).

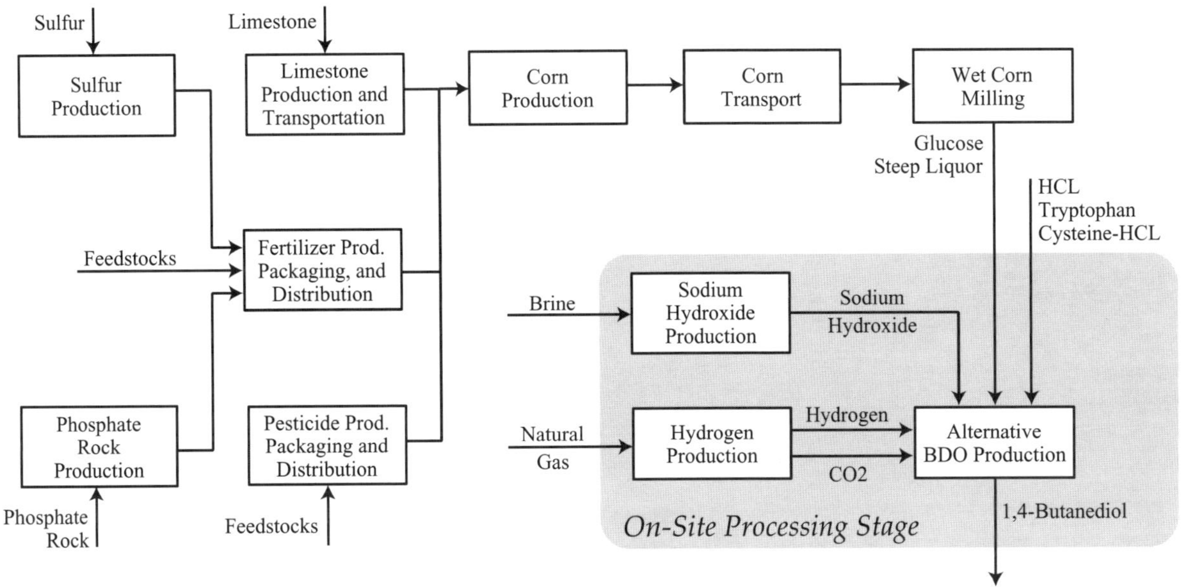

Figure 3.3: Simplified life cycle flow chart for BDO derived from corn glucose (energy consumption and associated processes not shown). Adapted from U.S. EPA (1997).

from metrics based on deterministic (single estimate) approaches and are starting to apply techniques that yield a measure of the probability or likelihood of a particular result. Although not commonly reflected in current relative comparison practices, summary statistics like the median or 95th percentile can be selected from the resultant probability distributions to help make consistent comparisons between alternatives while providing insights into the potential certainty associated with a decision.

The importance of uncertainty and variability associated with parameters (model and input data) in a methodology is often addressed using techniques like Monte Carlo analysis. It is important, however, to remember how well the methodology used to derive a metric represents reality (model and scenario uncertainty). For example, a simplistic approach based on a limited number of parameters may have low uncertainty associated with the input data, thus giving a sense of confidence; however, a more data-intensive method may have lower structural or model uncertainty by better representing mechanisms and interactions within the environment. Unfortunately, the trade-off between data and model uncertainties is often unclear, unquantified, and difficult to evaluate. These uncertainty issues therefore lead to the preference of using more than one approach (Pennington and Bare, 1999) or perspective (Hofstetter, 1998) to gain parallel insights. For clarity and brevity, uncertainty, variability, and perspective are not considered in the illustrative case studies presented in this chapter.

NONCARCINOGENIC TOXICOLOGICAL IMPACTS TO HUMAN HEALTH

A large number of methodologies have been proposed to derive relative comparison merits that help screen and rank chemicals or emissions in the context of noncarcinogenic toxicological impacts (Swanson and Socha, 1997; Pennington and Yue, 2000). Chemical-specific acute and worker health impacts are controlled to a large extent. Therefore, the focus in this chapter is directed toward comparison in the context of long-term toxicological impacts on a regional or global scale (where intermedia transport and degradation, not only dispersion, can control exposure to a large extent).

As expressed in the introduction, and unlike typical risk assessment, the metrics are often based on implicit concerns or "burdens" and do not provide measures of actual (as opposed to potential) impacts. Particularly in an LCA, the resultant aggregated scores are based on releases that occur in different locations, at different background concentrations, and can reflect a number of types of effect endpoints that differ in terms of severity (e.g., reproductive, developmental, neurotoxicological, and immunotoxicological effects). Although approaches like QALYs and DALYs have been proposed to help account for differences in endpoint severity (Hofstetter, 1998), often no distinction is made. All effects are inherently assumed equally important, an assumption that is sometimes debated. Population densities are similarly not always addressed, although consideration at the population level can help take into account differences in scale (e.g., number of people affected by global warming vs. exposure to toxic chemicals with a low long-range transport capability).

Pennington and Yue (2000) discussed available metrics and their limitations in a hierarchical framework that provides a distinction in the context of long-term toxicological impacts. The hierarchy of approaches was based on the degree of representation and level of sophistication of associated environmental mechanisms (model structure). It was therefore stated that the suitability of an approach in the hierarchy also depends heavily on the relative environmental behavior of the chemicals considered, the quality and consistency of available data, the comprehensiveness of the model, and hence the ability of more resource-intensive techniques to actually provide an improvement in discrimination.

Approaches like the Waste Minimization Prioritization Tool (WMPT) (U.S. EPA, 1998; U.S. EPA, 1997; Pennington and Ralston, 1999) in Figure 3.4

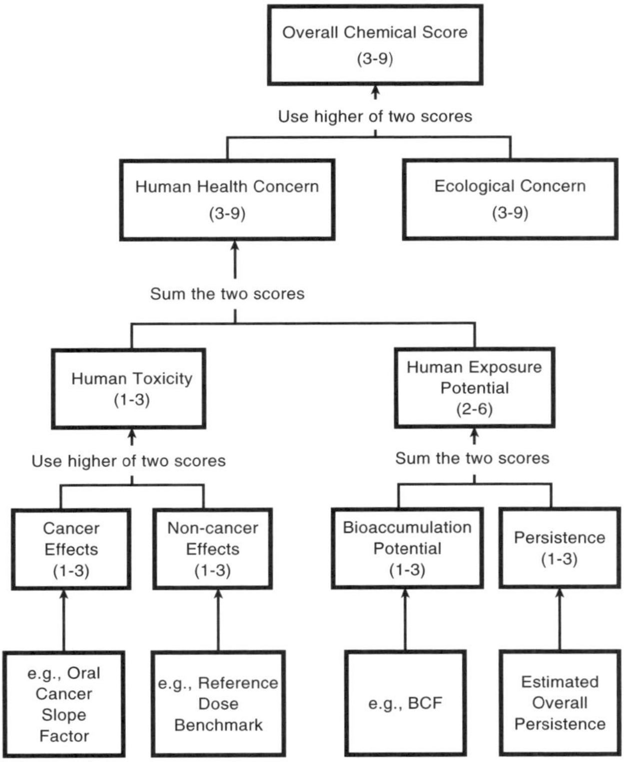

Figure 3.4: Schematic summary of the WMPT approach for human health (Pennington and Ralston, 1999).

can provide categorical insights into the relative concern associated with chemicals and possibly emissions (Pennington and Bare, 1999). Implicit comparison measures considered in the derivation of a score often include degradation half-lives in specific media, overall persistence (a prediction of the overall half-life of a chemical in the environment that accounts for intermedia transport), the potential to bioaccumulate, and hazard (considered here to be a measure of toxicological potency). These implicit measures are largely based on inherent chemical properties, like the octanol-water partitioning coefficient and vapor pressure. The resultant broad statements are sometimes considered more defensible than attempts to more precisely define relative concern, particularly when faced with undefined levels of model uncertainty and comprehensiveness; however, "binning" (categorical classification as high, medium, and low concern based on "cut-off" or "fenceline" criteria) is used in some straightforward approaches like the WMPT. Among other methodological issues, binning can result in misleading distinctions between similar chemicals (the "border-effect" that results in distinction between chemicals close to the cut-off criteria) and even possible misclassification (a chemical that marginally fails a cut-off criteria using low quality data).

There is some debate whether categorical scores should be used in the comparison of emissions. This debate hinges on whether the categorical scale is in the form of interval variables (numerical values quantitatively associated with each categorical level, e.g., discrete intervals of temperature) or ordinal variables (a qualitative ordering without a numerical value associated with each categorical level, e.g., car sizes like compact and regular). If the categories are ordinal, then they should not be directly combined with emissions data, although this obstacle may be overcome for certain applications, like national prioritization by categorically distinguishing between release volumes and combining this information with the other factors. If the categories are interval variables, however, then they can be directly combined with emissions data. Modified versions of the WMPT may provide interval variables that can be combined with logarithmic emissions data (Pennington and Bare, 1999).

To help avoid some of the limitations associated with categorical scales, a significant number of approaches have been presented that provide continuous metrics. Many continuous scales are still dependent on implicit measures combined by using subjective algorithms, commonly termed *scoring and ranking approaches* (Swanson and Socha, 1997; Pennington and Yue, 2000). Generic multimedia models that combine fate, exposure, and toxicological considerations, as in Figure 3.5, have been commonly adopted to derive quantitative metrics with a perception of greater confidence than these scoring and ranking approaches, due to a more explicit representation of environmental mechanisms (Hertwich et al., 1998; Hertwich, 1999; Guinee et al., 1996; Huijbregts, 1999); however, although the level of confidence in a comparison associated with parameter uncertainty can be calculated using probabilistic techniques, underlying model uncertainty is typically not specified. Caution, expert input, and further research are still required when deriving metrics with such models, particularly for metals, nonpolar chemicals, and

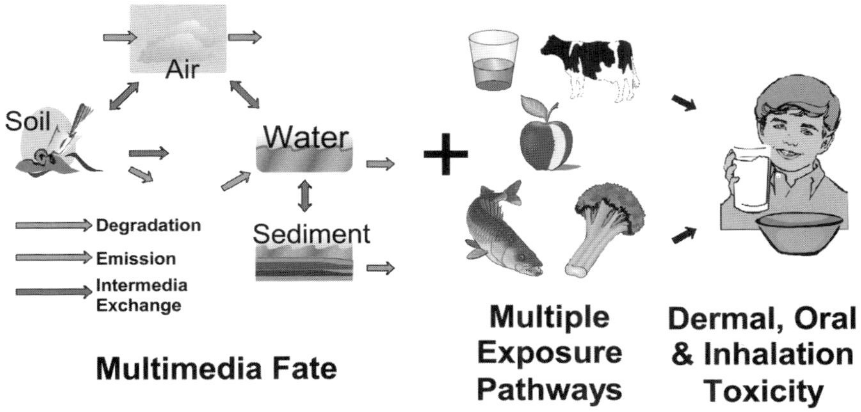

Figure 3.5: Summary of multimedia fate, exposure, and toxicity approach components.

compounds with high octanol-water partitioning coefficients (log $K_{ow} > 5$). It is recommended that different types of approach be used to provide complimentary metrics and insights.

CASE STUDY

The BDO inventory data were multiplied by Toxic Equivalency Potentials (TEPs) (Hertwich et al., 1998; Hertwich 1999) to facilitate comparison in the context of noncarcinogenic burdens (see Table 3.1). TEPs are one example of continuous metrics derived using a multimedia fate, exposure, and toxicity model. Results are presented for the top-ranking chemicals (28 chemicals with scores less than 1×10^{-3} toluene equivalents in all process options are not presented). Scores were not calculated for approximately 1 % of the emissions due to agglomeration in the supplied data set (e.g., as "metals" or "hydrocarbons") however the significance of these agglomerations could be evaluated using conservative estimates. Similarly, using expert judgment, a number of chemicals were considered here to be of no direct concern in the context of direct, noncarcinogenic impacts to human health (e.g., carbon dioxide, iron, chlorides, potassium, sodium, hydrogen chloride, and phosphorus).

As TEPs are limited in availability in the literature, additional values were calculated for 17 chemicals. Nonmeasured data were obtained using the estimation techniques and the data preference hierarchies of the WMPT (U.S. EPA, 1998). In the absence of model-based predictions, best-guess values were used to estimate the toxicological benchmarks [Acceptable Daily Intakes (ADIs)] used to calculate the TEPs. These estimates were based on WMPT noncarcinogenic score equivalents (1 = 0.06 mg/kg/day,

TABLE 3.1
Noncarcinogenic toxicological impact assessment results for the BDO options

	Noncarcinogenic TEPs		*Natural gas-based*		*Bio-based*	
Component	*Air*	*Water*	*Process*	*Life cycle*	*Process*	*Life cycle*
Lead	1,300,000		0.4	0.7	5.6	6.1
Copper compounds	670,000	690,000	2.4	2.5		
Sulfuric acid*	75	126		0.03		0.3
Cadmium		3,600,000		0.08		
Methane*	629		0.02	0.04	5E-03	9E-03
Mercury	220,000	1,200,000		6E-04	0.02	0.02
Nickel compounds	9,400	13,000	0.01	0.01		
Turbufos*	355	27,657				0.01
Chlorpyrifos*	807	6,301				8E-03
Cyanazine	160	110				5E-03
Ammonia	3	0.04	7E-08	2E-05	1E-06	1E-03
Totals			**2.9**	**3.3**	**5.7**	**6.5**

* Estimated using WMPT data and expert judgment.

2 = 0.03 mg/kg/day, and 3 = 0.0006 mg/kg/day). As the range for a WMPT toxicity score of 3 is associated with benchmarks >0.0006 mg/kg/day, the resultant estimate may be optimistic (significantly underestimate the true implicit toxicological potential). Where best-estimate ADIs were used is indicated in Table 3.1.

The totals in toluene-equivalents in Table 3.1 indicate that there is no significant difference between the two BDO life cycles (the scores are within an order of magnitude). A small number of chemicals in each life cycle and the associated processing stage (on-site and energy-related emissions) account for the majority of the impact potential (e.g., lead and copper compounds in the gas-based option). For both options, the consideration of the processing stage alone would not have resulted in the identification of all the trade-offs between the BDO alternatives. For example, lead (associated with energy use), mercury (water emissions from crude oil extraction and air emissions from mercury cells in the production of sodium hydroxide), and turbufos (from corn production) represent the most significant emissions in the bio-based BDO life cycle; however, the results for metals should also be adopted with caution. The use of high TEP values (the illustrated metric) for metals reflects high inherent concerns that may be a facet of current modeling limitations.

HUMAN HEALTH CARCINOGENIC IMPACTS

Carcinogenic impacts are the focus of considerable public attention and concern. Available metrics used to compare process alternatives in terms of potential contributions to carcinogenic impacts to human health and the associated methodologies are generally analogous to those of noncarcinogenic impacts discussed in the previous section. The main difference with noncarcinogenic impacts is the type of toxicity data used.

Preliminary screening to identify chemicals of concern in a carcinogenic context is often performed using qualitative likelihood measures like the IARC (International Agency for Research on Cancer) or the U.S. EPA weights-of-evidence (WOE). As these WOEs are only compiled for a limited number of chemicals, Quantitative Structure Activity Relationships (QSARs) can be used to provide estimates of the probability of other chemicals being a carcinogen on a 0 to 1 scale; however, the cut-off criteria required to reflect different levels of concern are arbitrary and strongly influenced by perspective (Hofstetter, 1998).

The relative potency of carcinogens usually is represented by slope factors, TD_{50}s, or some similar measure. Available slope factors typically represent the plausible upper bound of increased cancer risk from a lifetime exposure per unit dose in mg/kg/day. Providing the statistical basis is consistent, this is suitable for use in relative comparison applications. TD_{50}s are defined as the chronic dose rate in mg/kg body weight per day that is expected to induce tumors in half the test animals at the end of a standard life span for the species (Gold and Zeiger, 1997). The potency basis selected in a study should be consistent and may influence the result, although, for example, TD_{50}s are strongly correlated to the inverse of the slope factor (Krewski et al., 1993).

Due to the lack of prediction models for carcinogenic potency data, surrogate values based on expert judgment can be adopted for suspected carcinogens in the absence of test data. In this and analogous cases, it is beneficial to perform a sensitivity analysis to identify the significance of the associated result. WOE should not be used to imply potency, as this is only a measure of the probability that a chemical is a carcinogen of concern to humans.

It is clear from models that incorporate time-dependent dosing that the time at which exposure to carcinogenic substances occurs can be an important determinant of the level of risk (Krewski et al., 1989). Similarly, not all carcinogenic risks are combinatory and they can differ significantly in terms of severity. Nevertheless, these factors are not currently captured in most available comparison applications. For these and the reasons previously described, the agglomerated results do not indicate the degree of actual risk and are only suitable for use in a relative comparison, possibly precautionary, context.

CASE STUDY

The inventory data for the BDO alternatives was multiplied by TEPs (Hertwich et al., 1998; Hertwich, 1999) that facilitate comparison in the

TABLE 3.2
Carcinogenic impact assessment results for the BDO alternatives

	Carcinogenic TEPs		*Natural gas-based*		*Bio-based*	
Component	*Air*	*Water*	*Process*	*Life cycle*	*Process*	*Life cycle*
Cyanazine	11	9.2				3E-04
Lead	15		5E-06	8E-06	7E-05	7E-05
Atrazine	0.47	0.013				3E-05
Kerosene	1		2E-07	4E-07	3E-06	4E-06
Arsenic		950		2E-06		
Formaldehyde	0.003	0.00029	2E-07	2E-07		
Benzene	1	1		1E-07		
Cadmium		1.7		4E-08		
Nickel Compounds	0.83		9E-09	3E-08		
Totals			**5E-06**	**1E-05**	**7E-05**	**5E-04**

context of carcinogenic impacts (see Table 3.2). Only chemicals having an IARC WOE of 2A, 2B, or 1, or a U.S. EPA WOE of B2, B1, or A were considered (a value judgment strongly influenced by perspective (Hofstetter, 1998)). Less than 0.5% of the mass in each inventory set was excluded due to data agglomeration (e.g., reported as "metals," "hydrocarbons"). The results for the gas- and bio-based BDO life cycles suggest approximately two orders of magnitude difference in relative carcinogenic impact potential; however, the potential differs between the processing stages by less than an order of magnitude. As these values represent estimates without confidence limits from validated models, less than an order of magnitude differences can be considered insignificant. The majority of the potential carcinogenic concern in the bio-based option is associated with corn production (cyanazine and atrazine) and grid energy use (lead). Nevertheless, for both BDO alternatives, the results for metals should be adopted with caution.

ECOLOGICAL TOXICOLOGICAL IMPACTS

Public awareness is strongly driving efforts to protect ecosystems, historically through the control of the emissions of specific chemicals but now also in terms of contributions to potential or unforeseen long-term problems (the doctrines of the precautionary principle); however, our ability to generically compare process alternatives in the context of potential ecotoxicological impacts is limited by the availability of defendable models and required input data.

A multimedia fate, exposure, and toxicity model approach was demonstrated by Guinee et al. (1996) to facilitate comparison in terms of impacts to aquatic ecosystems, soil ecosystems, and top predators (fish and worm eating birds and mammals). This is somewhat analogous to the approach presented for human health in the previous two sections. The effect potency measure differs and is typically defined for a multiple group of species in a given medium as the concentration below which a certain percentage of species are protected.

The underlying approaches used to derive the effect potency measures for ecosystems are often adopted directly from regulatory screening applications and can result in significant inconsistency in a relative comparison context. The measures can incorporate unknown and inconsistent degrees of conservatism introduced during data extrapolation (e.g., estimating a chronic measure from acute data and across species). When sufficient chronic data are not available, probabilistic extrapolation factors and allometric correlations should therefore be used to help provide consistent potency estimates with associated confidence limits (Jager et al., 1997). QSARs, like those found in the ECOSAR (Meylan et al., 1998) and ASTER (Russom, 1994) suites, can similarly be used to fill in data gaps, but these estimates should not be multiplied by "safety factors" to account for uncertainty in data quality. Uncertainty in a relative comparison should be addressed independently, qualitatively, or quantitatively to provide a confidence indicator associated with the comparison basis. It should be noted that the common practice of comparison in terms of acute data alone due to its availability can provide consistency, but the inherent (model and scenario) uncertainty of the relative results in a chronic context is typically high and not quantified.

Combined fate, exposure, and toxicity modeling efforts are generally considered to represent the state of the art (see human health impact sections) and the future in terms of deriving ecological toxicological comparison metrics; however the current levels of sophistication, comprehensiveness, and consistency of these models are the source of debate and the driver of significant research efforts. As a result, the need remains to use less complex screening approaches with a potentially broader scope to provide parallel insights.

The WMPT (U.S. EPA, 1998) generates two scores, one for human health and one for aquatic ecosystems (see Fig. 3.4) and the previous noncarcinogenic human health section. The aquatic score algorithm mimics typical expert judgment adopted widely in chemical screening efforts and is based on the separate consideration of persistence, bioaccumulation, and aquatic toxicity (PBT). Data are selected in terms of defined preference hierarchies with associated guidelines to derive scores based on each data type. It has been demonstrated for human health that the WMPT may provide interval variables (numerical values quantitatively associated with each categorical level, e.g., discrete intervals of temperature) that logarithmically reflect the relative concern associated with chemical emissions (Pennington and Bare, 1999). The WMPT is likely to provide a similar logarithmic (base 10) scale for aquatic ecosystems and hence can be used to provide parallel insights (checks) to approaches perceived to be more sophisticated.

CASE STUDY

The BDO inventory data are evaluated here in terms of inherent concern in the context of toxicological impacts to aquatic ecosystems (see Tables 3.3 and 3.4). The illustrated approach is predominantly based on the 1998 WMPT methodology (U.S. EPA 1998). Scores are presented for each of the three key parameters (P, B, and T) for organic chemicals scoring a total of 5 or higher (implies a 2 in at least two categories or a 3 in at least one). This subjective screening basis eliminated 13 chemicals scoring 3 or 4 from requiring further consideration. Model-based approaches that should be considered

TABLE 3.3
Organic chemical inventory data (lb/lb BDO) ranked by WMPT PBT score

Component	*P Score*	*B Score*	*T Score*	*Aquatic score*	*Natural gas-based*		*Bio-based*	
					Process	*Life cycle*	*Process*	*Life cycle*
Chlorpyrifos	3	3	3	9				8E-06
Atrazine	3	1	3	7				6E-05
Turbufos	2	2	3	7				1E-05
Fonofos	3	1	3	7				5E-06
Phenol	3	1	3	7	8E-09	1E-08	1E-07	1E-07
Octane	1	3	3	7		5E-06		
Alachlor	2	1	3	6				7E-05
Cyanazine	2	1	3	6				3E-05
Benzene	2	1	3	6		1E-07		
Carbon monoxide	3	1	1	5	3E-03	3E-03	1E-02	2E-02
Phosphorus	1	1	3	5				1E-04
MEA	2	1	2	5	1E-05	1E-05	3E-05	3E-05
Methane	3	1	1	5	3E-05	6E-05	8E-06	1E-05
Kerosene	2	1	2	5	2E-07	4E-07	3E-06	4E-06
Chlorine	1	1	3	5			8E-07	2E-06
Formaldehyde	1	1	3	5	9E-05	1E-04		
Acetone	3	1	1	5	3E-05	3E-05		
Acetylene	2	1	2	5	6E-05	6E-05		
Toluene	1	1	3	5	8E-05	8E-05		
Hexane	1	1	3	5		6E-06		
Heptane	1	2	2	5		8E-06		
Paraffins (C5-C10)	1	1	3	5		1E-06		
Totals					**3E-03**	**4E-03**	**1E-02**	**2E-02**

TABLE 3.4
Metals inventory data (lb/lb BDO) ranked by toxicity potency score

Component	Toxicity score	*Natural gas-based*		*Bio-based*	
		Processing	*Life cycle*	*Processing*	*Life cycle*
Mercury	3		5E-10	1E-07	1E-07
Thallium	3		5E-10		
Arsenic	3		3E-09		
Cadmium	3		2E-08		
Iron	2	5E-04	8E-04	7E-03	7E-03
Lead	2	3E-07	5E-07	4E-06	5E-06
Boron	2		1E-06		
Zinc	2		3E-08		
Chromium compounds	1–3		2E-09		
Copper compounds	1–2	4E-06	4E-06		
Nickel compounds	1–2	8E-07	8E-07		
"Metals"	1–3	8E-09	1E-08	1E-07	1E-07
Totals		**5E-04**	**8E-04**	**7E-03**	**7E-03**

in parallel for preliminary screening, like Guinee et al. (1996), were illustrated in the previous sections in the context of human health. Metals and metallic compounds were considered in terms of toxic potency alone, as the WMPT PBT algorithm is sometimes considered to be less applicable for these substances.

Less than 2% of each inventory set was not scored due to data agglomeration into groups like "hydrocarbons" and "oil and grease"; however, agglomerations can be considered using conservative surrogates to estimate their significance. This is demonstrated for groups of metals, which can be assigned a toxicity score of 3 in Table 3.4. Compounds and groups considered here to be of no direct toxicological concern to ecosystems like acids, chlorides, and carbon dioxide were not addressed in this impact category.

Chlorpyrifos is released to the environment during the corn production stage of the bio-derived BDO process and is identified as the organic chemical of highest inherent concern (score = 9); however, the metric for carbon monoxide is 5 (due to its high persistence) but the emission accounts for the majority of the inventory for the bio-based BDO life cycle. A similar analysis can be performed for the gas-based process, although fewer chemicals of intrinsically high concern were reported. For both BDO alternatives, on-site processing (which includes emissions associated with energy consumption) accounts for a large proportion of the releases, but the majority of implicitly high concern chemicals (scores 6 to 9) would not be captured considering this stage alone.

The metals are considered to be predominantly of high inherent toxicological concern. The gas-based BDO option again has lower emissions from both a life cycle and a processing stage perspective. In addition to bioaccumulation concerns, mercury dominates the high toxicity category for both processes, but the associated releases only account for a small fraction of the total metal emissions. The use of more sophisticated models for the comparison of metals may yield different insights, however.

ACIDIFICATION

The emission of acidifying substances can have a variety of impacts on aquatic and terrestrial ecosystems through multiple pathways (see Fig. 3.1). After atmospheric reactions and transport, impacts begin with either wet or dry deposition of sulfur or nitrogen ions on leaves, soil, or water. Leaf exposure has been linked to tree stress and forest die-back. Direct acid deposition to water can lower the pH impacting biota and fauna directly and indirectly (e.g., mobilization of metals that result in a toxic effect). Sulfur ion deposition to soils is generally leached by rainwater, whereas deposited nitrogen may be retained in the soil (up to a point and depending on a variety of factors). This leaching of acidifying substances further contributes to the potential lowering of receiving water pH.

The basis for the commonly adopted acidification metrics reported in Heijungs et al. (1992) and Hauschild and Wenzel (1997) is the number of hydrogen ions that can theoretically be deposited somewhere per unit mass of pollutant X released, a midpoint metric (see Table 3.5). The metrics are developed relative to one of the acidifying substances, sulfur dioxide (SO_2 equivalents). This number is given by taking into account the stoichiometric coefficient ν in the chemical reaction equation

$$X + \cdots \rightarrow \nu H^+ + \cdots .$$

From this basis it is assumed that one mole of SO_2 will produce two moles of H^+; one mole nitrogen oxide compounds (NO_x) will produce one mole of H^+; and one mole reduced nitrogen compound (NH_x) will produce one mole H^+. As pollutant releases are specified in mass of emissions rather than moles, the coefficient ν must be divided by the molecular weight of the pollutant.

This *location-independent ion release* approach was cited by the SETAC Europe Working Group on LCIA (Udo de Haes et al., 1999a, 1999b) as possibly the best available practice for acidification characterization in the short term. Of course, a variety of factors are neglected that either influence or are influenced by ion formation (Fig. 3.1). Atmospheric chemistry and transport upstream of deposition govern which receptor (leaf, soil, freshwater, or ocean) receives the ion and in what form. Downstream of deposition are factors such as the acid-neutralizing capacity and current acidification

TABLE 3.5
Acidification characterization factors and results comparison

Method: Flow	*Comparison metrics* A	B	C	D	*Characterization results (Bio-based minus natural gas-based)* *Processing stage* A	B	C	D	*Life cycle (cradle-to-gate)* A	B	C	D
NO_x	0.7	0.7	0.35	0	5E-3	5E-3	3E-3	0	8E-3	8E-3	4E-3	0
HCl	0.88	0.88	0.88	0.88	0	0	0	0	1E-7	1E-7	1E-7	1E-7
Ammonia	1.88	1.88	0.94	0	6E-7	6E-7	3E-7	0	8E-4	8E-4	4E-4	0
Sulfuric acid	0	0.65	0.65	0.65	0	0	0	0	0	5E-7	5E-7	5E-7
SO_x	1	1	1	1	0.02	0.02	0.02	0.02	0.03	0.03	0.03	0.03
NO	1.07	1.07	0.535	0	0	0	0	0	2E-3	2E-3	1E-3	0
Nitric acid	0	0.51	0.255	0	0	0	0	0	0	3E-6	1E-6	0
Totals					**0.029**	**0.029**	**0.026**	**0.024**	**0.038**	**0.038**	**0.032**	**0.027**

Method A: (Heijungs et al., 1992); Method B: (Hauschild and Wenzel, 1997); Method C: (Hauschild and Wenzel, 1997), with N influence at 50%; Method D: (Hauschild and Wenzel, 1997), with N influence at 0%.

stress level of the receiving environment, the possible soil retention of nitrogen, the process of metals mobilization, and the responses of trees, fish, and other ecology to pH and metals. Several more sophisticated methods, which attempt to take into account one or more of these causal links, have been demonstrated and are under development (Potting and Hauschild, 1997; Potting et al., 1998). These more sophisticated approaches require information regarding the location of emissions within the life cycle and so are not demonstrated in the present case study example.

CASE STUDY

In the present case study we demonstrate the use of four different sets of comparison metrics (Table 3.5). The first two correspond to location-independent ion release approaches. To address the issue of soil nitrogen saturation and incomplete leaching of nitrogen anions, Finnveden et al. (1992) suggested that minimum and maximum scenarios of N deposition effects be characterized simultaneously in a sensitivity analysis to bracket actual results. The maximum N-influence characterization analysis uses the default (first two) metrics reported in Table 3.5, whereas minimum N-influence metrics are zero for nitrogen compounds (no leaching of nitrogen). Seppala (1997) applied alternative N leaching percentages of 20% and 100% to bracket acidification results based on reported N leaching percentages ranging from 15% to 40% in Scandinavia and central Europe.

Here we illustrate the use of N leaching percentages of 0% and 50%, together with the 100% leaching assumption implicit in the default factors from Heijungs et al. (1992) and Hauschild and Wenzel (1997).

Table 3.5 displays the acidification characterization results for bio-derived BDO minus the characterization results for natural gas-based BDO for all four characterization methods. Results for processing stage emissions are distinguished from those of the (cradle-to-gate) life cycle. Values greater than 0 imply that the bio-derived BDO has greater expected acidification impacts, which is the consistent result regardless of method or basis selected. Sulfur oxide emissions are consistently seen to have the greatest influence on the results for acidification in this example.

EUTROPHICATION AND NUTRIFICATION

Nutrification can occur when mineral nutrients are added to soil or water, resulting in increased biomass growth. An important endpoint in the cause-effect chain is oxygen depletion, resulting in eutrophication associated with the decomposition of dead algae. A similar cause of oxygen depletion is the release of organic matter, which is degraded through biotic (biodegradation) and abiotic (hydrolysis and photolysis) processes.

Heijungs et al. (1992) considered nitrogen and phosphorus as the two major limiting nutrients of importance in the context of eutrophication. Carbon and trace elements, such as magnesium and silicon, were reported to be rarely the constraining nutrients. The potential role of potassium was cited as unclear and therefore not included in commonly adopted eutrophication characterization metrics. The approach of Heijungs et al. (1992) is based on the average composition of algae ($C_{106}H_{263}O_{110}N_{16}P$). Any nutrient X containing either phosphorus (P) or nitrogen (N) is assumed to provide N or P needed for the formation of biomass on the basis of the relationship

$$X + \cdots \rightarrow \nu(C_{106}H_{263}O_{110}N_{16}P) + \cdots .$$

To provide units in terms of mass, the stoichiometric coefficient ν has to be divided by the molecular weight of the substance concerned. Phosphate (PO_4^{3-}) was selected as a reference substance (yielding phosphate equivalents). Microorganism cell composition (e.g., $C_{60}H_{87}O_{23}N_{12}P$) (Tchobanoglous and Burton, 1991) is not addressed, for example.

The link between mineral and organic loading is commonly based on Chemical Oxygen Demand ((COD)—derived using experimental oxidation techniques) and the assumption that 138 moles of O_2 is required in the complete breakdown of one mole of algal biomass ($C_{106}H_{263}O_{110}N_{16}P$) (Heijungs et al., 1992; Lindfors et al., 1995). This yields a characterization metric in terms of COD of 0.022 (see Table 3.6). Direct organic loading, however, can be quantified in terms of theoretical oxygen demand [(ThOD)—calculated from the chemical structure and stoichiometry for complete degradation] and biochemical or biological oxygen demand [(BOD_5)—experimentally measured; 5 denoting the test period in days], all

TABLE 3.6
Eutrophication and nutrification metrics and results comparison

					Characterization results (Bio-based minus natural gas-based)							
	Comparison metrics				*BDO processing*				*Life cycle (cradle-to-gate)*			
Method: Flow	*A*	*B*	*C*	*D*	*A*	*B*	*C*	*D*	*A*	*B*	*C*	*D*
NOx	0.13	0.13	0.13	0	1E-3	1E-3	1E-3	0	1E-3	1E-3	1E-3	0
Ammonia	0	0.35	0.35	0	0	1E-7	1E-7	0	0	2E-4	2E-4	0
NO	0.2	0.2	0.2	0	0	0	0	0	5E-4	5E-4	5E-4	0
BOD_5 -w*	**0.022**	**0.022**	**0.022**	**0.022**	**3E-5**	**3E-5**	**3E-5**	**3E-5**	**4E-5**	**4E-5**	**4E-5**	**4E-5**
Phosphorus -w	3.06	3.06	0	3.06	0	0	0	0	4E-4	4E-4	0	4E-4
Ammonia -w	0.35	0.35	0.35	0	0	0	0	0	3E-6	3E-6	3E-6	0
Nitrates (as N) -w	0	0.1	0.1	0	0	0	0	0	0	2E-4	2E-4	0
COD -w	0.022	0.022	0.022	0.022	3E-8	3E-8	3E-8	3E-8	3E-8	3E-8	3E-8	3E-8
Totals					**1E-3**	**1E-3**	**1E-3**	**2E-5**	**2E-3**	**3E-3**	**2E-3**	**4E-4**

* -w denotes releases to water.
Method A: (Heijungs et al., 1992); Method B: (Lindfors et al., 1992); Method C: (Lindfors et al., 1992) to N-limited environments; Method D: (Lindfors et al., 1995) to P-limited environments.

expressed in mass units of O_2. COD is an experimentally obtained estimate for ThOD, whereas BOD_5 is a measure of the amount of oxygen biochemically consumed in tests over a 5-day period (e.g., BOD_5 is expected to be approximately 0.5 COD for a chemical with a 5-day half-life). The differences in measurement basis raises the important issue of which should be used. BOD_5 is limited to a 5-day measurement, so it may underestimate the ultimate oxygen demand. On the other hand, ThOD and COD can overestimate the oxygen demand experienced within a given system (e.g., a river) if the degradation rate is low and advective removal high.

Some report BOD_5 whereas others report COD. This reporting forms the basis for life cycle inventories and LCI databases. Thus, the total life cycle inventory for a system of many unit processes is likely to report both.* Summing the BOD_5 and COD of the effluent stream for the same unit process would be double-counting. In the presented analysis we include default characterization factors for both BOD_5 and COD, as in many cases only one is reported for each unit process.

Other issues surrounding eutrophication and nutrification equivalency analysis include (1) differences among regions in terms of which nutrients are indeed limiting and/or eutrophying, (2) differences in the sensitivity of the receiving environment, (3) chemical fate, and (4) the consideration of

*This is the case, for example, in the LCI data for United States energy processes from the Franklin Associates databases, which have been used as the basis for several U.S. EPA studies and reports.

microorganisms. Samuelsson (1993) called for a type of multiscenario analysis in the context of aquatic eutrophication, where different sets of comparison metrics address different potential circumstances. For example, in P-limited aquatic environments (typically freshwater), the metrics for N-related compounds should all be zero (Samuelsson, 1993). Likewise, in N-limited environments (typically brackish/saltwater), the metrics for P-related compounds should be zero. Lindfors et al. (1995) suggested further sensitivity analyses or scaling factors for the N-related air emissions to take into account that some, but not all, of the nitrogen released to air will reach aquatic systems, either by direct deposition or after leaching from terrestrial systems. Similar arguments and modifications can be made to account for the fraction of organic emissions remaining or entering the aquatic environment (e.g., possibly using multimedia fate models). The original eutrophication metrics of Heijungs et al. (1992) are therefore considered "maximum impact" factors, as they assume that all N emissions to air reach water, as well as that both N and P are limiting.

CASE STUDY

In the present case study we demonstrate the use of four different sets of metrics for eutrophication: (A) the "maximal impact" factors of Heijungs et al. (1992); (B) the expanded set of "maximal impact" factors of Lindfors et al. (1995); and two limiting case scenarios: (C) an N-limited receiving environment (e.g., saltwater) and (D) a P-limited receiving environment (e.g., freshwater). The results in Table 3.6 first demonstrate that using the maximum impact methods, NO_x emissions comprise the bulk of the relative differences in total eutrophication impacts, whether processing only is examined or the life cycle. The results also illustrate that reducing the influence of the emissions to air (NO_x, ammonia, and NO) to account for incomplete transport to water would not change the final outcome that the bio-derived alternative has a higher eutrophication impact score.

Note that neither the addition of flows using the Lindfors et al. approach nor choosing the N-limiting case (zeroing the coefficient for phosphorus) has much effect on the relative score between natural gas-derived BDO and bio-derived BDO; however, the selection of a strictly P-limited environment (freshwater) has a *major* impact on their relative scores in this case study example (the score for the petroleum-based BDO becomes very close to zero). The ratio of the bio-based alternative's to the petroleum-based alternative's eutrophication result (an alternative representation of the differences that is not presented in Table 3.6) goes from roughly 3.0 with the first 3 methods to nearly 100,000 in the P-limited model.

PHOTOCHEMICAL OXIDANT FORMATION

The midpoint of comparison metrics in the cause-effect chain associated with photochemical oxidant (smog) respiratory impacts to human health is the formation of ozone molecules (O_3) in the troposphere. Rates of ozone formation in the troposphere are governed by complex chemical reactions, which are influenced by ambient concentrations of oxides of nitrogen (NO_x), concentrations of volatile organic compounds (VOCs), the type of VOCs, temperature, sunlight, and advective flow rates.

There are over 100 different types of VOCs emitted to the atmosphere, and they can differ by up to a factor of 200 in terms of their estimated relative influence on photochemical oxidant formation (Carter 1994). Beginning in the 1980s, atmospheric chemistry computer simulation models have been used in attempts to characterize the differences among VOCs in their potential to lead to photochemical oxidant formation. Derwent and Jenkin (1991) simulated three different trajectories (Germany to Ireland, France to Sweden, and within the UK) over multiple days. The potential contribution of a given VOC to ozone formation was gauged by comparing simulated ozone production with and without the presence of each test VOC. Using these results, Heijungs et al. (1992) presented calculated average POCP (photochemical oxidant creation potential) values for 23 alkanes, 7 halogenated hydrocarbons, 11 alcohols, 3 ketones, 5 esters, 2 ethers, 11 olefins, 14 aromatics, 8 aldehydes, and acetylene. Average values for each of these 10 chemical classes were similarly presented. The formula for the POCP of each VOC, relative to ethylene as the reference VOC, is

$$\mathrm{POCP}_i = (a_i/b_i)/(a_{\mathrm{ethylene}}/b_{\mathrm{ethylene}}),$$

where a_i is the simulated change in ozone concentration due to a change in the emission of VOC_i and b_i is the integrated (cumulative) emission of VOC_i up to that time; however, Derwent and Jenkin's (1991) analysis was recently updated (Derwent et al., 1998).

Carter 1994 (1998) developed and continues to refine sets of "ozone reactivities" under various bounding NO_x conditions in model scenarios representing 39 different urban areas in the United States. Carter's ozone reactivities are somewhat analogous to the POCPs developed by Derwent and Jenkin (1991) but with some notable differences in their derivation. They are based on simulated effects of incremental changes in the level of the test VOC rather than the presence or absolute absence of a test VOC, as in the POCP development. The reactivities effectively estimate the partial derivative of ozone concentration with respect to emissions of a particular VOC: the change in ozone caused by adding an arbitrarily small amount of the test VOC to the emissions in the episode, divided by the amount of test VOC added. These reactivity estimates are available for three different measures of ozone concentration: peak hourly, integrated (cumulative) during the simulated day, and integrated ozone concentrations above 90 ppb. It should be noted, however, that integrated daily ozone is an indicator related closely to the 8-hour average, which is the new basis for U.S. EPA ozone air quality regulations. The 8-hour average will replace peak hourly concentrations on the basis of research showing the 8-hour average as a better predictor of human respiratory impacts (Elston, 1998).

Reactivity estimates are similarly available for three NO_x availability scenarios for each of 39 U.S. urban regional scenarios (Edmond and Whitten, 1987). The three NO_x availability regimes are

- Maximum incremental reactivity (MIR) scenario, which reflects the upper-end NO_x availability level at which VOC influence on ozone concentrations are maximized and NO_x influence is near its minimum. MIR reactivities have been adopted by the California Air Resources Board for use in vehicle emissions regulations.
- Maximum ozone reactivity (MOR) scenario, a NO_x availability at which ozone levels are maximized and beyond which increases in NO_x begin to reduce ozone concentrations.
- Equal benefit incremental reactivity (EBIR) scenario, a relatively low-NO_x availability at which VOC and NO_x changes are equally influential on ozone concentrations.

Critically, neither the POCPs nor the MIRs address the influence of NO_x in smog formation relative to the influence of VOCs, neither individually nor collectively, even though a considerable body of research exists that indicates that regional-scale O_3 photochemical production is limited by the availability of NO_x (Cardelino and Chameides, 1995). Analyses in the United States, such as Cardelino and Chameides (1995), have characterized the relative influence of percentage reductions in NO_x or NMVOCs (nonmethane VOCs) upon area-weighted O_3 concentrations; however, emission inventories were not available from these observation-based studies. Hence, they cannot provide the measure of relative influence of *equal mass* changes in NO_x or NMVOCs upon O_3 concentrations, which are required to derive characterization metrics.

The results of simulations using the EMEP ozone model for various European scenarios by multiple researchers (Simpson, 1993; Heyes and Schopp, 1995) have repeatedly been used in the LCA literature for the purpose of estimating the relative influence of 1 kg of NO_x versus 1 kg of NMVOCs on regional O_3 concentrations. For example, Seppala (1997) drew on the results of Heyes and Schopp (1995) together with emissions modeling for Finland and Scandinavia to develop a ratio for the estimated O_3 concentration change due to a 1-ton change in NO_x divided by the estimated O_3 concentration change due to a 1-ton change in NMVOCs. This ratio was found to range between values of approximately 2 and 5, depending on the modeling scenario and the percentage reductions in NO_x and VOCs considered.

The results of Simpson (1993) were interpreted by Rabl and Eyre (1997) to derive estimated European average responses of daily 6-hour peak concentrations in O_3, averaged over 6 summer months and averaged over all countries of Europe. He found that the average ratio of NO_x-based changes to NMVOC-based changes in this O_3 metric ranged between 1.2 and 1.4, depending on whether simultaneous reductions of 50% in both pollutant categories were considered or not.

CASE STUDY

In the presented analysis we adopt Carter's best-estimate MIR factors (1998). We test the relative influence of the NO_x/NMVOC ratios of 1.2, 2.0, and 5.0 to span the full range of values cited in the photochemical oxidant formation metrics literature to date.

The results in Table 3.7 show that different VOCs contribute different amounts to the final summary score for photochemical oxidant (PCO) formation for the two different alternatives. The bio-based BDO has a total score for PCO almost twice that of the petroleum-based alternative. For both alternatives, "unspeciated hydrocarbons" constitute the greatest contribution. This

TABLE 3.7
Species-specific VOC equivalency analysis for photochemical oxidant formation

Flow	*Comparison metrics*	*Characterization results (Bio-based minus natural gas-based)*	
		BDO processing	*Life cycle (cradle-to-gate)*
CO	0.07	5.165E-4	9.563E-4
Methane	0.01	−2.563E-7	−4.328E-7
Hydrocarbons*	3.93	2.665E-3	2.969E-3
Aldehydes*	9.12	2.792E-6	1.415E-4
Acetylene	1.23	−5.670E-5	−5.670E-5
Formaldehyde	9.12	−5.198E-4	−6.603E-4
Butyl alcohol	3.53	−1.094E-4	−1.094E-4
Propionaldehyde	8.3	−5.428E-5	−5.428E-5
Acetone	0.48	−1.291E-5	−1.327E-5
Toluene	4.19	−3.377E-4	−3.377E-4
Methanol	0.99	0	−5.752E-5
Hexane	1.69	0	−9.920E-6
Heptane	1.43	0	−1.081E-5
Octane	1.24	0	−6.262E-6
Pentane	1.74	0	−6.368E-6
Ethane	0.35	0	−1.467E-6
Propane	0.64	0	−4.205E-6
n-Butane	1.44	0	−7.488E-6
iso-Butane	1.56	0	−4.040E-7
Benzene	1	0	−6.490E-8
Totals		**2.093E-3**	**2.730E-3**

* Unspeciated.

highlights a common, potential problem in LCA for photochemical oxidant formation analysis, namely the lack of VOC speciation in the inventory data.

To more fully compare the two BDO alternatives, we must also include the influence of NO_x emissions upon photochemical oxidant formation and test the sensitivity of our conclusions upon the factor for the relative influence of NO_x and NMVOCs. For this portion of the analysis, the inventory results for the different VOCs were summed and used together with the characterization factor for NMVOCs in general (see Fig. 3.6). The results show that for the full range of values for the relative influence of NO_x versus VOCs, NO_x is the dominant contributor to the final PCO score. This conclusion holds true both for the total life cycle results and for the results when only the BDO processing stage emissions are considered.

OZONE DEPLETION

Ozone is constantly being created and destroyed through chemical reactions within the stratosphere. Recent anthropogenic emissions of chlorfluorocarbons, halons, and other ozone-depleting substances (ODSs) are believed to be causing an acceleration of these destructive chemical reactions, resulting in lower ozone levels and ozone holes in certain locations. These reductions in the level of ozone in the stratosphere lead to increasing UVB radiation reaching the earth. Increasing UVB can cause additional cases of skin cancer and cataracts. UVB can also have a deleterious effect on crops, plastics, and marine life. The Montreal Protocol is an international agreement banning the use and emission of many of the most potent ozone-depleting substances, and limiting the use of less potent substances. Implementation of the Montreal Protocol has significantly helped address the ozone depletion problem (Solomon and Wuebbles, 1995).

There is a general consensus on the use of ODPs, a midpoint metric proposed by the World Meteorological Organization (WMO) (Solomon et al., 1992) for considering the relative importance of chlorofluorocarbons, hydrochlorofluorocarbons, and halons expected to contribute significantly to the breakdown of the ozone layer. ODPs are based on time-integrated calculations that account for atmospheric lifetime and the reaction rate with elemental oxygen in the stratosphere. Chemicals are characterized relative to CFC-11 and are calculated based on 8 time frames ranging from 5-year ODPs to 500-year ODPs (Solomon and Wuebbles, 1995; Solomon et al., 1992).

Comparisons in terms of ozone depletion can be made using the following equation:

$$\text{Ozone Depletion Index} = \sum_i \text{ODP}_i^* \text{m}_i,$$

where m_i = mass of compound (i).

A few ODSs, such as nitrous oxide (N_2O), are expected to exhibit significant effects but do not have calculated ODPs. Impact assessment modeling

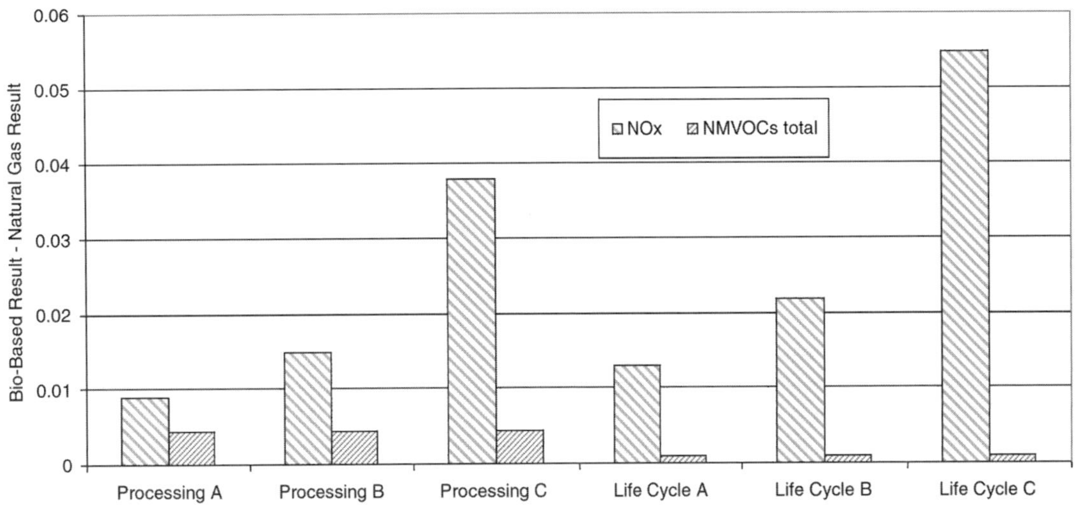

Figure 3.6: Tropospheric ozone characterization factors and results comparison. Method A: NO_x influence = 1.2 times that of nonmethane VOCs; Method B: NO_x influence = 2 times that of nonmethane VOCs; Method C: NO_x influence = 5 times that of nonmethane VOCs.

TABLE 3.8
Ozone depletion impact assessment results for the BDO alternatives

		Natural gas-based option		*Bio-based option*	
Component	*ODPs*	*Process*	*Life cycle*	*Process*	*Life cycle*
N_2O	N/A	0	0	7E-04	2E-03

conducted to derive the ODPs takes into account fate, effect on ozone, and anthropogenic background. However, current models and metrics do not attempt to represent endpoints like the number of cataracts. As the location of the release is assumed to have no bearing on the potential for global effects, the location of the emissions is similarly not considered important in this impact category; however, for some chemicals, it may be necessary to account for the fraction that is expected to partition to the atmosphere, and this may be location specific (possibly requiring the use of multimedia fate models).

CASE STUDY

No ODSs were reported for the life cycle inventory of the natural gas-based BDO process (see Table 3.8). For the bio-based process there were N_2O emissions in the BDO production energy stages and other energy life cycle stages totaling 2E-03 pounds of N_2O per pound of BDO produced. As N_2O was the only ODS reported and the current methodologies do not calculate a potency factor for this chemical, N_2O mass is used directly to represent changes in the ozone depletion potential. Conducting a process-specific approach and a life cycle approach would have shown the natural gas-based option to be potentially superior to the bio-based option in both cases.

GLOBAL WARMING

Global warming (or climate change) refers to potential changes in the earth's climate caused by the build-up of chemicals ("greenhouse gases" like water vapor, ozone, and carbon dioxide) that trap heat from reflected infrared radiation, which would have otherwise passed out of the earth's atmosphere. Whereas sinks exist for greenhouse gases (e.g., oceans absorb carbon dioxide), the rate of emissions in the industrial age is presumed to be exceeding the rate of absorption. Simulations are currently being conducted to try to quantify the potential effects of these exceedences, including increased droughts, floods, loss of polar ice caps, sea level rise, soil moisture loss, forest loss, change in wind and ocean patterns, changes in agricultural production, decreased biodiversity, and increasing occurrences of extreme weather events. Some simulations have predicted the severity of these potential effects to be so devastating that many countries have signed the Kyoto Protocol—a United Nations agreement to reduce the emissions of greenhouse gases (Environment Canada, 1998).

There is general consensus on the use of global warming potentials (GWPs), a midpoint metric proposed by the Intergovernmental Panel on Climate Change (IPCC) (Houghton et al., 1992) for the calculation of the potency of greenhouse gases relative to carbon dioxide. Calculations using these metrics in relative comparisons are made based on the following equation:

$$\text{Global Warming Index} = \sum_i \text{GWP}_i \times m_i,$$

where m_i = mass of compound (i).

GWPs are based on time-integrated radiative forcing predictions for the instantaneous release of 1 kg of gas relative to 1 kg of CO_2. These calculations take into account the atmospheric lifetime and infrared energy absorption properties of the gases. Degradation products are not usually considered, however (e.g., all organic emissions can degrade to form CO_2 and other compounds).

GWPs are available for time horizons of 20, 100, and 500 years. As with ozone depletion, the endpoint effects, like an increased number of malaria cases, are not quantified, and it is not considered necessary to consider the location of a release; however, chemical fate should be taken into account for some chemicals to estimate the fraction that reaches or remains in the atmosphere (possibly using multimedia fate models). This calculation can be location specific.

CASE STUDY

Carbon dioxide and methane were the only two global warmers inventoried for the BDO alternative case study. Calculations are shown in Table 3.9 for 20-year, 100-year, and 500-year time frames. For all cases, carbon dioxide from the BDO process energy dominated the calculations, and the methane emissions were relatively insignificant. The global warming results for the natural gas-based and bio-based processes were within an order of magnitude and are not expected to be significantly different. Conducting a process-specific approach, including energy-related emissions, and a life cycle approach would have drawn the same conclusions in this case study (i.e., no significant difference between the natural gas-based and bio-based options).

RESOURCE DEPLETION

Resource depletion refers to the loss, diminishment, or impairment of natural resources (air, water, land, and biomass) such that the resource is no longer available as an input into the system under consideration (SETAC, 1993). It

TABLE 3.9
Global warming impact assessment results for the BDO alternatives

		Natural gas-based option		*Bio-based option*	
Component	*20-yr GWPs*	*Process*	*Life cycle*	*Process*	*Life cycle*
Carbon dioxide	1	3	3	4	5
Methane	56	2E-03	3E-03	5E-04	8E-04
Totals		**3**	**3**	**4**	**5**
	100-yr GWPs	*Process*	*Life cycle*	*Process*	*Life cycle*
Carbon dioxide	1	3	3	4	5
Methane	21	7E-04	1E-03	2E-04	3E-04
Totals		**3**	**3**	**4**	**5**
	500-yr GWPs	*Process*	*Life cycle*	*Process*	*Life cycle*
Carbon dioxide	1	3	3	4	5
Methane	7	2E-04	4E-04	6E-05	1E-04
Totals		**3**	**3**	**4**	**5**

does not include the loss or impairment of resources as a result of emissions, as is the case with other impact categories like eutrophication, for example.

Resources are typically divided into two categories: stock and flow. Stock resources are those considered to be nonrenewable (finite), such as land, primary energy sources (e.g., natural gas, petroleum, coal), and minerals. Flow resources are renewable, although renewability can depend on several factors, including rate of use, economic factors influencing consumption, and whether reserves are actually available for use (SETAC, 1993).

Time-matrix models are generally used to characterize resource depletion. Examples include the Resource Consumption Ratio and the Resource Depletion Matrix (U.S. EPA, 1995). The Resource Consumption Ratio characterizes the depletion of natural resources by comparing consumption to available supplies or reserves. It is expressed as

$$\frac{\text{consumption per unit of use per unit time}}{\text{supply per unit time}}$$

and can be considered the inverse of "sustainability." The "unit of use" depends on the resource and objective of the study. For trees it may be board feet, and for coal it may be tons or BTU equivalents, for example.

The Resource Depletion Matrix characterizes depletion of stock or flow resources in local or global terms (see Table 3.10). For stock resources, the equation yields the measure of time until the resource is depleted. For flow resources, a dimensionless "depletion index" is produced and can be used to determine the sustainability of the resource use (U.S. EPA, 1995).

TABLE 3.10
Resource depletion ratio (SETAC, 1993)

	Stock	*Flow*
Local	$S_L/C_L = U_L$	$C_L/R_L = D_L$ if $D_L < 1$, then $C_L - R_L = E_L$ and $Q_L/E_L = U_L$
Global	$S_G/C_G = U_G$	$C_G/R_G = D_G$ if $D_G < 1$, then $C_G - R_G = E_G$ and $Q_G/E_G = U_G$

D = depletion index; E = excess consumption rate (amount/unit of time); G = global; C = consumption rate (amount/unit of time); L = local; Q = standing quantity (amount); R = replenishment rate (amount/unit of time); S = stock (amount); U = use-years (time).

Resource depletion also includes land use and water use, although these are sometimes treated as separate impact categories. The land use category is generally applied to depletion of land within a specific scope. For example, land use for solid waste disposal entails the depletion of landfill space, and a measurement of lost volume is appropriate. Land use for resource extraction or production, on the other hand, may be measured in terms of land use "intensity," such as the average annual acres of land (acre-years) that are unavailable for other uses due to production or extraction of the resource. For this calculation, the space occupied and the time of occupancy are relevant. This is demonstrated in the case study.

Water can be either a stock or flow resource, depending on its use. If it is an underground reservoir that is not recharged within a human lifetime, it is considered a stock resource; if it is withdrawn from a rechargeable reserve, it is a flow resource (SETAC, 1993). The location of the water source is also a consideration due to regional differences in hydrology and the high cost of transporting water over long distances. For this, regional scaling methods may be used to reflect the additional impact associated with water use within dry regions.

Existing or proposed approaches for addressing natural resources and land use impacts include the following examples. For the extraction of abiotic resources (most studied within the LCA context): (1) aggregating resources on the basis of mass; (2) aggregating at least energetic resources in energetic terms; (3) use of entropy or exergy as a measure of aggregation; (4) translating depletion of resources into future impacts within one or more other impact categories. For the extraction of biotic resources, proposed methods consider (1) size of the population used in conjunction with the net depletion rate; (2) impacts on biodiversity and life-support function; (3) qualitative scoring on the basis of lists of endangered species or ecosystems; (4) consideration under the ecosystem health impact category. Included in the land use category are (1) distinguish qualities of land use and consider changes from one use to another; (2) ecosystem and landscape degradation; and (3) disrupted land surface and its effect on ecosystem health. An analysis of these methods was presented by Heijungs et al. (1997). An emerging methodology is the life cycle Stressor-Effects Assessment (LCSEA) approach

(Rhodes and Kommonen, 1999). The LCSEA approach takes into account the inherent spatial and temporal components of resources, although additional data that have not been typically collected to date in LCA are required.

As with all impact categories, resource depletion is continuously being investigated and the reader is advised to consult the current literature for the most recent changes and trends. As an example, SETAC is currently considering two general categories of impacts, input related and output related (Udo de Haes et al., 1999a, 1999b). The proposed input-related categories would include extraction of abiotic resources, extraction of biotic resources, and land use; with land use being subdivided into increase in land competition, degradation of life support functions, and biodiversity degradation. A possible fourth category of land use would involve degradation of cultural values, such as landscapes. This categorization scheme is not final but is presented here to illustrate the dynamic nature of this area of study.

CASE STUDY

Four resource depletion categories are illustrated here: (1) fossil fuel resource depletion, (2) land use for solid waste disposal, (3) land use for resource extraction and production, and (4) water use.

Fossil Fuel Resource Depletion

The basis for the comparison metric for resource depletion of fossil fuels is the inverse of sustainability (which is essentially the resource consumption ratio discussed above), not taking into account potential technological advancements for economically locating or mining resources, nor the influence of recycling or reducing demand. The results in Table 3.11 indicate that the natural gas-based option has a potential for greater impacts in this category. From a life cycle perspective, the natural gas-based option has roughly twice the potential impact on fossil fuel resource depletion compared to the

TABLE 3.11
Resource depletion impacts in BTUs (N.B. The results here are given in terms of BTUs to satisfy the energy-related objectives of the study)

Fossil fuel	*Comparison metric*	*Natural gas-based*		*Bio-based*	
		Processing	*Life cycle*	*Processing*	*Life cycle*
Coal	0.0034	3.0	5.0	47	56
Natural gas	0.0151	320.0	599.0	170	203
Petroleum	0.0214	0.2	3.3	2	34
Total		**323**	**607**	**219**	**293**

TABLE 3.12
Land used for solid waste disposal in cubic yards

Component	*Comparison metric*	*Natural gas-based*		*Bio-based*	
		Processing	*Life cycle*	*Processing*	*Life cycle*
FGD solids	1.67	2E-06	3E-06	3E-05	3E-05
Fly ash	1.67	4E-06	7E-06	6E-05	7E-05
Slag	1.04	4E-07	6E-07	5E-06	5E-06
Depleted uranium	0.17	3E-08	6E-08	5E-07	5E-07
Other solids	2.22	0	0	4E-04	4E-04
Total	**N/A**	**5E-06**	**8E-06**	**7E-05**	**4E-04**

bio-based process. Within the processing stage alone, the difference between the natural gas-based and bio-based is only slight.

Land Use for Solid Waste Disposal

The comparison metric for land use for solid waste disposal was based on the estimated volume of solid waste using the specific gravity of each type of solid waste (the LCA inventory for solid waste was in weight per functional unit and was converted to a volume by multiplying the weight by the inverse of the specific gravity of the wastes). As indicated in Table 3.12, the impacts of the natural gas-based options are greater than the bio-based and are significantly greater on a life cycle basis.

Land Use for Resource Extraction/Production

Land use for resource extraction and production was based on the "intensity" of land use for the extraction of oil and gas and the production of corn (the fuel sources considered in the case study). The calculation assumed a U.S. average well productivity of 156,000 ft^3/day for natural gas wells and 27,130 ft^3/day for crude oil wells. It was also assumed that a natural gas well occupies an average of 0.5 acres per well for a typical well life of 30 years. For corn production, it was assumed that the land produces one crop per year over a growing season of 0.5 years. As presented in Table 3.13, no impacts occurred in the processing stage for either the natural gas-based or bio-based option; however, impacts are much greater for the bio-based option than for the natural gas-based in the life cycle context.

Water Use

For water use, regional scaling factors were used to account for the fact that the impacts would be greater in areas where water is scarce. This analysis

TABLE 3.13
Land use for resource extraction/production in acre-years per pound of BDO

	Natural gas-based		*Bio-based*	
Metrics	*Processing*	*Life cycle*	*Processing*	*Life cycle*
N/A	0	6.31E-07	0	1.30E-04

was based on maps showing areas of inadequate surface water supply for in-stream use and groundwater overdraft (withdrawing water from a source that cannot be renewed or withdrawing it more quickly than it can be recharged). A scoring mechanism was adopted that considered the percentage of water depleted in an average or dry year and the gallons per day of groundwater "overdraft." Additional consideration was given for consumptive water use within the state in question. For water use, the natural gas-based option was shown to use almost half as much water as the bio-based option on a processing-only basis (see Table 3.14). On a life cycle basis, the natural gas-based option was still favored but to a greater extent.

SUMMARY

In this chapter we briefly outlined many considerations associated with current state-of-the-art metrics that are used in environmental comparison applications and illustrated the importance of comparing design alternatives from a life cycle as well as site-specific perspective. It is clear that our ability to readily compare alternatives in Life Cycle Assessment (LCA), as well as in the context of more restricted scopes like the direct comparison of on-site emissions associated with process design alternatives, remains limited. Despite these limitations and the significant debates that ensue, decisions must ultimately be made to help focus our finite resources (time and money) in an effort to continually reduce the potential for negative consequences associated with chemical emissions and the depletion of natural resources.

It is important to note that the methodologies and the metrics in the presented case study are commonly used but may not necessarily be considered the state of the art. The case study represents a "snapshot in time," and all

TABLE 3.14
Water use in pounds of water per pound of BDO

	Natural gas-based		*Bio-based*	
Metrics	*Processing*	*Life cycle*	*Processing*	*Life cycle*
N/A	4.96E+01	4.97E+01	7.19E+01	8.34E+01

the stages are subject to change, particularly in the developing bio-based life cycle. Updates and more detailed information can be found in the attached references and through the ongoing research activities of organizations like SETAC.

DISCLAIMER

Although the U.S. EPA has funded the research described in this article wholly (or in part), it does not constitute agency policy.

REFERENCES

Cardelino, C. A., and Chameides, W. L. (1995). An observation-based model for analyzing ozone precursor relationships in the urban atmosphere. *J. Air Waste Manage. Assoc.* 45:161–180.

Carter, W. (1994). Development of ozone reactivity scales for volatile organic compounds. *J. Air Waste Manage. Assoc.* 44:881–899.

Carter, W. (1998). Updated maximum incremental reactivity scale for regulatory applications. California Air Resources Board.

Derwent, R. G., and Jenkin, M. E. (1991). Hydrocarbons and the long range transport of ozone and PAN across Europe. *Atmospheric Environment* 25A:277–279.

Derwent, R. G., Jenkin, M. E., Saunders, S. M., and Pilling, M. J. (1998). Photochemical ozone creation potentials for organic compounds in northwest Europe calculated with a master chemical mechanism. *Atmospheric Environment* 32:2429–2441.

Edmond, G. M. R., and Whitten, G. (1987). Tropospheric ultraviolet radiation: Assessment of existing data and effects on ozone formation. *Final Report* EPA-600/3-87-047.

Elston, J. (1998). Personal communication. New Jersey Dept. of Env. Protection.

Environment Canada, Canada signs Kyoto protocol on Climate Change, Environment Canada News Release, 29 April, 1998.

Finnveden, G., Andersson-Skold, Y., Samuelsson, M.-O., Zetterberg, L., and Lindfors, L.-G. (1992). Classification (Impact Analysis) in connection with life cycle assessments: A preliminary Study, IVL—Draft prepared for the Nordic Council of Ministers.

Gameson, T., editor. (1998). Private sector methods for weighting environmental indicators, Institute for Prospective Technological Studies, Joint Research Centre, European Commission.

Gold, L. S., and Zeiger, E. (1997). *Handbook of Carcinogenic Potency and Genotoxicity Databases*, CRC Press, Boca Raton, FL.

Guinee, J., Heijungs, R., van Oers, L., van de Mennt, D., Vermeire, T., and Rikken, M. (1996). LCA impact assessment of toxic releases: Generic modelling of fate, exposure and effect for ecosystems and human beings with data for about 100 chemicals, Centre of Environmental Science, Leiden University (CML), Leiden, The Netherlands, and National Institute of Public Health and Environmental Protection (RIVM), Bilthoven, The Netherlands.

Hauschild, M. Z., and Wenzel, H. (1997). Scientific background for environmental assessment of industrial products, Hauschild, M. Z., Ed., Chapman & Hall, London, U.K.

Heijungs, R., Guinee, J., and Huppes, G. (1997). Impact categories for natural resources and land use, CML Report 138, Centre of Envionmental Science, Leiden University, Leiden, The Netherlands.

Heijungs, R., Guinee, J. B., Huppes, G., Lankreijer, R. M., Udo de Haes, H. A., Wegener Sleeswijk, A., Ansems, A. M. M., Eggels, P. G., van Duin, R., and de Goede, H. P. (1992). *Environmental Life Cycle Assessment of Products*, Vol. 1 and 2, Center for Environmental Studies (CML), Leiden University, Leiden, The Netherlands.

Hertwich, E. G., Pease, W. S., and McKone, T. E. (1998). Evaluating toxic impact assessment methods: What works best? *Env. Sci. Tech.* 32:138A–145A.

Hertwich, E. G. (1999). Toxic equivalency: Addressing human health in lifecycle assessment, Ph.D. thesis, University of Berkeley, California.

Heyes, C., and Schopp, W. (1995). Towards a simplified model to describe ozone formation in Europe. International Institute for Applied Systems Analysis, Laxenbourg, Austria, Working Paper WP-95-34.

Hofstetter, P. (1998). *Perspectives in Life Cycle Impact Assessment: A Structured Approach to Combine Models of the Technosphere, Ecosphere and Valuesphere*. Kluwer Academic Publishers.

Houghton, J. T., Callander, B. A., and Varney, S. K. (1992). Climate change. The supplementary report to the IPCC scientific assessment, Cambridge University Press, Cambridge, U.K.

Huijbregts, M. A. J. (1999). Priority assessment of toxic substances in the frame of LCA: Development and application of the multi-media fate, exposure and effect model USES-LCA, University of Amsterdam, The Netherlands.

ISO (1997). Environmental management, life cycle assessment: Principles and framework, ISO 14040.

Jager, T., Rikken, M. G. J., and van der Poel, P. (1997). Uncertainty Analysis of EUSES: Improving Risk Management by Probabilistic Risk Assessment, National Institute of Public Health and The Environment, Bilthoven, The Netherlands.

Krewski, D., Gaylor, D. W., Soms, A. P., and Sszyszkowicz, M. (1993). An overview of the report: Correlaton between carcinogenic potency and the maximum tolerated dose: Implications for risk assessment. *Risk Anal.* 13:383–398.

Krewski, D., Murdoch, D., and Withey, J. R. (1989). Recent developments in carcinogenic risk assessment. *Health Physics* 57(Suppl. 1):313–325.

Lindfors, L.-G., Christiansen, K., Hoffman, L., Virtanen, Y., Juntilla, V., Hanssen, O.-J., Roenning, A., and Ekvall, T. (1995). Nordic guidelines on life-cycle assessment, report Nord 1995:20, Nordic Council of Ministers, Copenhagen.

Meylan, W., and Howard, P. (1998). User's guide for the ECOSAR class program, prepared for Nabholz, J., and Cash, G., U.S. EPA, Washington, Syracuse Research Corporation.

Pennington, D. W., and Yue, P. L. (2000). Options for comparison of process design alternatives in terms of regional environmental impacts. *J. Cleaner Production* 8:1–9.

Pennington, D. W., and Bare, J. C. (1999). Comparison of human health screening and ranking approaches: The Waste Minimization Prioritization Tool (WMPT) and Toxic Equivalency Potentials (TEPs), submitted to *Risk Analysis*.

Pennington, D. W., and Ralston, M. (1999). Multimedia persistence and EPA's Waste Minimization Prioritization Tool, SETAC News, 19:30.

Potting, J., and Hauschild, M. (1997). Predicted environmental impact and expected occurrence of actual environmental impact, Part 2: Spatial differentiation in life-cycle assessment by site-dependent characterization of environmental impact from emissions. *Int. J. LCA* 2:209–216.

Potting, J., Schopp, W., Blok, K., and Hauschild, M. (1998). Site-dependent life-cycle impact assessment of acidification. *J. Industrial Ecology* 2:63–87.

Rabl, A., and Eyre, N. (1997). An estimate of regional and global O_3 damage from precursor NO_x and VOC emissions, Unpublished manuscript, Ecole des Mines, Paris.

Rhodes, S., and Kommonen, F. (1999). Life-cycle stressor-effects assessment (LCSEA), A practitioner's manual, Scientific Certification Systems, Inc. and Soil and Water, Ltd.

Russom, C. (1994). Aster tools for the evaluation of risk: Version 1.0, National Health and Environmental Effects Research Laboratory, U.S. EPA.

Samuelsson, M. O. (1993). Life cycle assessments and eutrophication: A concept for calculation of the potential effects of nitrogen and phosphorous. IVL Report B1119, IVL, Stockholm, Sweden.

Seppala, J. (1997). Decision analysis as a tool for life cycle impact assessment, Finnish Environment Institute, Helsinki.

SETAC (1993). *A Conceptual Framework for Lifecycle Impact Assessment*, Fava, J., Consoli, F., Denison, R., Kickson, K., Mohin, T., and Vigon, B., Eds., Society of Environmental Toxicology and Chemistry and SETAC Foundation for Environmental Education, Inc., Pensacola, FL.

SETAC (1996). *Towards a Methodology for Life Cycle Impact Assessment*. Udo de Haes, H. A., Ed., Brussels, Belgium.

Simpson, D. (1993). Photochemical model calculations over Europe for two extended summer periods: 1985 and 1989, Model results and comparison with observations. *Atmospheric Environment* 27A:921–943.

Solomon, S., and Albritton, D. L. (1992). Time dependent ozone depletion potentials for short and long-term forecasts. *Nature* 357:33–37.

Solomon, S., and Wuebbles, D. (1997). World Meteorological Organization (WMO) Document, 1995, Streamlined life-cycle assessment of 1,4-butanediol produced from petroleum feedstocks versus bio-derived feedstocks, prepared by Battelle Columbus Division, Pacific Northwest National Laboratory, U.S. EPA, Systems Analysis Branch, Cincinnati, OH.

Swanson, M. B., and Socha, A. C., editors, (1997). Chemical ranking and scoring: Guidelines for relative assessments of chemicals, *Proceedings of the Pellston Workshop on Chemical Ranking and Scoring*, February 12–16, 1995, SETAC Press, Florida.

Tchobanoglous, G., and Burton, F. L., eds. (1991). *Wastewater Engineering: Treatment, Disposal and Reuse*, 3rd edition, Metcalf & Eddy Inc., McGraw-Hill, New York.

U.S. EPA (1995). Life-cycle impact assessment: A conceptual framework, key issues, and summary of existing methods, 452/R-95-002.

U.S. EPA (1998). Notice of availability of draft RCRA waste minimization PBT chemical list, EPA 530-Z-98-009, *Federal Register*, 63, 216.

U.S. EPA (1997). Notice of availability of waste minimization software and documents, EPA530-C-97-003, *Federal Register*, 62, 33868.

Udo de Haes, H. A., Jolliet, O., Finnveden, G., Hauschild, M., Krewitt, W., and Mueller-Wenk, R. (1999a). Best available practice regarding impact categories and category indicators in life cycle impact assessment: Part 1. *International Journal of LCA* 4:66–74.

Udo de Haes, H. A., Jolliet, O., Finnveden, G., Hauschild, M., Krewitt, W., and Mueller-Wenk, R. (1999b). Best available practice regarding impact categories and category indicators in life cycle impact assessment: Part 2. *International Journal of LCA* 4:167–174.

Part Three

PROCESS INTEGRATION FOR THE ENVIRONMENT

CHAPTER 4

Challenges for Applying Process Integration Design Methodologies for Waste Reduction within Chemical Process Companies

Russell F. Dunn
SOLUTIA, Gonzalez, FL 32560

OVERVIEW OF PROCESS INTEGRATION DESIGN METHODOLOGIES

There has been considerable activity over the past decade toward the development of new environmental process design methodologies and tools which have been used in a large variety of chemical process industries/companies (e.g., petroleum companies, pulp and paper companies, pharmaceutical companies, food products companies, plastics producing companies, etc.). These developments have been driven primarily by increasingly stringent environmental regulations and the high cost associated with pollution prevention processes. These new design methodologies and tools are generally targeting the systematic identification of the minimum cost strategy (process designs) for a company to achieve environmental compliance. The cost of environmental compliance becomes even more critical as companies strive to remain financially competitive in a global business and as companies strive to solidify their long-term sustainability. To address these issues, several novel design methodologies have been developed over the past decade that have been collectively grouped under the broad classification of *process integration technology*. Process integration technology is "a holistic approach to process design, retrofitting, and operation which emphasizes the unity of the process," as opposed to a design approach that optimizes at the unit operation level (El-Halwagi, 1997). Process integration technology generally is used at the macroscopic level of process design, as indicated in Figure 4.1, and tools within this technology are used to determine the optimal allocation of mass and energy within an overall process and/or site to minimize waste, increase capacity, conserve energy, etc.

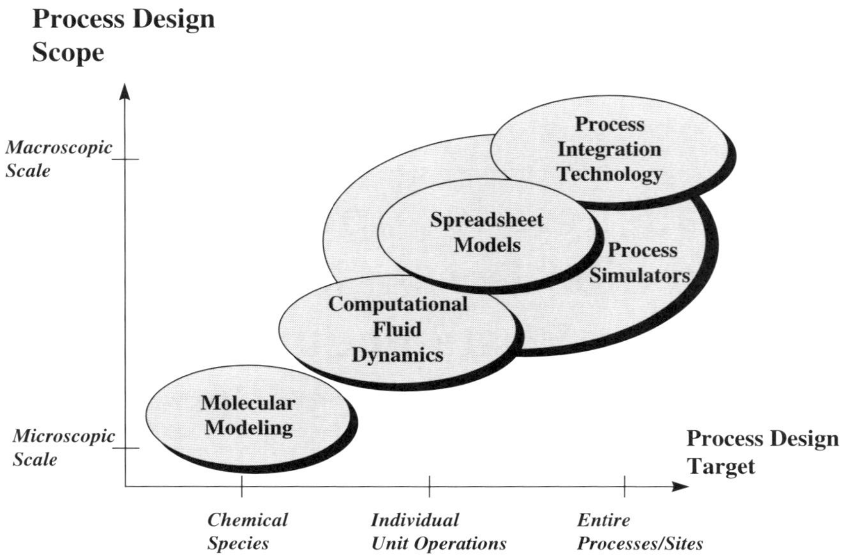

Figure 4.1: A schematic representation of process integration technology in relation to other process design tools.

The initial efforts in the field of process integration were focused toward the energy dimension of a process. This resulted in the development of significant design methodologies during the late 1970s and 1980s, specifically the formation of heat pinch analysis and the development of optimal heat exchange networks (HENs) for energy recovery, and these collective efforts are referred to as the field of heat integration technology (Linnhoff and Hindmarsh, 1983; Papoulias and Grossmann, 1983; Cerda et al., 1983; Gunderson and Naess, 1988; Shenoy, 1995). Over the last decade, the field of process integration technology has been expanded to address the mass dimension of a process. This has resulted in the development of significant design methodologies, specifically the formation of mass pinch analysis and other tools for waste recovery process design, and these collective efforts are referred to as the field of mass integration technology (El-Halwagi and Manousiouthakis, 1989b; El-Halwagi, 1997). More recently, design methodologies have been developed to simultaneously address both the mass and energy dimensions of the process (Dunn and El-Halwagi, 1994a, 1994b; El-Halwagi et al., 1995; Dunn et al., 1995; Dunn and Srinivas, 1997; Dunn et al., 1999).

There have been major breakthroughs within the process integration technology category targeting the development of environmental process design methodologies and tools. Specifically, they are the development of *end-of-pipe design methodologies and tools* and the development of *source reduction design methodologies and tools*. The *end-of-pipe* design methodologies and tools introduced the novel concepts of the design of mass exchange networks (MENs), reactive mass exchange networks (REAMENs), heat-induced separation networks (HISENs), energy-induced separation networks (EISENs), membrane-based separation systems, and several other design methodologies (El-Halwagi and Manousiouthakis, 1989b, 1990a, 1990b; El-Halwagi and Srinivas, 1992; Srinivas and El-Halwagi, 1994a, 1994b; Gupta and Manousiouthakis, 1994; Papalexandari et al., 1994; Papalexandari and Pistikopoulos, 1994; Zhu and El-Halwagi, 1995; Warren et al., 1995; Kiperstock and Sharratt, 1995; Dunn and El-Halwagi, 1994a, 1994b; El-Halwagi et al., 1995; Richburg and El-Halwagi, 1995; Crabtree et al., 1998; Dunn and Dobson, 1998; El-Halwagi, 1992; Srinivas and El-Halwagi, 1993). A brief description of some of the end-of-pipe methodologies and the technologies (unit operations) covered by these methodologies is provided in Table 4.1.

TABLE 4.1
Summary of some end-of-pipe methodologies for environmental process design

Design methodology	*Description*	*Example technologies targeted*
Mass exchange networks (MENs)	A network of process units that removes pollutants from end-of-pipe streams via the use of physical, direct-contact mass separating agents (MSAs).	Adsorption Absorption Liquid-liquid extraction Ion exchange
Reactive mass exchange networks (REAMENs)	A network of process units that removes pollutants from end-of-pipe streams via the use of reactive, direct-contact, MSAs.	Adsorption Absorption Liquid-liquid extraction Ion exchange
Heat-induced separation networks (HISENs)	A network of process units that removes pollutants from end-of-pipe streams via the use of indirect-contact energy separating agents (ESAs).	Condensation Evaporation Drying Crystallization
Energy-induced separation networks (EISENs)	A network of process units that removes pollutants from end-of-pipe streams via the use of indirect-contact ESAs, including stream pressurization and/or depressurization.	Condensation Evaporation Drying Crystallization Compressors Vacuum pumps
Membrane separation networks	A network of process units that removes pollutants from end-of-pipe streams via the use of membranes and stream pressurization and/or depressurization.	Reverse osmosis Pervaporation

The *source reduction* design methodologies and tools have generally been focused toward three major strategies:

1. If the undesirable species are generated in the process (via reaction), find an alternative, environmentally acceptable reaction path (Crabtree and El-Halwagi, 1994).
2. If the undesirable species are not generated in the process, replace it with a more environmentally benign material (Hamad and El-Halwagi, 1998; Dunn et al., 1997; Jobac 1995; Constantinou et al., 1995).
3. If the emissions are associated with in-plant utility systems (e.g., blowdowns, greenhouse gas emissions, etc.), implement in-plant heat integration modifications to minimize the amount of emissions from thermal pollution or to reduce the amount of emissions associated with utility systems (Linnhoff, 1994; Linnhoff, 1995, Dhole, 1995).
4. If the undesirable species is not generated in the process and cannot be replaced with a more environmentally benign material, implement in-plant modifications to recycle streams with the undesirable species. This approach has resulted in the introduction of the novel concepts of the design of wastewater minimization systems, waste interception and allocation networks (WINs), heat-induced waste minimization networks (HIWAMINs), and energy-induced waste minimization networks (EIWAMINs) (Wang and Smith, 1994; Dunn and Dobson, 1999; El-Halwagi et al., 1996; Dunn and Srinivas, 1997; Dunn et al., 1999). The HIWAMIN and EIWAMIN design approaches simultaneously address waste reduction and energy conservation process design to further improve economics.

A brief description of some of the source reduction methodologies and the technologies (unit operations) covered by the various methodologies is provided in Table 4.2.

The major benefit derived from new environmental design tools occurs when the tools are applied to real chemical process problems to reduce costs, reduce emissions, debottleneck processes, increase capacity, etc. In this light, often there are significant challenges that are encountered when applying process integration design technology in industrial practice. Some of these hurdles are as follows:

- ***Management-Based Issues:*** Does management support new design technology from both a monetary and time allotment perspective? What are the benefits and drawbacks of employing technical resources within the industry to use new design technology versus using consultants, etc.?
- ***Personnel-Based Issues:*** Who should be a member of the overall design team? What are the responsibilities of the team members, etc.?
- ***Technology-Based Issues:*** Which design approaches should be used and for what types of design tasks? Is the technical approach acceptable for the size of the design task, etc.?

TABLE 4.2
Summary of some source reduction methodologies for environmental process design

Design methodology	*Description*	*Example technologies targeted*
Environmentally acceptable reactions (EARs)	The development of an alternative reaction path that will no longer result in the generation of the pollutant.	Reactions
Solvent selection systems	The identification of alternative solvents and/or solvent combinations that can be used to provide a desired effect (solubility, material transport, separating agent, etc.).	Substitute refrigerants Substitute coating constituents Simultaneous solvent and separation unit selection
Heat integration systems	The identification of heat recovery devices that minimize environmental emissions resulting from utility generation systems.	Heat exchangers Heat pumps Boilers/cooling towers
Wastewater minimization systems	A design strategy for reuse, regeneration reuse, and regeneration recycling of wastewater streams that minimizes water usage and minimizes wastewater discharge.	Direct recycle opportunities Regeneration reuse and recycling opportunities
Waste interception and allocation networks (WINs)	A network of process units that removes pollutants from in-plant streams via the use of physical or reactive direct-contact MSAs and/or rerouting of in-plant process streams.	Direct recycle opportunities Adsorption Absorption Liquid-liquid extraction Ion exchange
Heat-induced waste minimization networks (HIWAMINs)	A network of process units that removes pollutants from in-plant streams via the use of indirect-contact ESAs and/or rerouting of in-plant process streams. Full site heat integration is simultaneously addressed by this technique.	Direct recycle opportunities Heat exchange/heat integration Condensation Evaporation Drying Crystallization

(Continued)

TABLE 4.2 (*Continued*)

Design methodology	*Description*	*Example technologies targeted*
Energy-induced waste minimization networks (EIWAMINs)	A network of process units that removes pollutants from in-plant streams via the use of indirect-contact ESAs with stream pressurization and/or depressurization and/or rerouting of in-plant process streams. Full site heat integration is simultaneously addressed by this technique.	Direct recycle opportunities Heat exchange/heat integration Condensation Evaporation Drying Crystallization Compressors Vacuum pumps Heat pumps
Membrane separation networks	A network of process units that removes pollutants from streams via the use of membranes and stream pressurization and/or depressurization.	Reverse osmosis Pervaporation

Explorations of these issues are provided in the following sections. Because all environmental design tasks that are tackled using process integration technology are unique, it may be necessary to modify the management-based issues, personnel-based issues, and technology-based issues provided below to fit a company-specific design task.

MANAGEMENT-BASED ISSUES FOR APPLYING PROCESS INTEGRATION ENVIRONMENTAL DESIGN TOOLS WITHIN CHEMICAL PROCESS COMPANIES

Several key management-based issues that must be addressed when applying process integration technology for environmental design tasks are

- the defining of the design task deliverables;
- the establishment of a management support structure for the design effort;
- the selection of the process integration technology team;
- the realization that no waste reduction improvements may be identified.

The initial management issue that must be addressed is the defining of the design task deliverables. Key parameters such as the level of waste reduction requirements (flowrates and/or compositions), the consideration of new environmental process technologies during the design process, financial and personnel allocations for the project, and the time schedule for project implementation to meet business and/or environmental regulation requirements are examples of the deliverables that can be initially defined. Additional deliverables tailored to the specific design task can be added.

The next management issue that must be addressed is whether middle and upper management fully supports the funding and time allotment required to complete the environmental design task. The key issue is whether management will accept the first design that works or whether sufficient time and resources (personnel, finances, etc.) will be allocated to identify the most cost effective design option. As a general rule, the majority of the cost of a new process is established during the basic or preliminary design phase and Research & Development phase of the project (Hertz, 1995). This indicates that the majority of the project environmental cost reductions are achieved during the early design phases of the project. In addition, developing the technical feasibility and preliminary cost estimates for a large number of design options may require substantial time and technical effort, even though the use of process integration design tools dramatically reduces that time.

The third management issue that must be addressed is where the company will acquire process integration trained resources to address the environmental design tasks. It is a common approach for medium- to large-size chemical industries to develop a group whose sole function is environmental process design, and within this group some level of process integration technology expertise should reside. If this approach is taken, it is imperative that a group is formed whose background includes a diverse knowledge of new design tools, combined with an acceptable level of industrial experience concerning environmental process design. This group most often functions from a corporate center and provides design services for the entire chemical company enterprise. It is also recommended that members of this design group remain updated on new environmental design methodologies. This can be accomplished via a variety of environmental conferences, environmental design short courses, periodic technical journal searches, and research collaborations with academic institutions that are active in developing new environmental design tools.

It may be difficult to develop a good mixture of talent for this design group, and this often leads to an alternative approach, which is the use of environmental consulting companies. This presents additional complexities, as there are several consulting companies who "specialize" in process integration-based environmental process design. The ability to identify a consulting company that does possess expertise in the variety of new design techniques, specifically process integration design tools, may be difficult. In addition, there is often less accountability for an environmental consulting company to achieve a solution that fits the long-term strategy of the company. A summary of the benefits and drawbacks of maintaining an internal corporate environmental design group, with process integration expertise, versus the use of external consultants is provided in Table 4.3.

A potential solution to this issue is to establish environmental process design responsibilities within a larger design group. Establishing the

TABLE 4.3
Selecting the process integration environmental design team

	An internal corporate environmental design group	*An external environmental consulting group*
General benefits	Better utilization of company personnel and monetary resources for medium to large chemical process companies Closer alignment with current corporate goals and long-term corporate strategy Better access to company data and personnel resources Better long-term accountability for the design implementation and success	Greater diversity of design experience Better utilization of company personnel and monetary resources for small to medium chemical process companies Greater flexibility of resource levels that can be tailored to the size and complexity of the specific environmental design task
General drawbacks	Represents high overhead costs for the company May not possess a diversity of environmental design experience May create issues due to the lack of resource flexibility	Greater likelihood of specilization in a specific environmental technology (e.g., biotreatment, membrane technology, etc.). This reduces the likelihood of a adequate assessment of environmental technologies that are technically feasible but outside of the consultant's area of expertise Less long-term accountability for the design implementation and success Multiple consultants to chose from

environmental design responsibilities as a subset of a larger corporate group can allow flexibility of resources to fit the timing and technical needs for a particular task. For example, the entire group could be used for a "crisis" design situation that must be addressed as rapidly as possible. In addition, these resources can focus on other corporate tasks when environmental design tasks are minimal. This approach can also result in a more optimal assignment of overhead costs to the company to cover environmental design requirements. A second solution is to establish a long-term partnership with a consulting company that can provide the process integration expertise. This is often a positive strategy for both the chemical company and the consulting company.

The final management issue that must be addressed is the realization that no waste reduction improvements may be identified. Obviously, some data are needed in order to apply process integration design tools. Much effort may be required to collect these data, especially in those cases where the data do not already exist. For example, measurement of process stream flowrates and compositions may be required. After the sufficient plant/site data have been collected, process integration design tools can then be used to evaluate the plant/site. After the analysis, the design team *may* determine that the process is indeed optimal regarding its allocation of energy and mass to the extent that environmental emissions are already minimal. This determination could only occur after the data collection and analysis phases of the study have been completed, thus resulting in what may be viewed as an inefficient use of time and resources. While this scenario may occur, it is *highly* unlikely that an existing plant is already optimal regarding its waste generation and/or emissions.

PERSONNEL-BASED ISSUES FOR APPLYING PROCESS INTEGRATION ENVIRONMENTAL DESIGN TOOLS WITHIN CHEMICAL PROCESS COMPANIES

Several key personnel-based issues that must be addressed when applying process integration technology for environmental design tasks are

- the formation of the overall design team (three subteams);
- the establishment of subteam responsibilities;
- the assignment of design implementation responsibilities.

One of the key steps of the environmental design task is to form the overall design team. This team should contain the management sponsors for the design task, the process integration environmental technology design team responsible for the actual design/economic assessment, and the engineering/operations team responsible for data collection and design feasibility assessments. Figure 4.2 is a schematic representation of this team. The engineering/operations team should consist of key personnel from the operating units that are involved in the generation of the emission streams. In general, this may be a small operating unit; however, the environmental process design solutions are more cost effective when the entire chemical site is included in the analysis. Economically attractive solutions often involve the integration of key mass and energy streams across process unit/plant boundaries.

The second key initiative in forming the team is to clearly define the responsibilities of the subteams. Table 4.4 includes a generic listing of the responsibilities of each of the subteams. As indicated previously, the responsibilities may be modified to be more closely aligned with a specific environmental design task.

TABLE 4.4
Responsibilities of the members of the environmental design team

Subteams	*Responsibilities*
Management Subteam	Provide personnel and financial support for the technical design team and the engineering operations team Provide management and financial support for the implementation of solution projects Provide commitment for the design task procedure so that all subteams are unified toward a common goal
Engineering/Operations Subteam	Collect site mass and energy data (flow rates, compositions, heat duties, etc.) Evaluate results (process designs) provided by the Process Integration Technology Team to consider practical feasibility issues (safety, unit locations, etc.) Lead in the implementation of selected process designs
Process integration Technology Subteam	Utilize process integration analysis methodologies highlighted in Tables 4.1 and 4.2 to identify cost effective environmental design solutions Provide general process flow diagrams for proposed solutions with estimated annual operating costs and total capital costs Estimate the overall profitability for the environmental process designs Iterate the proposed solutions with the Engineering/ Operations Team to assess practical feasibility issues

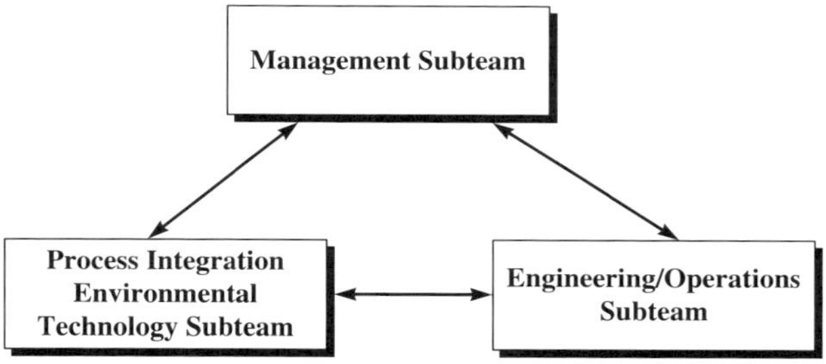

Figure 4.2: Schematic of the overall environmental design team.

A final key personnel-based issue involves the implementation of those projects (environmental process designs) that meet the financial goals, thermodynamic feasibility screening, and practical feasibility screening. Those projects must have assigned engineering "champions" who are dedicated to the procurement of project funding, detailed project engineering, project implementation, and project start-up. In general, this activity is most effectively addressed using personnel from the engineering/operations subteam. The assignment of the project implementation responsibility is a crucial step which is often overlooked.

TECHNOLOGY-BASED ISSUES FOR APPLYING PROCESS INTEGRATION ENVIRONMENTAL DESIGN TOOLS WITHIN CHEMICAL PROCESS COMPANIES

Several key technology-based issues that must be addressed when applying process integration technology for environmental design tasks are

- the establishment of an environmental design team with process integration expertise;
- the selection of an overall technical design strategy;
- the testing of the selected designs via lab scale and/or pilot scale units.

As mentioned earlier, it is possible to establish process integration expertise within a chemical company or through the use of external consultants. Regardless of the source of this expertise, this team must remain up to date on the developments in the field of process integration technology. Again, as mentioned earlier, this can be accomplished via a variety of environmental conferences, environmental design short courses, periodic technical journal searches, and research collaborations with academic institutions that are active in developing new environmental design tools based on process integration approaches.

The second issue is the selection of a systematic approach for managing the use of the large variety of design tools available. An example of a systematic design strategy for wastewater reduction design has been provided via the introduction of the Water Allocation Design and Engineering (WADE) approach (Dunn and Dobson, 1999). The WADE approach uses a variety of environmental design tools (water management tools, systems analysis tools, interception technology tools), as indicated in Figure 4.3, to arrive at the most cost-effective solution in a systematic manner, and a detailed explanation of these tools has been provided in literature. Another example of a systematic approach for wastewater minimization has been developed based on a water pinch graphical methodology (Wang and Smith, 1994). Both methodologies fall in the category of process integration design tools.

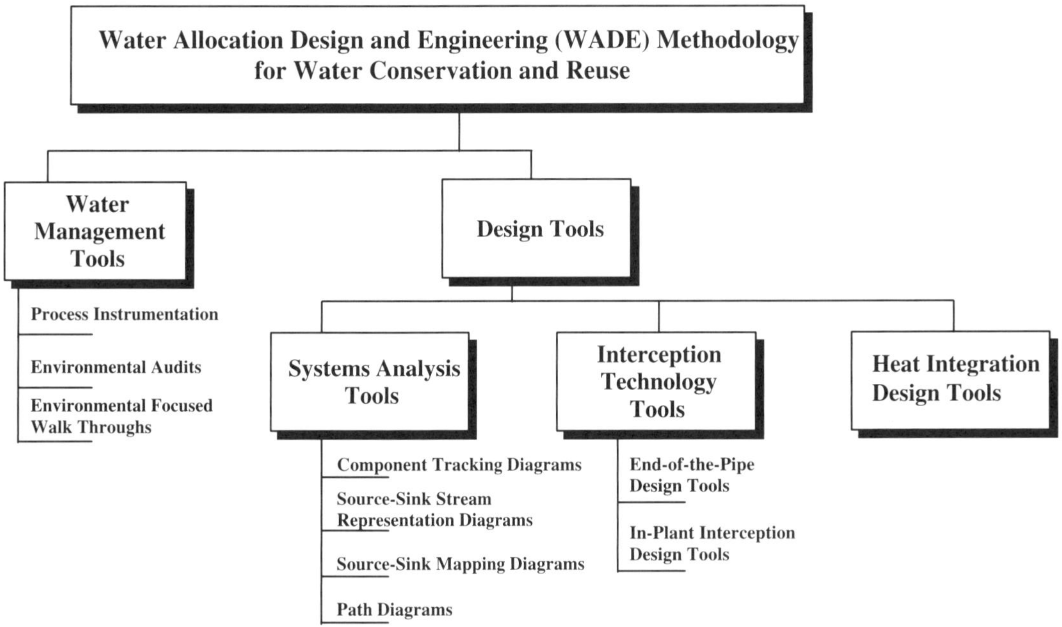

Figure 4.3: Summary of the design tools used in the WADE methodology.

Within the WADE methodology, the interception technology tools are the most technically complex and time consuming to utilize; however, it is often possible to achieve the desired environmental emission reductions via the water management tools and the systems analysis tools. In general, these tools cover process designs that involve general water conservation and direct recycle designs for wastewater streams. When the direct recycle designs are insufficient to achieve the desired reductions in environmental emissions, it is necessary to identify cost-effective interception technologies for key streams. As previously indicated in Tables 4.1 and 4.2, several design methodologies have been developed to evaluate a host of interception technology options in a systematic and time effective manner.

In summary, the WADE methodology and the accompanying tools have been applied at numerous large chemical process companies and most recently at two chemical process facilities whose results have been documented (Dunn, 1996; Dunn et al., 1997; Dunn and Dobson, 1999). The WADE systematic approach for environmental process design can also be adapted to address process design for gaseous emissions or the process integration subteam can develop its own overall strategy for using the large variety of tools. The important point is that some overall strategy for environmental design tool management and use should be employed.

The final technology issue that must be addressed is the lab scale or pilot scale testing of the selected technologies. This step may or may not be necessary, depending on many issues surrounding the final designs, such as complexity, experience with the technology selected, controllability, etc. A pilot scale operation is generally established to address these issues and to allow refinement of the economic assessment of the designs.

CONCLUDING REMARKS

Process integration technology is emerging as one of the most innovative design approaches to address environmental process design. While the process integration design methodologies are valuable for addressing the design tasks, there are a number of challenges in the form of management-based issues, personnel-based issues, and technology-based issues surrounding the use of this technology. Guidelines for addressing these challenges have been provided in this chapter so that this new technology can be more readily applied to solve environmental process design tasks within chemical process companies.

ACKNOWLEDGMENTS

The author wishes to acknowledge the many individuals who have contributed to his efforts in developing and applying process integration design tools within chemical process companies, especially the continued personal encouragement and technical suggestions provided by Dr. Mahmoud El-Halwagi, Auburn University. The author also wishes to acknowledge God for guiding his path and for allowing him to use his knowledge and experience to be a better steward of the environmental resources that He has created. We are all commanded by God to govern the earth's resources responsibly.

"The Lord God took the man and put him in the Garden of Eden to work it and take care of it." Genesis 2:15

REFERENCES

Cerda, J., Westerberg, A. W., Mason, D., and Linnhoff, B. (1983). Minimum utility usage in heat exchanger network synthesis—a transportation problem. *Comp. Chem. Eng.* 38:373–387.

Constantinou, L., Jacksland, C., Bagherpour, K., Gani, R., and Bogle, L. (1995). Application of group contribution approach to tackle environmentally-related problems. *AIChE Symp. Ser.* 90:105–116.

Crabtree, E. W., and El-Halwagi, M. M. (1995). Synthesis of environmentally acceptable reactions. *AIChE Symp. Ser.* 990:117–127.

Crabtree E. W., Dunn, R. F., and El-Halwagi, M. M. (1998). Synthesis of hybrid gas permeation membrane/condensation systems for pollution prevention. *Journal of the Air and Waste Management Association* 48:616–626.

Dhole, V. (1995). Total site integration for reducing process emissions. *Waste Minimization Through Process Design* (A. P. Rossiter), pp. 79–92. McGraw-Hill, New York.

Dobson, A. M. (1998). Shortcut design methodologies for identifying cost-effective heat-induced and energy-induced separation networks for condensation-hybrid processes. Masters Thesis, Auburn University.

Dunn, R. F., and El-Halwagi, M. M. (1994a). Selection of optimal VOC-condensation systems. *Waste Mgt.* 14:103–113.

Dunn, R. F., and El-Halwagi, M. M. (1994b). Optimal design of multi-component VOC-condensation systems. *J. Haz. Mtls.* 38:187–206.

Dunn, R. F., Zhu, M., Srinivas, B. K., and El-Halwagi, M. M. (1995). Optimal design of energy-induced separation systems for VOC recovery. *AIChE Symp. Ser.* 90:74–85.

Dunn, R. F. (1996). Solutia Special Report ICP 14142, Pensacola Site Water Conservation Probe, July, company confidential.

Dunn, R. F., Dobson, A. M., and El-Halwagi, M. M. (1997). Optimal design of environmentally-acceptable solvent blends for coatings. *Advances in Environmental Research* 1:115–134.

Dunn, R. F., Dobson, A. M., and Stonecipher, J. (1997). Solutia Special Report ICP 14173, Chocolate Bayou Wastewater Minimization Study, December, company confidential.

Dunn, R. F., and Srinivas, B. K. (1997). Synthesis of heat-induced waste minimization networks (HIWAMINs). *Advances in Environmental Research* 1:275–301.

Dunn, R. F., and Dobson, A. M. (1998). A spreadsheet based approach to identify cost-effective heat-induced and energy-induced separation networks for condensation-hybrid processes. *Advances in Environmental Research* 2:269–290.

Dunn, R. F., and Dobson, A. M. (1999). Water Conservation & Reuse at a Major CPI Facility, National Association of Corrosion Engineers Annual Conference Proceedings, paper 317, pp. 1–18.

Dunn, R. F., Hamad, A. A., and Dobson, A. M. (1999). Synthesis of energy-induced waste minimization networks (EIWAMINs) for simultaneous waste reduction and heat integration. *Clean Products and Processes*. 1:91–106.

Dye, S. R., Berry, D. A., and Ng, K. M. (1995). Synthesis of crystallization-based separation schemes. *AIChE Symp. Ser.* 91:238–241.

El-Halwagi, M. M., and Manousiouthakis, V. (1989a). Synthesis of mass-exchange networks. *AIChE J.* 35:1233–1244.

El-Halwagi, M. M., and Manousiouthakis, V. (1989b). Design and analysis of multicomponent mass-exchange networks. AIChE Annual Meeting, San Francisco.

El-Halwagi, M. M., and Manousiouthakis, V. (1990a). Automatic synthesis of mass-exchange networks with single component targets. *Chem. Eng. Sci.* 45:2813–2831.

El-Halwagi, M. M., and Manousiouthakis, V. (1990b). Simultaneous synthesis of mass-exchange and regeneration networks. *AIChE J.* 36:1209–1219.

El-Halwagi, M. M. (1992). Synthesis of optimal reverse-osmosis networks for waste reduction. *AIChE J.* 38:1185–1198.

El-Halwagi, M. M., and Srinivas, B. K. (1992). Synthesis of reactive mass-exchange networks. *Chem. Eng. Sci.* 47:2113–2119.

El-Halwagi, M. M., Srinivas, B. K., and Dunn, R. F. (1995). Synthesis of heat-induced separation networks. *Chem. Eng. Sci.* 50:81–97.

El-Halwagi, M. M., Hamad, A. A., and Garrison, G. W. (1996). Synthesis of waste interception and allocation networks. *AIChE J.* 42:3087–3101.

El-Halwagi, M. M. (1997). *Pollution prevention through process integration: Systematic design tools*. Academic Press, San Diego.

Gunderson, T., and Naess, L. (1988). The synthesis of cost optimal heat exchanger networks: An industrial review of the state of the art. *Comp. Chem. Eng.* 12:503–530.

Gupta, A., and Manousiouthakis, V. (1994). Waste reduction through multicomponent mass exchange network synthesis. *Comp. Chem. Eng.* 18:S585.

Hamad, A. A., and El-Halwagi, M. M. (1998). Simultaneous synthesis of mass separating agents and interception networks. *Trans. IChemE* 76:376–388.

Hertz, D. W. (1995). Managing the design process. *Waste Minimization Through Process Design* (A. P. Rossiter), pp. 265–288. McGraw-Hill, New York.

Jobak, K. G. (1995). Solvent substitution for pollution prevention. *AIChE Symp. Ser.* 90:98–104.

Kiperstok, A., and Sharratt, P. N. (1995). On the optimization of mass-exchange networks for fixed removal of pollutants. *Trans. IChemE* 73B:271–277.

Linnhoff, B., and Hindmarsh, E. (1983). The pinch design method for heat exchange networks, *Chem. Eng. Sci.* 38:745–763.

Linnhoff, B. (1994). Use pinch analysis to knock down capital costs and emissions, *Chem. Eng. Prog.* 90:32–57.

Linnhoff, B. (1995). Pinch analysis in pollution prevention. *Waste Minimization Through Process Design* (A. P. Rossiter), pp. 53–68. McGraw-Hill, New York.

Papalexandri, K. P., and Pistikopoulos, E. N. (1994). A multiperiod MINLP model for the synthesis of heat and mass exchange networks.*Comp. Chem. Eng.* 18:1125–1139.

Papalexandri, K. P., Floudas, C., and Pistikopoulos, E. N. (1994). Mass exchange networks for waste minimization: A simultaneous approach. *Trans. IChemE* 72: 279–294.

Papoulias, S. A., and Grossmann, I. E. (1983). A structural optimization approach in process synthesis. II: Heat recovery networks, *Comp. Chem. Eng.* 7:707.

Richburg, A., and El-Halwagi, M. M. (1995). A graphical approach to the optimal design of heat-induced separation networks for VOC recovery. *AIChE Symp. Ser.* 91:256–259.

Shenoy, U. V. (1995). *Heat Exchanger Network Synthesis*. Gulf Publishing Company, Houston.

Srinivas, B. K., and El-Halwagi, M. M. (1994a). Synthesis of reactive mass-exchange networks with general nonlinear equilibrium functions. *AIChE J.* 40:463–472.

Srinivas, B. K., and El-Halwagi, M. M. (1994b). Synthesis of combined heat and reactive mass-exchange networks. *Chem. Eng. Sci.* 45:2059–2074.

Srinivas, B. K., and El-Halwagi, M. M. (1993). Optimal design of pervaporation systems for waste reduction. *Comp. Chem. Eng.* 17:957–970.

Wang, Y. P., and Smith, R. (1994). Wastewater minimization. *Chem. Eng. Sci.* 49:981–1006.

CHAPTER 5

Pollution Prevention and Process Integration: Two Complementary Philosophies

Alan P. Rossiter

Process Improvement Consultant, 4421 Darsey Street, Bellaire, TX 77401

Abstract

This chapter describes the various classes of process integration techniques—pinch analysis, knowledge-based approaches, and numerical/graphical optimization approaches—and explores their applications in pollution prevention. Pollution prevention aims to improve overall processes to reduce emissions, rather than resolving emission problems with add-on, end-of-the-pipe solutions. In a similar way, process integration seeks to improve process designs by considering complete systems rather than individual unit operations. The two concepts are therefore highly compatible, and process integration has become an important tool in the quest for designs that minimize environmental impact.

Keywords: process improvement, emissions, systematic design methods

INTRODUCTION

Pollution prevention and process integration are two complementary philosophies. Pollution prevention seeks to eliminate pollution problems by not creating potential pollutants at all or, if their creation cannot be avoided, by containing them within the process that makes them. Process integration is an evolving discipline encompassing a number of interrelated methodologies for designing and revamping industrial processes based on a philosophy that recognizes the unity of the whole process.

From their earliest applications, process integration techniques have been used in ways that are inherently conservation oriented (i.e., enhancing process efficiency by minimizing the use and/or maximizing the recovery of energy and materials). Moreover, because process integration techniques view processes in their entirety, they can be readily employed to achieve pollution prevention goals by minimizing emissions from the entire process. This chapter discusses the various classes of methodologies under the process integration umbrella and illustrates how these have been developed and applied to pollution prevention problems.

POLLUTION PREVENTION AND THE POLLUTION PREVENTION HIERARCHY

Pollution has always been a part of industrial progress. At the time of the industrial revolution it seems that it was considered an inevitable part of the price of progress. However, over time the severity of the environmental impact of our industrial activities has become apparent. This has led to growing pressure to develop "greener" processes. It is no longer socially, legally, or politically acceptable to pollute.

Until recent times, the emphasis has primarily been on "pollution control" (i.e., finding ways to prevent undesirable substances from reaching the environment). Typically this involves providing "end-of-pipe" solutions to environmental problems (e.g., scrubbing or incinerating to remove toxic vapors from exhaust air streams or neutralizing acidic or alkaline wastewaters before discharging them).

This approach has certain attractions. It provides well-defined technical solutions to specific problems. These solutions are readily transferable from one location to another, and companies have developed methods to perfect and market the technologies. Furthermore, end-of-pipe solutions are easy to legislate and regulate. You simply define the type of technology that has to be used in certain applications, together with some key performance parameters, and require all operators of that type of process to install the appropriate equipment and comply with the performance standards.

Unfortunately, many pollution control techniques simply transfer pollution problems from one medium to another. For example, absorption of undesirable vapors may remove an air emission problem, but it often creates a water discharge problem. This situation can also be reversed: air stripping is sometimes used to remove volatile contaminants from water. Other techniques simply disguise real problems. Historically it was a common "pollution control" practice for operators of processes with gaseous and aqueous emissions to dilute their discharges to comply with mandated concentration limits. "Dilution is the solution to pollution!"

In recent years there has been a philosophical shift away from "pollution control" toward "pollution prevention." The difference is basically this: Pollution control aims to prevent the release of undesirable materials and relies heavily on end-of-pipe technologies to achieve this objective. Pollution prevention, on the other hand, seeks to eliminate pollution problems entirely by either not creating potential pollutants at all or, if their creation cannot be avoided, by keeping them as far as possible from the end of the pipe.

The "pollution prevention hierarchy" (Freeman, 1990) ranks the basic means of reducing emissions in terms of global desirability. Several versions of the hierarchy have been proposed, of which the following (showing the ranking, from most to least desirable) is one:

1. Source Reduction (e.g., use cleaner feed material; eliminate leaks).
2. Reuse and Recycle (e.g., reuse waste from one unit as a feed to another; recycle contaminated material to the unit that generated it, with a small purge to maintain acceptable contaminant levels).
3. Treatment (e.g., acid neutralization; incineration).
4. Safe Disposal (e.g., landfill). It is important to note that even under the pollution prevention philosophy it is recognized that some end-of-pipe discharges may be inevitable; however, they are to be minimized, both in terms of quantity and potential harmfulness.

Any pollution prevention option can, in principle, be classified under one of these categories, although in practice the classifications are not always clear cut. For example, "treatment" options could be used to make a stream suitable for reuse or recycle (these are commonly known as "regeneration" options), or they could be precursors to safe disposal (end-of-pipe treatment).

Pollution prevention has several clear advantages over pollution control. Any pollutant that reaches the end of the pipe and has to be destroyed represents a loss of raw material. As pollution prevention seeks to minimize the amount of material that suffers this fate, it is a way of conserving natural resources, which is good stewardship. This also results in savings in raw material and operating costs for the process operator, and thus tends to improve process economics.

There are also some difficulties inherent in the pollution prevention concept. From the perspective of the regulators, it is far more problematic to legislate and monitor. Unlike conventional pollution control, pollution prevention is not usually accomplished simply by placing a piece of equipment immediately ahead of a discharge point. The solutions are less clear cut and well defined, and may not always be transferable, even between outwardly similar industrial facilities.

Pollution prevention also poses a unique challenge for those engaged in the design and improvement of industrial processes. As already noted, pollution prevention usually is not accomplished simply by placing a piece of equipment immediately ahead of a discharge point. Rather, it depends on overall process designs that are intrinsically friendly to the environment. This calls for a philosophy of process design that recognizes the unity of the whole process. Such a design philosophy has been evolving for several years within the process engineering community, and it is known as "process integration."

PROCESS INTEGRATION METHODS

Several different systematic approaches to the design and retrofit of industrial processes have developed under the banner of process integration, and they are fundamentally changing the way in which process design and retrofit activities are carried out. Process integration is a term used for integrated and systems-based design approaches that are geared towards enhancing the performance of new and retrofitted processes.

Typically the procedures start with an overview of the process as a whole, rather than focusing initially on individual unit operations or pieces of equipment. In this way, a "correct" structure can be developed for the overall plant, with individual items of equipment being fitted into this structure. This approach has been successfully applied for many years in cost minimization activities, with the emphasis on the trade-off between capital and operating costs; however, the same basic methods can also be used to explore the three-way capital cost/operating cost/environmental trade-off. Process integration has proven to be extremely useful in the overall management of mass and energy (El-Halwagi, 1997; Linnhoff, 1994).

PROCESS INTEGRATION OR PROCESS SIMULATION?

Process integration is used to identify the most appropriate flow sheet structure and optimize process parameters such as flow rates, temperatures, and pressures. Process simulation, on the other hand, is used to develop accurate heat and material balances and physical properties data for a given flow sheet. Process simulation is used primarily to understand the operating characteristics of a single design, whereas process integration is used to develop new and better designs.

Process integration activities can often be carried out with very limited data about the process under consideration. In contrast, to carry out a process simulation, it is generally necessary to have sufficient information to define all major process flows, temperatures, pressures, and compositions, and to characterize the performance of all major items of equipment.

Process integration and process simulation are both essential to good process design. They are closely related and may take place in parallel within the overall design process. Moreover, the distinction between them is somewhat blurred, especially in the area of parametric optimizations. Process simulation software packages are becoming more powerful, enabling them to carry out this task effectively in many instances. We could view this as using process simulation tools to assist in process integration work, or alternatively, we could see it as changing the scope of process simulation to include parametric optimization.

CLASSES OF PROCESS INTEGRATION METHODS

Process integration methodologies can be classified under three major headings (Gunderson and Naess, 1987):

- Numerical modeling
- Knowledge-based approaches
- Pinch analysis

These three classes of process integration methodologies are outlined below.

NUMERICAL MODELING

Numerical modeling includes a variety of approaches, ranging from the use of simplified spreadsheet models of a process to sophisticated mathematical programming methods. These are generally combined with cost equations to quantify the impact of design decisions on process economics. The results of this type of analysis are often presented in graphical form, for ease of interpretation.

One simple graphical format that has been used in pollution prevention studies is the "cost-versus-emission limit curve" (Rossiter et al., 1992). This can be developed with the aid of a spreadsheet, and it allows engineers to explore the impact of process changes on cost and emission levels. At the same time, it serves to define the most cost-effective means of achieving a given emissions target. No complicated mathematics are involved here; optimization is achieved by purely graphical means.

An example of this type of curve is shown in Figure 5.1 (Rossiter and Kumana, 1994). This is the "minimum average control cost vs. emission limit curve" for NO_x emissions from an oil refinery. Each point on the graph represents the performance and costeffectiveness of that combination of pollution prevention and pollution control equipment that provides the least-cost combination of measures to achieve a given emissions target. In this example, the following technologies are considered: heat integration (HI), low excess air burners (LEA), staged burners (SB), flue gas recirculation (FGR), selective noncatalytic reduction (SNCR), and selective catalytic reduction (SCR).

The average cost ($ thousands per ton) of reducing NO_x emissions to any given level is shown on the vertical axis. This cost includes both capital and operating cost elements, which are combined to give a "total annualized cost" (TAC, $ thousands/year) as follows:

$$\text{TAC} = \text{Operating Cost} + b(\text{Capital Cost}),$$

where b is the "annualization factor." Dividing the TAC by the reduction in emissions from the base case to any specified level (ton/year) yields the average cost for reducing the emissions to that level ($ thousands/ton).

Interval algebra has been used recently to develop targeting techniques for process integration (Noureldin and El-Halwagi, 1999). The key concept is

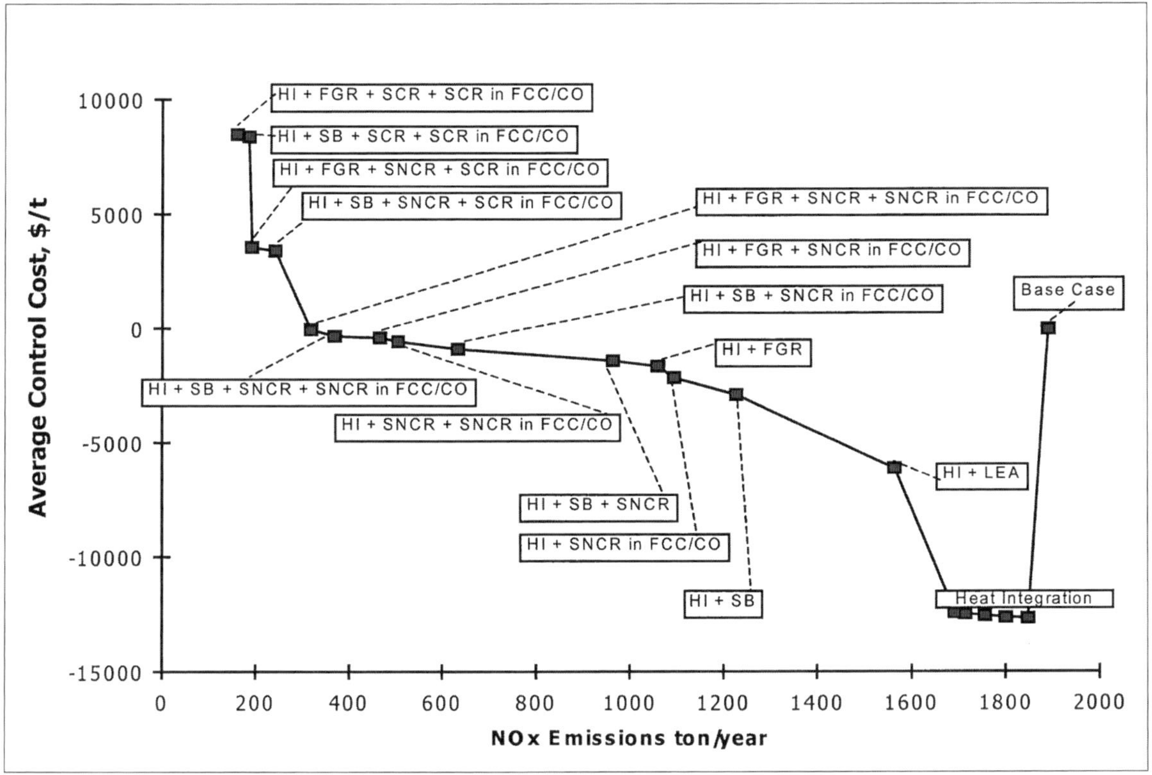

Figure 5.1: Minimum average control cost vs. NO_x emissions plot. This graph shows the cost of achieving any given level of NO_x emissions from a refinery, starting from a given base case.

to bound process performance by integrating all possible changes in process structure as well as design and operating parameters. Any potential variation is described by a bounding interval. Interestingly enough, very few data and computations are needed to establish these targets.

Linear and nonlinear programming techniques offer a powerful mathematical approach and have been used in many different applications (El-Halwagi, 1997). In general, the process or other system under investigation is modeled mathematically, and a suitable optimization algorithm is then used to determine the preferred solution to the problem. The level of detail and accuracy of the model can vary considerably. Data are often linearized to simplify problems and facilitate convergence, but with some loss of rigor (Nath and Kumana, 1992). However, powerful linear optimization techniques are also available, so more precise linear formulations can also be used (Srinivas and El-Halwagi, 1994; El-Halwagi et al., 1995).

In the field of pollution prevention, this approach has been used for many types of studies, including wastewater minimization (Rossiter and Nath, 1995), waste minimization in pulp and paper operations (Dunn and El-Halwagi, 1993), and synthesis of reverse osmosis networks for waste minimization (El-Halwagi, 1992). Much attention has also been devoted to minimization of VOC emission using these methods (Sung et al., 1992; Rossiter et al., 1992; Dunn and El-Halwagi, 1994; Stanley and El-Halwagi, 1995; Dunn and Dobson, 1998). Similar approaches have also been used for real-time optimization to reduce certain emissions, such as NO_x (Nath and Kumana, 1992).

KNOWLEDGE-BASED APPROACHES

The term "knowledge-based system" (or "expert system") is used to describe a class of artificial intelligence applications that embodies a system of rules based on an area of expert knowledge. In the present discussion, the term "knowledge-based approaches" is used in a broader sense, describing those process integration methods that build on an accumulated "knowledge base" of proven ideas for waste minimization. The underlying basis is the recognition that there are similarities even between processes that are outwardly very different. Therefore, solutions that work in any given process should be transferable to others.

The most basic form of this process integration class is simply a list of specific pollution prevention ideas that can be transferred directly from one application to others (Nelson, 1990). The list can be developed into a database and managed in such a way that potential solutions are matched to specific problems. The UO-WM (*U*nit *O*perations *W*aste *M*inimization) software system from the University of South Carolina (Smith and Khan, 1995) is one example of this approach. In this case, applicable solutions to a given problem are located and related through common unit operations between industrial categories. CPAS (Clean Process Advisory System) is a more comprehensive suite of programs being developed as a collaborative effort by numerous organizations in industry, academia, and government (Hertz, 1995).

The system includes design tools containing new technology performance data; experience-based information or expert guidance; case studies; simulation modules; health, environmental, safety, and risk data; and federal, state, and local regulatory information.

Hierarchical design and review procedures provide another structured knowledge-based approach to process integration. In this case, the logical sequence of flow sheet evolution provides a framework for identifying and evaluating waste minimization options (Douglas, 1992; Rossiter et al., 1993). The basic premise is that process design proceeds through a series of decisions that involve progressively greater levels of detail. By identifying the key decisions and implementing them in the correct order, one can identify appropriate design or retrofit options to minimize pollution with a minimum of effort and rework. At each stage there are specific questions to address, heuristics to apply, and trade-offs to consider. This is illustrated below for the earliest steps in the design procedure.

The most basic questions are whether a process is to be operated continuously or in batch mode, which product or products will be made, and which feed or feeds will be used. At this stage a simple input/output diagram (Fig. 5.2) is a useful tool for gaining an overview of the material transformations that are taking place, and identifying any major waste streams and how they are generated. It also helps in formulating and addressing key questions, such as

- Can the feed quality be improved economically?
- Should impurities be removed before processing or should they be allowed to pass through the process?
- Can any input streams be eliminated?
- Can any "waste" output stream be used beneficially within the process plant or sold to an external customer?

Based on the answers to these questions, an initial view can be taken on the economic viability of the process. If it looks attractive, the material conversion aspects of the process and recycle structure must now be considered (Fig. 5.3). At this stage, key questions typically include

- Do any "waste" output streams contain feed or product material that can be recovered and recycled?

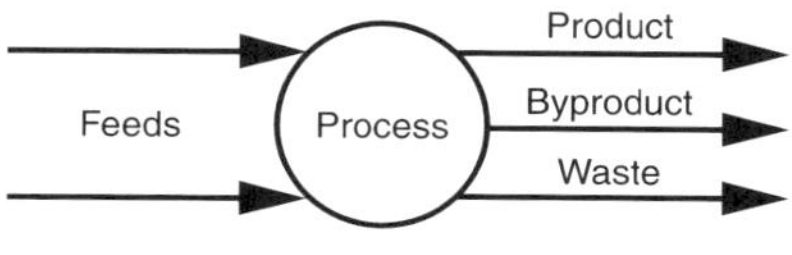

Figure 5.2: The input/output diagram is used to address fundamental feed and product questions.

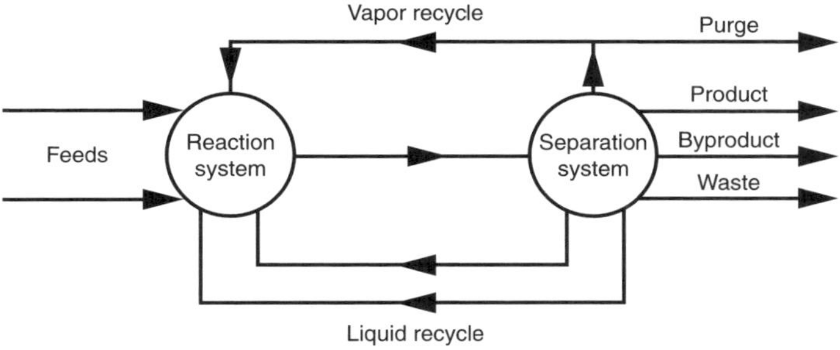

Figure 5.3: The reaction and recycle structure diagram helps to define material conversion and recovery issues.

- Can reaction conditions be altered to minimize formation of "waste" by-products?
- Can "waste" by-products be recycled to extinction?

Additional details, such as separation systems and heat integration, can then be added. More complete structural diagrams are developed, key questions are addressed, and economic viability is calculated at each level. Only after these conceptual decisions have been made is it appropriate to consider the selection and design of individual items of equipment.

The most sophisticated approach within this process integration class is artificial intelligence. In this case, computer programs use a combination of first-principles knowledge and heuristic rules to develop "clean" process designs. Numerous problems have been addressed using systems of this type [e.g., minimization of hazardous species and waste energy (Huang and Fan, 1994), fault detection in the operation of a hazardous waste incineration system (Huang and Fan, 1993), and minimization of waste generation and energy consumption in a petroleum refinery (Huang and Edgar, 1995)].

The knowledge-based process integration approaches described above can either be used for developing new designs or for identifying retrofit options, often starting with minimal data. In new plant design, use of this type of procedure generally results in one or more "good" design(s) for the process (i.e., cost-effective, with low emissions). In retrofits, they typically generate a list of potential process improvements for use in revamp projects.

PINCH ANALYSIS

Pinch analysis (Linnhoff et al., 1992) is a systematic technique, based on fundamental thermodynamic principles, for analyzing the flow of heat and mass through industrial processes. The basic concept for thermal or heat

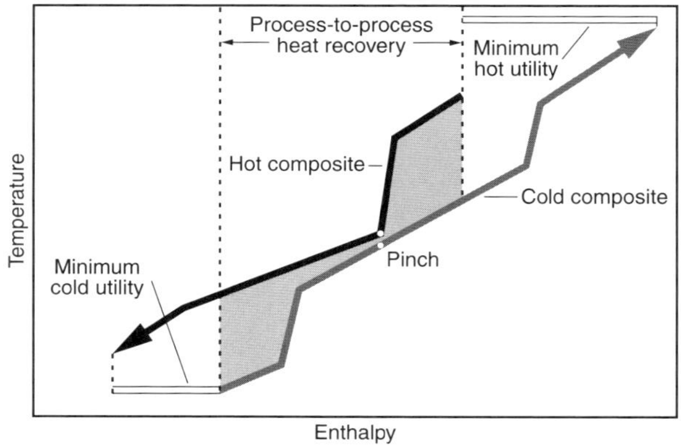

Figure 5.4: Composite curves represent the heat sources and sinks within a process. They are used to show where heat flow is restricted (the "pinch").

transfer pinch analysis is illustrated in Figure 5.4, which is a set of "composite curves" for an arbitrary process. The hot composite curve represents the sum of all heat sources (i.e., streams that require cooling) within the process in temperature-enthalpy terms. The cold composite curve similarly represents the sum of all heat sinks (i.e., streams that require heating). When the curves are shown together on a single temperature-enthalpy plot, most processes display a pinch, a region where the curves come very close together. This divides the process into two distinct sections—above the pinch it has a net heat deficit and requires an external utility heat source; below the pinch it has a net heat surplus and requires an external utility heat sink.

The decomposition between net source and net sink regions is a key characteristic of the pinch approach, and forms the starting point for process optimization using the pinch method. It is widely used to determine the scope for energy savings in industrial operations and to define possible process changes to reduce intrinsic energy consumption. The trade-off between energy consumption and capital investment can also be assessed, as well as the pressure drop implications of heat recovery, distillation column optimization, total site analysis, and other aspects of energy conservation (Linnhoff, 1994).

In the environmental context, thermal pinch analysis is useful primarily as a means of determining the extent to which energy consumption can be reduced, with the attendant reductions in NO_x, SO_x, and CO_2 (Sung et al., 1995); however, it also provides some guidance in the generation of design options for the reduction of other, process-related emissions (Rossiter et al., 1991).

The pinch method has also been adapted for analyzing direct contact mass transfer-based separation process design (El-Halwagi and Manousiouthakis, 1989). In this case, composite curves are developed for all rich (waste) and all lean (mass separating agent) process streams. Further developments include heat-induced pinch diagrams and energy-induced pinch diagrams. Both of these couple heat integration and separation objectives by developing composite curves of all hot process streams, cold process streams, and waste streams that could be separated via heat-induced separation processes. In the case of energy-induced pinch diagrams, these are adjusted via the addition of pressurization/depressurization units (Dunn and El-Halwagi, 1996). These diagrams are used for waste separation process design coupled with heat integration process design. They provide a basis for simultaneously assessing heat and mass integration to achieve environmental objectives, and have been used to solve a number of in-plant separation problems (El-Halwagi et al., 1996; Dunn and Srinivas, 1997; Dobson et al., 1999).

Dilute aqueous systems represent another special case: water pinch (Wang and Smith, 1994; Buehner and Rossiter, 1996). Here, the mass transfer system is represented graphically in terms of mass flow and concentration. This approach has been used to analyze wastewater systems and identify opportunities for improvements in numerous processes.

CONCLUSION

To achieve the objective of pollution prevention, it is necessary to adopt a philosophically complementary approach to process design. Process integration provides that approach.

Process integration methodologies can be placed in three major categories:

- Numerical modeling
- Knowledge-based approaches
- Pinch analysis

The three different branches of process integration have somewhat different areas of application, and tend to yield different, complementary results. The numerical modeling approach, for example, is best suited to situations where only a limited number of well-defined design options require evaluation. For complex processes with multiple variants, the modeling effort can become overwhelming, and the other approaches, especially knowledge-based methods, are needed to identify potentially attractive options and narrow down the scope of the problem. The pinch approach is good for identifying fundamental insights into heat transfer and mass transfer problems, and can result in step-change design improvements.

These approaches are continuously evolving. They have already been applied to a wide range of industrial processes, and the scope of their applicability, both in new design situations and in retrofits, is likely to increase in the future.

REFERENCES

Buehner, F. W., and Rossiter, A. P. (1996). Minimize waste by managing process design. *Chemtech* 26(4):64–72.

Dobson, A. M., Hamad, A. A., and Dunn, R. F. (1999). Design of energy-induced waste minimization networks. Process Systems Engineering Topical Conference Proceedings. AIChE 1999 Spring Meeting, Houston, pp.102–124.

Douglas, J. M. (1992). Process synthesis for waste minimization. *Ind. Eng. Chem. Res.* 31:238–243.

Dunn, R. F., and Dobson, A. M. (1998). A spreadsheet-based approach to identify cost-effective heat-induced and energy-induced separation networks for condensation-hybrid processes. *Advances in Environmental Research* 2:269–290.

Dunn, R. F., and El-Halwagi, M. M. (1993). Optimal recycle/reuse policies for minimizing the wastes of pulp and paper plants. *J. Environ. Sci. Health* A28:217–234.

Dunn, R. F., and El-Halwagi, M. M. (1994). Optimal design of multi-component VOC condensation systems. *Journal of Hazardous Materials* 38:187–206.

Dunn, R. F., and El-Halwagi, M. M. (1996). *Design of Cost-Effective VOC Recovery Systems*. Published by Tennessee Valley Authority Environmental Research Center/EPA Center for Waste Reduction, 82 pp., http://www.p2pays.org/ref/01/00034.htm.

Dunn, R. F., and Srinivas, B. K. (1997). Synthesis of heat-induced waste minimization networks HIWAMINs. *Advances in Environmental Research* 1:275–301.

El-Halwagi, M. M. (1992). Synthesis of reverse-osmosis networks for waste minimization. *AIChE J.* 38:1185–1198.

El-Halwagi, M. M. (1997). *Pollution Prevention through Process Integration: Systematic Design Tools*. Academic Press, San Diego.

El-Halwagi, M. M., Hamad, A. A., and Garrison, G. W. (1996). Synthesis of waste interception and allocation networks. *AIChE J.* 42:3087–3101.

El-Halwagi, M. M., and Manousiouthakis, V. (1989). Synthesis of mass-exchange networks. *AIChE J.* 35:1233–1244.

El-Halwagi, M. M., Srinivas, B. K., and Dunn, R. F. (1995). Synthesis of heat-induced separation networks. *Chem. Eng. Sci.* 50:81–97.

Freeman, H. M. (1990). *Hazardous Waste Minimization*, p. 7. McGraw-Hill, New York.

Gunderson, T., and Naess, L. (1987). The synthesis of cost optimal heat exchanger networks. XVIII EFCE Congress, Giardini Naxos, Italy.

Hertz, D. W. (1995). Managing the design process. *Waste Minimization through Process Design* (A. P. Rossiter), pp. 265–287. McGraw-Hill, New York.

Huang, Y. L., and Edgar, T. F. (1995). Knowledge-based design approach for the simultaneous minimization of waste generation and energy consumption in a petroleum refinery. *Waste Minimization through Process Design* (A. P. Rossiter), pp. 181–196. McGraw-Hill, New York.

Huang, Y. L., and Fan, L. T. (1993). A fuzzy-logic-based approach to building efficient fuzzy rule-based expert systems. *Comp. Chem. Eng.* 17:181–192.

Huang, Y. L., and Fan, L. T. (1994). HIDEN: A hybrid intelligent system for synthesizing highly controllable exchanger networks: Implementation of a distributed strategy for integrating process design and control. *IE&C Res.* 33:1174–1187.

Linnhoff, B. (1994). Pinch analysis: Building on a decade of progress. *Chem. Eng. Prog.* 90(8):32.

Linnhoff, B., Townsend, D. W., Boland, D., Hewitt, G. F., Thomas, B. E. A., Guy, A. R., and Marsland, R. H. (1982). *User Guide on Process Integration for the Efficient Use of Energy*. IChemE, Rugby, England.

Nath, R., and Kumana, J. D. (1992). NOx dispatching in plant utility systems-using existing tools. 14th Industrial Energy Technology Conference, Houston, Texas.

Nelson, K. E. (1990). Use these ideas to cut waste. *Hydroc. Proc.* 69(3):93–98.

Noureldin, M. B., and El-Halwagi, M. M. (1999). Interval-based targeting for pollution prevention via mass integration. *Comp. Chem. Eng.* 23:1527–1543.

Rossiter, A. P., Klee, H., Jr., and Sung, R. D. (1992). Process integration techniques for pollution prevention. American Chemical Society I&EC Special Symposium on Emerging Technologies for Hazardous Waste Management, Atlanta.

Rossiter, A. P., and Kumana, J. D. (1994). Rank pollution prevention and control options. *Chem. Eng. Prog.* 90(2):39.

Rossiter, A. P., and Nath, R. (1995). Wastewater minimization using nonlinear programming. *Waste Minimization through Process Design* (A. P. Rossiter), pp. 225–243. McGraw-Hill, New York.

Rossiter, A. P., Rutkowski, M. A., and McMullan, A. S. (1991). Pinch technology identifies process improvements. *Hydroc. Proc.* 70(1):63–66.

Rossiter, A. P., Spriggs, H. D., and Klee, H., Jr. (1993). Apply process integration to waste minimization. *Chem. Eng. Prog.* 89(1):30–36.

Smith, R. L., and Khan, J. (1995). Unit operations database for transferring waste minimization solutions. *Waste Minimization through Process Design* (A. P. Rossiter), pp. 133–148. McGraw-Hill, New York.

Srinivas, B. K., and El-Halwagi, M. M. (1994). Synthesis of reactive mass-exchange networks with general nonlinear equilibrium functions. *AIChE J.* 40:463–472.

Stanley, C., and El-Halwagi, M. M. (1995). Synthesis of mass exchange networks using linear programming techniques. *Waste Minimization through Process Design* (A. P. Rossiter), pp. 209–224. McGraw-Hill, New York.

Sung, R. D., Rossiter, A. P., and Kumana, J. D. (1992). Process integration and the environment. International Energy Agency Workshop on Process Integration. Gothenberg, Sweden.

Sung, R. D., Kumana, J. D., and Rossiter, A. P. (1995). Southern California Edison's CTAP: Saving energy and reducing pollution for industrial customers. *Waste Minimization through Process Design* (A. P. Rossiter), pp. 335–343. McGraw-Hill, New York.

Wang, Y. P., and Smith, R. (1994). Wastewater minimization. *Chem. Eng. Sci.* 49:981–1006.

CHAPTER 6

Process Integration, Synthesis, and Analysis for Cleaner Processes

A. Alva-Argáez, A. C. Kokossis, and R. Smith
Department of Process Integration, UMIST, Manchester M60 1QD, U.K.

1 INTRODUCTION

Process integration is technology to optimize the use of raw materials and reduce energy and waste. The benefits are not only cost savings, as the improvement of the process efficiency and the product quality is a key prerequisite for the development of a sustainable environment. The direct environmental benefits typically are linked with accomplishments in waste reduction (Rossiter et al., 1993) that subsequently translate into lower operating costs. The mutual gains in improving the efficiency and the environmental impact represent a typical feature that is better understood as the process integration track record expands. A process integration approach to pollution prevention provides a means to see how materials move through the production chain and venues through which waste is generated and accumulated. The analysis that is available with its relevant technologies identifies the major bottlenecks of the process and explains important steps for dramatic reductions in waste and costs (Linnhoff, 1993).

The major potential of process integration relies on its ability to review complex and highly integrated flowsheets. The simultaneous consideration of pollution prevention opportunities with process modifications, as well as redesign, retrofit, and reengineering opportunities, appears a challenging and difficult problem. Its solution, however, assumes a significance that goes beyond the typical benefits of local improvements, as the considerations simultaneously include upstream and downstream processes. Furthermore, the analysis helps to understand the cause of the problem and assess grounds on which improvements are made. The latter assessment is vital to pursuing the project with confidence and probably extends the benefits with further ideas not initially embedded in the options considered by the analysis. Although process integration fosters the process efficiency and the environment as part of its overall objectives, a special category of its technologies is dedicated to the development of environmentally conscious designs, including hierarchical methods, pinch analysis methods, and wastewater minimization techniques. The following section reviews the available technology, discusses advantages and limitations, and outlines the latest developments of the water-related methods—especially the ones integrated with mathematical programming—as a result of the growing momentum they have gained over the last years.

2 REVIEW OF TECHNOLOGIES FOR ENVIRONMENTALLY CONSCIOUS DESIGNS

2.1 HIERARCHICAL METHODS

Smith and Petela (1991a, 1991b, 1992a, 1992b, 1992c) report on waste reduction techniques that use the "onion" diagram (Fig. 6.1), a classification of wastes and a systematic evaluation scheme. The wastes are classified into two major groups: (i) process waste as it may originate from reaction processes or separation products and (ii) utility waste as it develops from the outer layers of the "onion" and relates to heat-exchange network (HEN) and utilities.

The principal sources of utility waste are identified with the hot and cold utilities. Furnaces, steam boilers, gas turbines, and diesel engines produce gaseous waste from the combustion. The waste streams they consider include gaseous contaminants (oxides of sulfur and nitrogen, carbon dioxide) and aqueous effluents generated in feedwater treatment units as boiler blowdown, lost condensate, and cooling tower blowdown. The authors

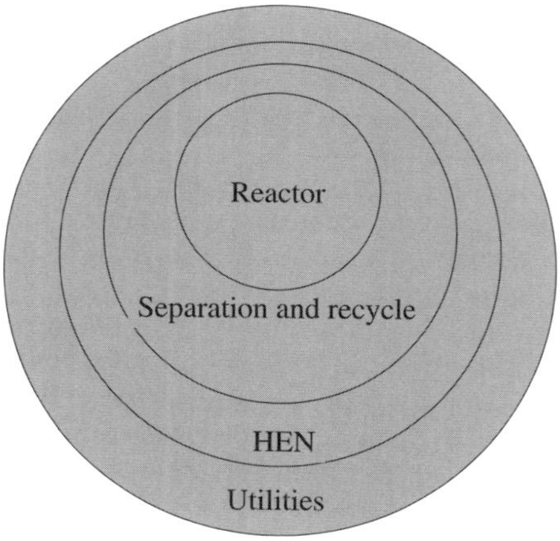

Figure 6.1: Decomposing design through the onion diagram.

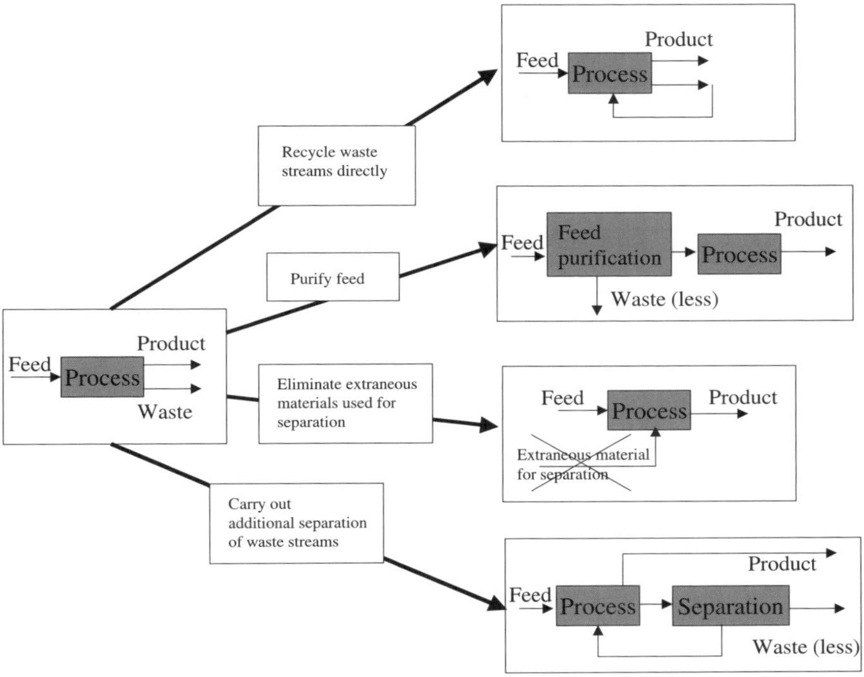

Figure 6.2: The main strategies in recycle and separation.

applaud practices with comprehensive audits as effective prerequisites for waste minimization efforts. They also recommend an improved heat recovery, as it results in an overall reduction of the utility demand and a further reduction of utility waste. Figure 6.2 outlines the main strategies proposed by Smith and Petela in dealing with separation and recycling systems.

The hierarchical method first defended by Douglas (1992) can be adapted to address issues related to waste minimization. Waste minimization opportunities are identified with design considerations that have an environmental impact on the performance. Pollution prevention at the source, rather than an end-of-pipe treatment, always yields favorable savings. Table 6.1 summarizes the overall procedure:

(a) A preliminary stage (level 1) processes input information about the reaction chemistry, the feed streams, and the data related to safety, toxicity, and the potential environmental impact of the components.

(b) An intermediate stage (level 2) evaluates the feed as well as the product and byproduct streams that enter and exit the plant. Decisions are made about the purity of the feeds, the recovery and the recycles of the reactants, the employment of purges, and the potential allocation

TABLE 6.1
Hierarchical procedure for process synthesis

Level 1	Input information: type of problem
Level 2	Input-output structure of the flowsheet
Level 3	Recycle structure of the flowsheet
Level 4	Specification of the separation system
Level 4a	General structure: phase splits
Level 4b	Vapor recovery system
Level 4c	Liquid recovery system
Level 4d	Solid recovery system
Level 5	Energy integration
Level 6	Evaluation of alternatives
Level 7	Flexibility and control
Level 8	Safety

of byproducts. The difficult problems are the ones originating from the nature of the chemical pathways. Typical cases involve the oxidation of air and specialty chemicals (agricultural chemicals, specialty polymers). The economic potential, EP_2, at the second stage is defined as

$$EP_2 = (\text{product value}) + (\text{byproduct value}) - (\text{raw material costs}).$$

Waste byproducts are assigned negative values. The sensitivity to these costs is claimed important, as unit prices may change rapidly.

(c) A final stage (levels 3 and 4) at which the streams (especially reactor recycles) are analyzed. The design decisions consider feed ratios, heat effects, possible favorable shifts from equilibrium (i.e., introduction of inerts), the potential participation of agents, and the integration of reaction with separation. The stage first identifies decisions about the separation system (where separating agents are considered) and is followed by considerations about the vapor recovery system (solvents, regeneration processes, wasted adsorbents), the liquid recovery system (entrainers, catalysts, purge streams), and the solid recovery system.

The classification of waste minimization problems as proposed by Douglas is presented in Table 6.2. Decisions differ on the basis of the magnitude of the changes they introduce and the types of streams they introduce to the flowsheet.

2.2 PINCH ANALYSIS

Pinch analysis is a thermodynamic approach that gained importance during the early 1980s as a tool for heat exchanger network design. A

TABLE 6.2
Waste minimization problems

Level 2. Input-output structure of the flowsheet	a. The reaciton chemistry: change the chemistry b. Air oxidation to NO_x: change to O_2 in recycle CO_2 oxidations c. Spent catalysts: regenerate the catalysts
Level 3. Recycle structure of the flowsheet	a. Adding reactor diluents to shift the product distribution or to shift the equilibrium conversion: change the diluent b. Adding heat carriers: change the heat carrier c. By adding reactor solvents: change the solvent
Level 4. Specification of the separation system	
Level 4a. General structure: phase splits Level 4b. Vapor recovery systems	a. Absorber solvents: change the solvent b. Regeneration of adsorption beds: change the bed stripping agent c. Removing spent adsorbents: change to absorption or condensation d. Use of reactive absorbers to remove toxic materials: change absorber solvent
Level 4c. Liquid recovery system	a. Stripping agents: change the agent b. Extraction solvents: change the solvent c. Crystallizer recycle and purge streams (almost pure water): reuse the purge water elsewhere in the process d. Crystallizer purge streams (not almost pure water): remove the contaminants and recycle the water or look for a different separation system e. Reactive crystallization byproducts: look for a different separation technique f. Spent adsorbents: regenerate the adsorbent
Level 4d. Solid recovery system	a. Cake washing: same as for crystallizer; filter mother liquor streams

comprehensive review of the design methods is available from Gundersen and Naess (1988). The thermodynamics enable a high-level approach with subsequent opportunities to review the synthesis problem, determine absolute performance targets, and help the engineer identify and remove bottlenecks. The conceptual background simplifies the challenging synthesis problem into tasks that are easy to target, understand, and analyze. The conceptual knowledge of pinch technology has been further exploited by conceptual programming in the form of customized optimization technology that automates the development of industrial heat exchanger networks (Briones and Kokossis, 1999a, 1999b, 1999c).

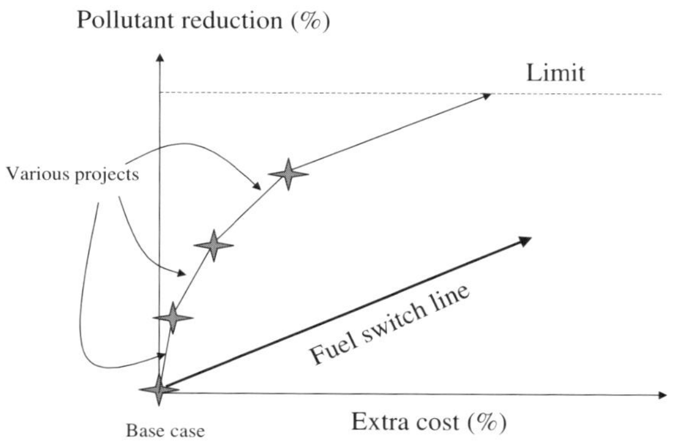

Figure 6.3: Pollution reduction versus additional cost.

Delaby and Smith (1995) proposed a pinch analysis method to minimize flue gas emissions in fuel combustion processes. In the case of new designs, their method first advocates the selection of utilities, as these choices usually have the major impact. Process modifications and further iterations with additional steam levels subsequently follow the utility selection. The method employs plots of the pollutant reduction versus the incremental cost (Fig. 6.3). Projects are reviewed and assessed on the basis of their relative efficiency, as reported from this graph. The base line is given by the option to switch fuel, and projects are rejected unless they feature smaller slopes.

As a given utility (e.g., a steam boiler) is modified to accommodate for a different fuel (e.g., switch from oil to natural gas with progressive modifications to the burners), the line of Figure 6.4 translates the fuel switch. In addition, the figure shows the regulatory limit for a particular pollutant.

The fuel switch is not enough to comply with limits, but, as shown in Figure 6.5, it complements other measures. Fuel switch lines separate the

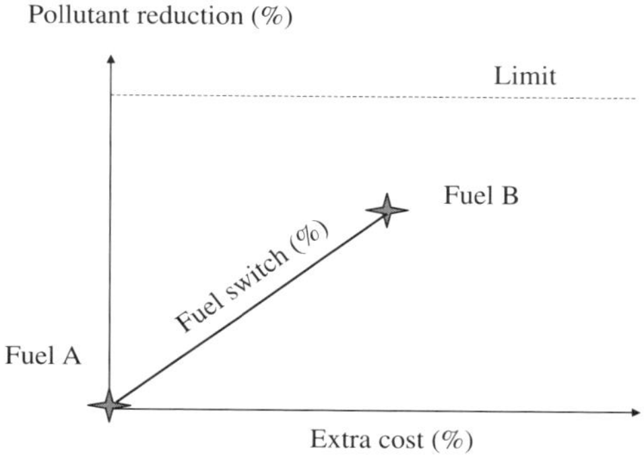

Figure 6.4: Effect of fuel switch.

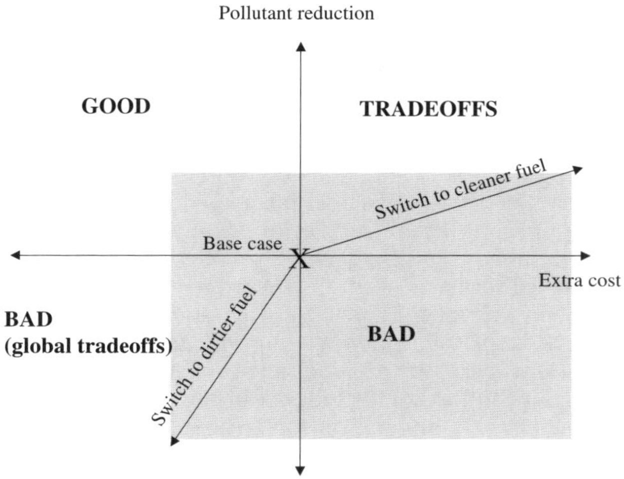

Figure 6.5: Classification of zones for reduction versus cost.

solution space into four regions. Scenarios below the line are not valid alternatives (as it is cheaper to perform the fuel switch). The development of global regions for emissions is consistent with life cycle analysis and sustainable development concepts. The latter puts an emphasis on the global picture, where regions can authorize the local increase in emissions to save benefits for the overall system. The global emissions, $G^{\text{emissions}}$, are defined as

$$G^{\text{emissions}} = G^{\text{site utilities}} + G^{\text{cent.power}} - S^{\text{cent.power}},$$

where

$G^{\text{site utilities}}$ are the emissions from on-site utilities.

$G^{\text{cent.power}}$ are the emissions from the central power station corresponding to the amount of electricity imported.

$S^{\text{cent.power}}$ are the emissions saved at the central power station corresponding to the amount of electricity exported from the site.

The emissions from a central power station are strongly dependent on the technology and type of fuel used. The plot introduced by Delaby and Smith can help the designer select combinations of reduction policies that may include

(a) Changes of the fuel and modifications on the utility system.

(b) Process changes and improved heat recovery projects.

(c) Chemical treatment (end-of-pipe) projects.

2.3 WASTEWATER MINIMIZATION AND THE DESIGN OF DISTRIBUTED EFFLUENT TREATMENT SYSTEMS

El-Halwagi and Manousiothakis (1989) have successfully extended the boundaries of pinch analysis developments by introducing the mass-exchange network problem. Following analogies from energy systems, mass transfer had replaced energy transfer and the development of mass exchanger networks had been systematically accomplished with the consideration of targets, the minimum number of transfer units, and driving forces based on concentration differences. This early work has been extended separately by each one of these authors. Furthermore, the more general concept of mass integration has been introduced by El-Halwagi and co-workers to provide a holistic framework for the allocation, generation, and interception of streams and species. El-Halwagi (1997) provides an excellent reference on the subject and the latest contributions of his group. El-Halwagi and Spriggs (1998) provide an overview of mass integration fundamentals and applications in their review paper. Other contributions (Bagajewicz et al., 1998; Gupta and Manousiouthakis, 1994; Bagajewicz and Manousiouthakis, 1992) explain the development of the state-space approach as a generalized tool for structural optimization of mass-exchange networks.

Smith and co-workers developed a targeting approach for wastewater minimization that is based on graphical developments. The approach was first reported by Wang and Smith (1994a, 1994b, 1995) and later improved by Kuo and Smith (1996, 1997, 1998). The technology addresses water and wastewater minimization problems and the design of distributed effluent treatment systems. Wang and Smith introduced a water use representation that consists of a concentration-mass load diagram (Fig. 6.6). In this diagram a process stream is plotted in countercurrent contact with a water stream. The contaminant concentration is reduced as mass is transferred to the water stream that picks up the contaminant.

The water-using operations are described by *limiting water profiles* defined by the maximum inlet and outlet concentrations (Fig. 6.7). Among

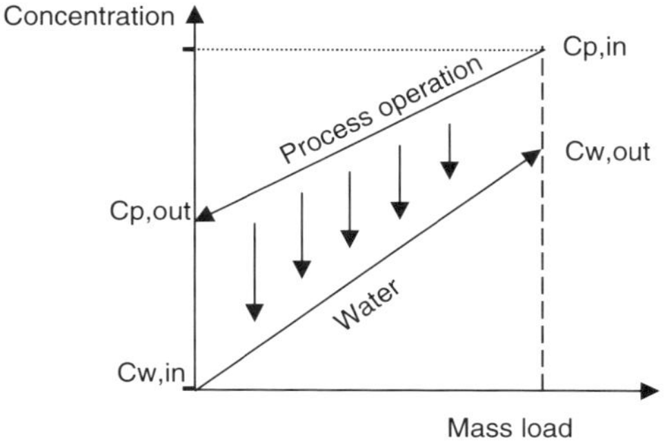

Figure 6.6: Composition versus load diagram.

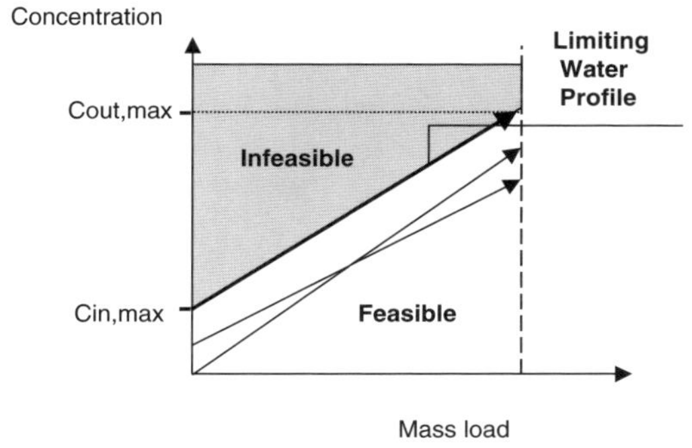

Figure 6.7: Limiting water profile.

others, available driving forces determine the limits for mass transfer, along with constraints such as the ones related to corrosion and fouling. The limiting water profiles are boundaries between feasible and infeasible operations.

The different inlet and outlet concentrations determine concentration intervals within which profiles can be added. The result is a *composite profile* that simultaneously accounts for all streams (Fig. 6.8).

The freshwater and wastewater are minimized by maximizing the outlet concentration of the water supply line. The point where the supply line touches the limiting water profile corresponds to the water pinch. The design target is achieved by decomposing the problem at the pinch. The water requirements are calculated independently above and below the pinch. Imaginary water "mains" available at different concentration levels provide water sources and sinks. The water-using operations are serviced by the mains with the use of a grid diagram that guides the development of a water network that achieves the targets (Fig. 6.9).

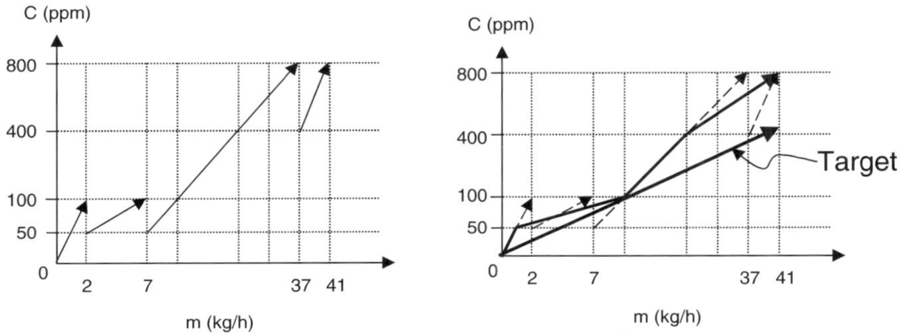

Figure 6.8: An example of a composite profile.

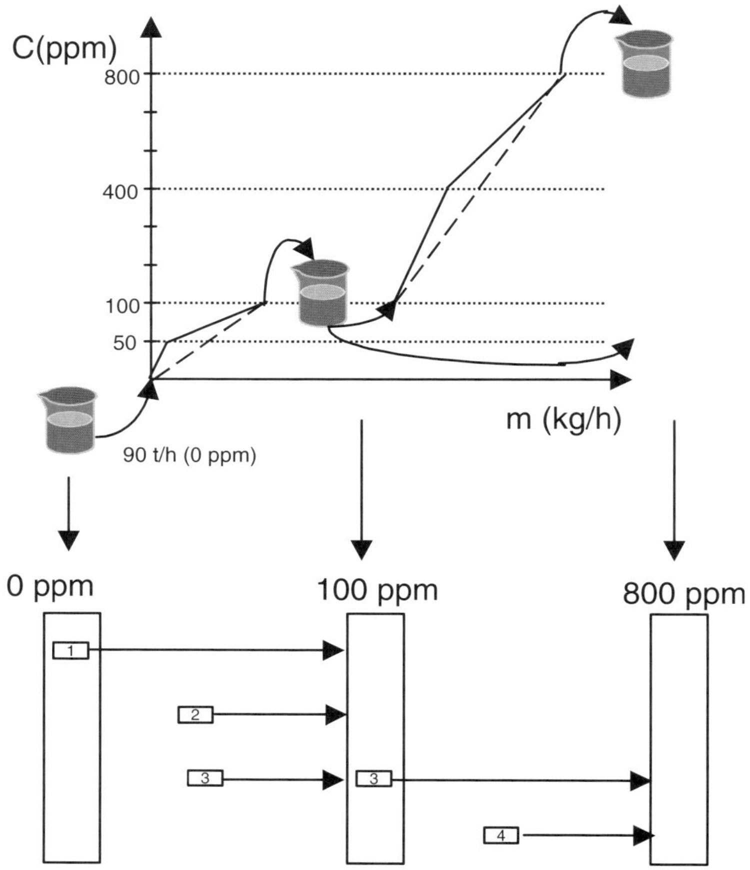

Figure 6.9: An example of the grid diagram.

In summary, the water pinch is useful in developing targets for single contaminant processes and insightful in visualizing the design problem. Water losses can also be considered by adjusting the water supply line at the contaminant concentration where the loss takes place; however, the approach is difficult to extend for problems with multiple contaminants, designs with process constraints, and problems with objectives other than the minimization of water. Another limitation is that water pinch assumes a fixed contaminant load and is unable to address different mass transfer models.

Kuo and Smith (1996) improved the earlier methods and provided important insights to study interactions within the water system. The water system is divided, the latter into three subsystems: the water-using subsystem, the regeneration subsystem, and the effluent treatment subsystem. The water-using subsystem is designed ahead of the effluent treatment subsystem. The interactions between the subsystems are therefore not explored by the decomposition. The target for maximum reuse is found from the limiting composite curves. The number of units is calculated by plotting boundary lines of the treatment units. Their sequence is defined by practical considerations. For multiple contaminants there exist different sequences, and the approach is inconclusive. The effluent composite positions a targeting curve. Different water-using systems lead to different curves and different treatment targets (even if the freshwater target remains the same). Based on the limiting composite curve, a lower and upper bound is found for the treatment line. The minimum effluent treatment is calculated by eliminating pockets from the composite. The water regeneration is considered with respect to freshwater and regenerated water. A "migration" procedure identifies optimal groups of components. The optimization makes use of rules driven from concepts.

The diagrams are useful in visualizing and analyzing changes in the performance of the regeneration units. Applied to each water user group, the approach provides the number of regeneration and effluent treatment units. Hallale and Fraser (1998) described a procedure to include capital cost targets in the water minimization problem. The method applies to single component problems. Following Ahmad and Smith (1989), they calculated targets for the minimum number of trays required to carry out predetermined separation tasks in a water network. The assumption of a fixed number of units is relaxed, as it is reported that the minimum total cost does not usually correspond to the minimum number of exchangers. Following El-Halwagi and Manousiouthakis (1989), a minimum allowable composition difference is employed to calculate trade-offs between capital and operating costs.

2.4 DESIGN OF EFFICIENT WATER SYSTEMS

A methodology for the design of water systems is described in this section. The approach uses insights from water pinch and combines them with mathematical programming tools. In working with complex systems, the understanding of the basic processes is critical in reducing the number of variables and the computational effort required. The nature of the problem dictates the need for focused rather than general purpose approaches (e.g., Mass Exchanger Network Synthesis). A major feature is that the largest water users are not mass transfer operations (e.g., cooling tower, steam system, etc.).

The method presented in this section is based on a superstructure model formulated as a Mixed-Integer Nonlinear Programming (MINLP) problem. The nonlinearities in the formulation are due to bilinear terms that appear in the mass balances (superstructure mixers and splitters), the nonlinear terms of the sizing equations, and the cost functions for the water-using operations. The difficulties associated with the solution of MINLP problems are well documented in the published literature (Floudas, 1995) along with algorithms to address bilinear terms (Quesada and Grossman, 1995; Al-Khayyal and Falk, 1983; Adjiman et al., 1996). In these algorithms, a branch and bound strategy is employed together with convex underestimators to arrive at the global optimum solution. Earlier work to tackle bilinear terms includes applications of recursive LP and Successive Linear Programming (SLP). Haverly (1977, 1978, 1979, 1980) applied a recursive linear programming approach on the "pooling" problem, but the results have proved very sensitive to initialization. Palacios-Gomez et al. (1982) and Zhang et al. (1985) report on theoretical properties of SLP algorithms. SLP (or successive linear programming) methods can tackle a general nonlinear programming problem, which is solved through a sequence of linear problems to approximate the optimal solution. Successful industrial applications of SLP have been published (Baker and Lasdon, 1985; Simon and Azma, 1983).

In the method described later, the optimization strategy employs a decomposition scheme and associates binary variables with the existence of different connections. The use of binary variables enables the approach to address complexity issues of the water network, as many practical constraints can be incorporated into the analysis. Among others, these include geographical constraints, flow rate constraints, and forbidden/compulsory matches. The formulation extends the domain of water pinch analysis with elements of capital cost to study the trade-offs between freshwater costs, mass exchanger costs, and the cost of the required pipework. Consequently, the optimization results in cost-efficient networks rather than networks featuring minimum freshwater consumption. Former limitations of the water pinch method to address multiple contaminants are now overcome, as there are, in principle, no limits on the number of components or the number of freshwater sources.

3 PROBLEM STATEMENT

The design of a water-using system involves water-related elements in the form of (i) freshwater sources available to satisfy the demand, and (ii) water-using operations described by loads of contaminants and concentration levels. Sources and operations are sketched in Figure 6.10. The design task is to find the network configuration to minimize the overall demand for freshwater (and consequently reduce wastewater volume) compatible with minimum total annual cost. The synthesis objective consequently combines terms for low freshwater consumption, suitable network topology for water reuse, and low investment cost. The assumptions are made that (i) all data for the limiting water profiles are available and certain; (ii) the number of water-using operations is fixed; (iii) heat integration is not allowed and thermal effects are not important; and (iv) the network operates under constant pressure. The investment cost of the network includes piping costs. The

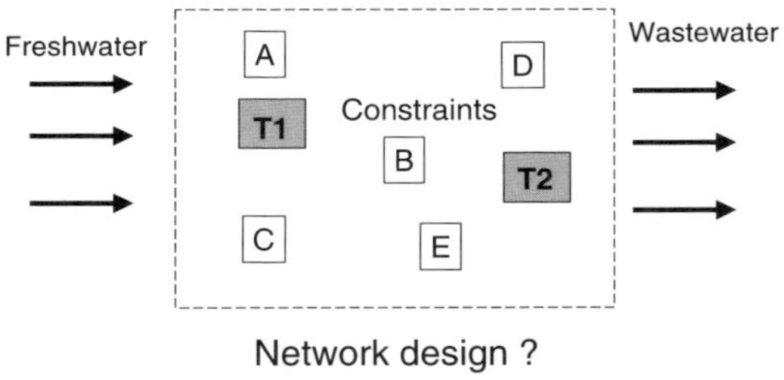

Figure 6.10: Schematic representation of problem statement.

approximate length of the pipe can be specified for each possible connection, together with the materials of construction. The cost of mass exchange units assumes thermodynamic parameters and equilibrium relationships between the process streams and the water streams for the key contaminants, as well as the corresponding design equations and cost functions.

In what follows, the water-using system synthesis problem is presented based on the optimization of a superstructure representation. The problem is initially formulated as an MINLP model. A natural decomposition of the problem is then developed based on the physical nature of the design problem. Engineering insights gained through water pinch analysis allow projection of the bilinear terms following an iterative procedure.

4 SYNTHESIS REPRESENTATION AND THE MATHEMATICAL MODEL

The method relies on the optimization of a superstructure model with all possible connections between (i) freshwater sources and operations and (ii) different water-using operations. Important features of the superstructure are as follows: (i) The process streams in the mass exchange operations are considered *implicitly* through the construction of the limiting water profiles; (ii) The number of units in the water network determines existing units, and discrete options account for connections between water sources and sinks. The superstructure is developed so that (i) each freshwater stream entering the network is split among the water-using operations; (ii) all the effluent streams generated from each operation are mixed in a final discharge point; (iii) each operation is preceded by a mixer fed by streams from the freshwater splitters and reuse streams emanating from the outlets of all other operations; and (iv) each operation is followed by a splitter that feeds the final mixer and the other operations in the system.The example of a superstructure with a single freshwater source and two water-using operations is shown in Figure 6.11. Unlike previous mass-exchange network (MEN)

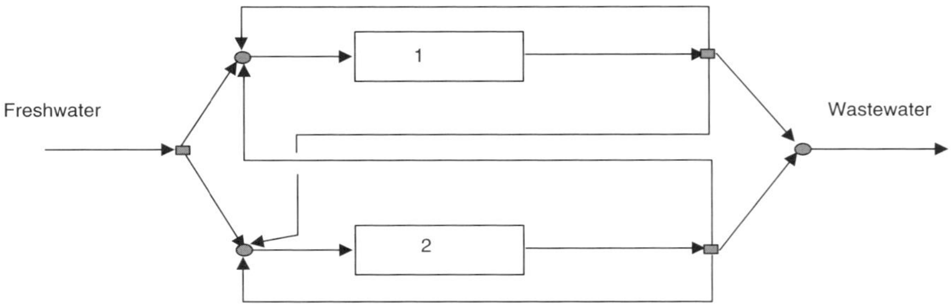

Figure 6.11: Superstructure representation of water usage.

developments (Papalexandri et al., 1994a), the superstructure provides options for water streams (lean streams in MEN terminology) to merge.

The formal description of the design problem is given based on a set of constraints and parameters. The following sets are defined:

SETS

I = $\{i \mid i$ is an operation involved in the water system$\}$, $i = 1, 2, \ldots, N_{OP}$

I_{ME} = $\{i \mid i$ is a water user and a mass exchanger$\} \subseteq I$, $i = 1, 2, \ldots, N_{ME}$

C = $\{c \mid c$ is a contaminant in the analysis$\}$, $c = 1, 2, \ldots, N_C$

J = $\{j \mid j$ is a freshwater source$\}$, $j = 1, 2, \ldots, N_S$

R = $\{r \mid r$ is a linear segment for the piping cost functions$\}$,
$r = 1, 2, \ldots, N_R$

PROBLEM VARIABLES AND PARAMETERS

Decision variables include continuous and binary variables. The continuous variables are associated with flow rates,

$F^{w}_{j,i}$ freshwater stream from source j to operation $i \in I$,

$F^{IP}_{i,i'}$ stream from operation i to operation i', $i, i' \in I$,

F^{OUT}_{i} stream discharged from operation $i \in I$,

F^{T}_{i} total through operation $i \in I$,

cross-sectional areas of the pipes associated to different connections,

$A^{Fw}_{j,i}$ between freshwater source $j \in J$ and operation $i \in I$,

$A^{IP}_{i,i'}$ between operation $i \in I$ and operation $i' \in I$,

A^{OUT}_{i} from operation i to discharge $i \in I$,

mass load and contaminant concentrations,

$M_{c,i}^{in}$ inlet load in operation $i \in I, c \in C$,

$M_{c,i}^{out}$ outlet load in operation $i \in I, c \in C$,

$C_{c,i}^{out}$ contaminant concentration from $i \in I, c \in C$,

cost terms associated with

$Cost_j^{supp}$ freshwater supply $j \in J$,

$Cost_{j,i}^{Fw,pipe}$ piping from source $j \in J$ to operation $i \in I$,

$Cost_{i,i'}^{IP,pipe}$ piping from operation $i \in I$ to operation $i' \in I$,

$Cost_i^{OUT,pipe}$ piping for discharge from operation $i \in I$,

$Cost_i^{ME}$ capital cost of mass exchanger $i \in I_{ME}$,

$N_{c,i}$ number of equilibrium stages in operation $i \in I$ based on key contaminant $c \in C$.

Binary variables are introduced to account for the

$Y_{j,i}^{Fw}$ stream from freshwater source $j \in J$ and operation $i \in I$,

$Y_{i,i'}^{IP}$ stream from operation $i \in I$ and operation $i' \in I$,

Y_i^{OUT} stream discharged from operation $i \in I$.

The parameters required to formulate the design problem include bounds on the

$C_{c,i}^{in,max}$ inlet concentration of operation $i \in I$, contaminant $c \in C$ (upper),

$C_{c,i}^{out,max}$ outlet concentration of operation $i \in I$, contaminant $c \in C$ (upper),

$F_i^{T,L}$ total flow through operation $i \in I$ (lower),

$F_i^{T,U}$ total flow through operation $i \in I$ (upper),

$F_{j,i}^{w,U}$ flow from source $j \in J$ to $i \in I$ (upper),

$F_{j,i}^{w,L}$ flow from source $j \in J$ to $i \in I$ (lower),

$W_j^{in,U}$ withdrawal from freshwater source $j \in J$ (upper),

$A_{j,i,r}^{Fw,UP}$ cross-sectional area of pipe connecting source $j \in J$ and $i \in I$ in segment r (upper),

$A_{i,i',r}^{IP,UP}$ cross-sectional area of pipe connecting $i' \in I$ and $i \in I$ in segment r (upper),

$A_{i,r}^{OUT,UP}$ cross-sectional area of discharge pipe from $i \in I$ in segment r (upper),

distances between sources and sinks of water,

$d_{j,i}^{Fw}$ from freshwater $j \in J$ to operation $i \in I$,
$d_{i,i'}^{IP}$ from operation $i \in I$ to operation $i' \in I$,
d_{i}^{OUT} from operation $i \in I$ to discharge,

flow velocities in the connections,

$\upsilon_{j,i}^{Fw}$ from freshwater $j \in J$ to operation $i \in I$,
$\upsilon_{i,i'}^{IP}$ from operation $i \in I$ to operation $i' \in I$,
υ_{i}^{OUT} from operation $i \in I$ to discharge,

materials of construction for the pipes,

$\xi_{j,i}^{Fw}$ from freshwater $j \in J$ to operation $i \in I$,
$\xi_{i,i'}^{IP}$ from operation $i \in I$ to operation $i' \in I$,
ξ_{i}^{OUT} from operation $i \in I$ to discharge,

regression parameters for the fixed charge for piping costs from

$b_{j,i}^{Fw,pipe}$ freshwater $j \in J$ to $i \in I$,
$b_{j,i}^{IP,pipe}$ operation $i \in I$ to $i' \in I$,
$b_{j,i}^{OUT,pipe}$ operation $i \in I$ to discharge,
$b_{j,i,r}^{Fw,pipe}$ freshwater $j \in J$ to $i \in I$ in segment $r \in R$,
$b_{i,i',r}^{IP,pipe}$ operation $i \in I$ to $i' \in I$ in segment $r \in R$,
$b_{i,r}^{OUT,pipe}$ operation $i \in I$ to discharge in segment $r \in R$,

and the variable charge for piping costs from

$a_{j,i}^{Fw,pipe}$ freshwater $j \in J$ to $i \in I$,
$a_{j,i}^{IP,pipe}$ operation $i \in I$ to $i' \in I$,
$a_{j,i}^{OUT,pipe}$ operation $i \in I$ to discharge,
$a_{j,i,r}^{Fw,pipe}$ freshwater $j \in J$ to $i \in I$ in segment $r \in R$,
$a_{i,i',r}^{IP,pipe}$ operation $i \in I$ to $i' \in I$ in segment $r \in R$,
$a_{i,r}^{OUT,pipe}$ operation $i \in I$ to discharge in segment $r \in R$.

Other parameters include the

$m_{c,i}^{L}$ mass load of contaminant $c \in C$ from user $i \in I_{OP}$,
$cw_{c,j}^{in}$ contaminant concentration in freshwater source $j \in J$, $c \in C$,
α_j cost per unit mass flow of $j \in J$,

$C_{c,i}^{out,PS*}$ equilibrium concentration of $c \in C$ with the water inlet of $i \in I$,
F_i^{loss} flow rate loss at the inlet of $i \in I$,
$m_{c,i}^{eq}$ equilibrium parameter for the distribution of $c \in C$ between the process and the water stream,
F_i^{PS} mass flow rate of process stream in $i \in I$,
$C_{c,i}^{in,PS}$ inlet concentration of $c \in C$ in the process stream of $i \in I$,
$C_{c,i}^{out,PS}$ outlet concentration of $c \in C$ in the process stream of $i \in I$,
M_i^{NS} maximum number of sources that may feed $i \in I$.

The limiting flow rate is related to the mass load of contaminants by

$$F_i^L = \max_c \frac{m_{c,i}^L}{C_{c,i}^{out,\max} - C_{c,i}^{in,\max}}. \tag{1}$$

CONSTRAINTS

Constraints include the following:
Overall mass balance:

$$\sum_{j \in J} \sum_{i \in I} F_{j,i}^w - \sum_{i \in I} F_i^{OUT} = \sum_{i \in I} F_i^{loss}; \tag{2}$$

Mass balances and an overall flow balance around each operation:

$$M_{ci}^{in} - M_{ci}^{out} + m_{ci}^L = 0 \qquad \forall c \in C, i \in I, \tag{3}$$

$$\sum_{i' \in I} F_{i',i}^{IP} + \sum_{j \in J} F_{j,i}^w - \sum_{i' \in I} F_{i,i'}^{IP} - F_i^{OUT} = F_i^{loss} \qquad \forall i \in I \tag{4}$$

along with

$$\left[\sum_{i' \in I} (F_{i',i}^{IP} \cdot C_{c,i'}^{out}) + \sum_{j \in J} (F_{j,i}^w \cdot cw_{cj}^{in}) \right] - M_{c,i}^{in} = 0 \qquad \forall c \in C, i \in I, \tag{5}$$

$$C_{ci}^{out} \cdot \left[\sum_{j \in J} F_{j,i}^w + \sum_{i' \in I} F_{i',i}^{IP} - F_i^{loss} \right] - M_{c,i}^{out} = 0 \qquad \forall c \in C, i \in I; \tag{6}$$

Feasibility constraints on the inlet concentration and the flow rates:

$$M_{ci}^{in} \le C_{c,i}^{in,\max} \cdot F_i^T \qquad \forall c \in C, i \in I, \tag{7}$$

$$\sum_{i \in I} F_{j,i}^w \le W_j^{in,U} \qquad \forall j \in J; \tag{8}$$

Flow balances for the total flow around each operation:

$$F_i^T - \left(\sum_{j \in J} F_{j,i}^w + \sum_{i' \in I} F_{i',i}^{IP} - F_i^{loss} \right) = 0 \qquad \forall i \in I. \tag{9}$$

Note that the implicit assumption in the above set of equations is that the water losses occur at the inlet of the operation.

Modeling constraints for costing

$$Cost_{ji}^{Fw,pipe} = d_{j,i}^{Fw,pipe} \cdot \left(a_{j,i}^{Fw,pipe} \cdot A_{j,i}^{Fw} + b_{j,i}^{Fw,pipe} \cdot Y_{j,i}^{Fw} \right) \qquad \forall j \in J, i \in I \tag{10}$$

or (through a more refined linearisation)

$$Cost_{j,i}^{Fw,pipe} = d_{j,i}^{Fw,pipe} \sum_r \left(a_{j,i,r}^{Fw,pipe} \cdot A_{j,i,r}^{Fw} + b_{j,i,r}^{Fw,pipe} \cdot Y_{j,i,r}^{Fw} \right) \tag{11}$$

coupled with constraints in the form

$$A_{j,i,r}^{Fw} - A_{j,i,r}^{Fw,UP} Y_{j,i,r}^{Fw} \leq 0, \tag{12}$$

$$\sum_r Y_{i,j,r}^{Fw} \leq 1, \tag{13}$$

and

$$\sum_r A_{j,i,r}^{Fw} = A_{j,i}^{Fw}. \tag{14}$$

The index r accounts for the degree of linearization. The cross-sectional area $A_{j,i}^{Fw}$ is related to the flows $F_{j,i}^w$ by

$$F_{ji}^w = A_{j,i}^{Fw} \cdot \mathbf{v}_{j,i}^{Fw,pipe} \qquad \forall j \in J, i \in I. \tag{15}$$

For water-using operations that include mass transfer (absorbers, strippers, extraction, etc.), the capital cost of the units is expressed as a function of the number of stages $N_{c^*,i}$,

$$Cost_i^{ME} = f(N_{c^*,i}), \tag{16}$$

where c^* is the key contaminant in the separation. Unless otherwise specified, the dependence of $N_{c^*,i}$ on the concentration and flow rates follows Kremser (King, 1980):

$$N_{c,i} = \frac{\ln\left\{\left[1-\left(m_{ci}^{eq} F_i^{PS} / F_i^T\right)\right]\left[\left(C_{c,i}^{in,PS} - C_{c,i}^{out,PS^*}\right) / \left(C_{c,i}^{out,PS} - C_{c,i}^{out,PS^*}\right)\right] + \left(m_{ci}^{eq} F_i^{PS} / F_i^T\right)\right\}}{\ln\left(F_i^T / m_{ci}^{eq} F_i^{PS}\right)} \qquad \forall i \in I_{ME}. \tag{17}$$

The cost of the mass exchanger will depend on the specific nature of the operation. For an absorption column, for example, additional constraints are needed to calculate the diameter and the height of the column, together with appropriate cost correlations.

The design of a water network with a specific flow pattern for water reuse defines the driving force profile for the mass exchange. The reuse of contaminated water implies a reduction in the driving forces available for mass transfer.

Logical constraints:

$$F^{w}_{j,i} - F^{w,U}_{j,i} \cdot Y^{Fw}_{j,i} \leq 0, \tag{18}$$

$$F^{w}_{ji} - F^{w,L}_{j,i} \cdot Y^{Fw}_{j,i} \geq 0, \tag{19}$$

$$\sum_{j\in J} Y^{Fw}_{j,i} + \sum_{i'\in I} Y^{IP}_{i',i} \leq M^{NS}_{i}. \tag{20}$$

Similar constraints can be generated for the remaining connections of the superstructure. In what follows, and for simplicity, only the sets of constraints corresponding to connections from sources j to operations i are described.

OBJECTIVE FUNCTION

The objective function includes terms for the operating cost of the network (freshwater costs on an annual basis), the capital cost for piping, and the investment required for the mass exchangers:

$$Cost^{tot} = Cost^{op} + Cost^{cap}, \tag{21}$$

where

$$Cost^{op} = N_t \cdot \sum_{j} \sum_{i} \left(F^{w}_{j,i} \cdot \alpha_j\right), \tag{22}$$

$$Cost^{cap} = b^{ann} \left[\begin{array}{l} \sum_{j\in J} \sum_{i\in I} Cost^{Fw,pipe}_{ij} + \sum_{i\in I} \sum_{i'\in I} Cost^{IP,pipe}_{ii'} \\ + \sum_{i\in I} Cost^{OUT,pipe}_{i} + \sum_{i\in I_{ME}} Cost^{ME}_{i} \end{array} \right], \tag{23}$$

where N_t is the operating time and b^{ann} is an annualization factor.

The above formulation is an MINLP problem. The nonlinearities appear in equations (5), (6), (16), and (17). Binary variables are used to identify all connections in the network through constraints (18) and (19) and for controlling the complexity of the resulting design through constraints (20). The minimum allowable composition difference used by MEN developments (El-Halwagi and Manousiouthakis, 1989, 1990a, 1990b; Papalexandri et al., 1994a, 1994b), to avoid infinite-sized mass exchangers, is inherently built into the limiting water profiles. In this approach, the given value is only a lower bound and it does not imply any preoptimized cost trade-offs between capital and operating costs.

The use of equations (18) and (19) provides immediate controls on the flow rates of existing connections and unrealistically small streams. To allow a certain degree of control over the complexity of the flow pattern, constraints related to the number of sources that can feed a particular operation [constraints (20)] are included in a similar way. As the binary variables are introduced to identify the connections, simple assignment constraints can exploit alternatives, as they typically abound on these problems. In other words, the water system design problem presents highly degenerate solutions (i.e., many alternative optimal solutions). In general, network problems are renowned for their degeneracy (Magnanti and Wong, 1984).

5 SOLUTION STRATEGY

Equation (1) describes the inverse relationship between the water flow in a particular operation and the outlet concentrations of the contaminants. Because the mass load is assumed fixed, the outlet concentrations must be maximized to reduce the water flow. Motivated by this observation, provided all outlet concentrations in the water-using operations are set to their maximum levels, $C_{c,i,}^{out,max}$, an upper bound, is obtained for the performance of the system. Furthermore, the mass balances become linear in the flow rates, and the nonlinearities, due to bilinear terms, are avoided.

The observation can be further exploited with the development of a *natural decomposition strategy*. The scheme decomposes the original MINLP problem into two related subproblems that are solved sequentially within an iterative procedure. The subproblems are identified as (P_1) and (P_2). Problem (P_1) is an MILP formulation that optimizes against an expanded economic objective. Problem (P_2) minimizes infeasibilities and is formulated as an LP problem. Both are explained in the remainder of this section, along with the solution algorithm to implement the decomposition.

PRIMAL PROBLEM (P_1)

Primal problem, (P_1), is developed by projecting the nonlinear constraints onto the concentration space so that

$$C_{c,i}^{out} = C_{c,i}^{out,\max}. \tag{24}$$

Let us assume a relaxed version of equation (3) in the form

$$M_{c,i}^{in} - M_{c,i}^{out} + m_{c,i}^{L} + s_{c,i}^{1} = 0 \qquad \forall c \in C, i \in I. \tag{3'}$$

Constraints (5) and (6) are projected onto the concentration space, with respect to the outlet concentrations, $C_{c,i}^{out}$, to yield the following linear expressions:

$$\left[\sum_{i'\in I}\left(F_{i',i}^{IP}\cdot C_{c,i'}^{out,k}\right)+\sum_{j\in J}\left(F_{ji}^{w}\cdot cw_{cj}^{in}\right)\right]-M_{c,i}^{in}=0 \qquad \forall c\in C, i\in I, \tag{5'}$$

$$C_{ci}^{out,k}\cdot\left[\sum_{j\in J}F_{ji}^{w}+\sum_{i'\in I}F_{i',i}^{IP}-F_{i}^{loss}\right]-M_{c,i}^{out}=0 \qquad \forall c\in C, i\in I, \tag{6'}$$

where

$C_{c,i}^{out,k}$ is the fixed concentration of c leaving operation $i \in I$, $c \in C$, at iteration k.

Problem (P_1) can be stated as follows:

$$\min Cost^{tot} = Cost^{op(l)} + Cost^{cap(l)}$$

subject to

constraints (2), (4), (7)–(10),

constraints (15)–(20), and

constraints (3'), (5'), and (6').

Nonnegativity constraints for all variables complete the formulation. With the exception of equations (16) and (17), all constraints in (P_1) are linear. The sizing equation for the mass exchange unit (17) is a function of the flow rate of water through the operation and of the inlet concentration of this water stream. That is,

$$N_{c,i} = \mathrm{f}\left(F_{i,}^{T}C_{c,i}^{out,PS^*}\right).$$

Provided the Kremser equation is used to calculate the number of equilibrium stages, it can be expressed as a univariate function of F_i^T and approximated by

$$N_{c,i}^{k} = \sum_{q=1}^{N_L} N_{c,i}^{q}\lambda_q \tag{25}$$

together with

$$A_{c,i} = \sum_{q=1}^{N_L} A_{c,i}^{q}\lambda_q \tag{26}$$

and

$$\sum_{q}\lambda_q = 1, \tag{27}$$

where

$$A_{c,i} = \frac{F_{Ti}}{w_{c,i}^{1}} \tag{28}$$

is the absorption factor.

$N_{c,i}^{k}$ is the approximate number of equilibrium stages calculated at some iteration k. Problem (P_1) results in an MILP problem that can be solved to global optimality using a reliable branch and bound algorithm.

PRIMAL PROBLEM (P_2)

Once the flow pattern corresponding to the optimal solution of (P_1) has been identified (i.e., $F_{ji}^{w^*}$, $F_{i,i'}^{IP^*}$, $F_{i}^{OUT^*}$), a second primal problem, (P_2), can be solved for the concentrations. This is an LP model and includes equations (5) and (6), projected now onto the flow rate space with respect to all flows in the network, $F_{j,i}^{w}$, $F_{i,i'}^{IP}$, andF_{i}^{OUT}. The exit concentrations of all contaminants are the variables of the problem.

Problem (P_2) is stated as follows:

$$z_2^k = \min \sum_{c} \sum_{i} s_{c,i}^{1}$$

subject to

$$\left[\sum_{i' \in I} \left(F_{i,i'}^{IP^*} \cdot C_{c,i'}^{out}\right) + \sum_{j \in J} \left(F_{ji}^{w^*} \cdot cw_{cj}^{in}\right)\right] - M_{c,i}^{in} = 0 \qquad \forall c \in C, i \in I, \tag{5''}$$

$$C_{ci}^{out} \cdot \left[\sum_{j \in J} F_{j,i}^{w^*} + \sum_{i' \in I} F_{i',i}^{IP^*} - F_{i}^{loss}\right] - M_{c,i}^{out} = 0 \qquad \forall c \in C, i \in I, \tag{6''}$$

$$M_{c,i}^{in} - M_{c,i}^{out} + m_{c,i}^{L} + s_{c,i}^{1} = 0 \qquad \forall c \in C, i \in I. \tag{3'}$$

The optimal solution to (P_2) results in an updated vector of exit concentrations for all operations and all contaminants,$C_{c,i}^{out^*}$, that can be used to set up a new (P_1). Let us denote the formulations of (P_1) and (P_2) at iteration k as $(P_1)^k$ and $(P_2)^k$.

Outline of the Decomposition Procedure

The overall iterative solution procedure is illustrated in Figure 6.12. The algorithm involves the following steps:

1. Set $k = 1$. Solve problem $(P_1)^k$ with $C_{c,i}^{out,\max}$ to obtain flow rates, $F_{j,i}^{w,k}$, $F_{i,i'}^{IP,k}$, $F_{i}^{out,k}$. Let the optimal value of the objective be z_1^k. Calculate $w_{c,i}^{1}$ and $w_{c,i}^{2,k}$ from equations (26) and (27).

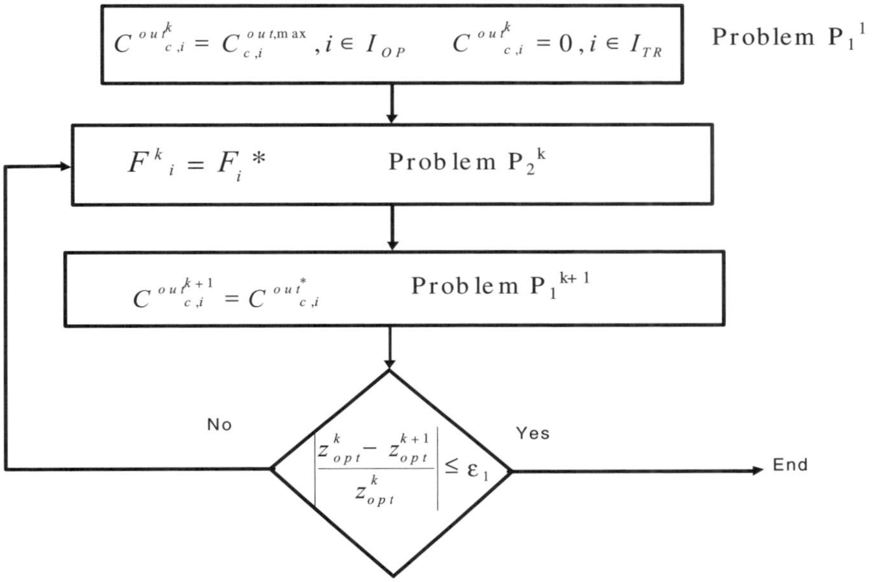

Figure 6.12: Iterative procedure.

2. Solve $(P_2)^k$ on $[F^{w,k}_{j,i}, F^{IP,k}_{i,i'}, F^{OUT,k}_i]$ to obtain new values for the outlet concentrations, $C^{out,k+1}_{c,i}$. Calculate $w^{2,k+1}_{c,i}$.

3. Solve problem $(P_1)^{k+1}$ on $C^{out,k+1}_{c,i}$, to obtain new flow rates; let the optimal objective be z_1^{k+1}.

4. If $|\frac{z_1^k - z_1^{k+1}}{z_1^k}| \le \varepsilon$, stop. Otherwise, set $k = k + 1$ and go to Step 2.

For the first iteration of the procedure ($k = 1$), it is assumed that clean water is used in all mass exchangers (i.e., $C^{out,PS*}_{c,i}$ is zero). In this way, the number of equilibrium stages needed for the separation is approximated. The linearization parameters are updated at every iteration after the corresponding inlet and outlet concentrations of all contaminants are known. Piecewise linearization of this function is performed using special ordered sets 2 (SOS2) variables (Williams, 1987), as this does not increase significantly the computational effort needed to solve the problem. In summary, the original MINLP formulation (P) has been addressed as an MILP-LP/NLP problem where NLP is only applied once and upon convergence of the MILP-LP decomposition. The initial point is always determined by the problem data (limiting water profiles), and the improvements gained by the NLP are marginal and due to the approximations behind the linearized costs.

TABLE 6.3
Process stream data for Example 3

Gas stream	*Flow rate (Gs) (kmol inert gas/hr)*	*Four gas streams Ys (kmol SO_2/ kmol inert gas)*	*Yt (kmol SO_2/ kmol inert gas)*	*Load removed (kmol SO_2/hr)*
1	50	0.01	0.004	0.3
2	60	0.01	0.005	0.3
3	40	0.02	0.005	0.6
4	30	0.02	0.015	0.15

6 ILLUSTRATIVE EXAMPLE

The example illustrates the employment of the above methodology and is taken from Hallale and Fraser (1998). The network consists of four absorbers, with the data given in Table 6.3.

A linear equilibrium relation for SO_2 is assumed in the form

$$y^* = 26.1x - 0.00326. \tag{29}$$

Each absorber is assigned to a separate gas stream, and the limiting water profile data are obtained assuming equilibrium restrictions. The minimum composition difference, $\varepsilon_{\min}$, is 5×10^{-6} kmol SO_2/kmol H_2O. The maximum inlet and outlet concentrations of each absorption column are given by

$$C^{in,\max} = \frac{Y_t - b^{eq}}{m^{eq}} - \varepsilon_{\min}, \tag{30}$$

$$C^{out,\max} = \frac{Y_s - b^{eq}}{m^{eq}} - \varepsilon_{\min}, \tag{31}$$

where m^{eq} and b^{eq} are equilibrium parameters. Table 6.4 shows data for the limiting water profiles.

TABLE 6.4
Limiting water profile data for Example 3

Gas stream	*Limiting flow rate (kmol H_2O/hr)*	*$C^{in,max}$ (kmol SO_2/kmol H_2O)*	*$C^{out,max}$ (kmol SO_2/kmol H_2O)*	*Load removed (kmol SO_2/hr)*
1	1305	0.000273	0.000503	0.3
2	1566	0.000311	0.000503	0.3
3	1044	0.000311	0.000886	0.6
4	783	0.000695	0.000886	0.15

TABLE 6.5
Cost data for Example 3

Water cost	£0.64/ton
Operating time	8600 h/y
Shell cost	£6400 $H^{0.95}$ $D^{0.6}$
Trays	£304 $e^{0.8D}$ per tray
Capital annualization factor	0.2

There is a single uncontaminated freshwater source available ($cw_{c,j}^{in} = 0$). Costing and operating parameters are given in Table 6.5. The cost expressions include an installation factor:

$$Cost^{shell} = 6{,}400 \cdot H_i \cdot D_i^{0.6}, \tag{32}$$

$$Cost^{tray} = 308e^{0.8D_i}, \tag{33}$$

$$Cost^{col} = Cost^{shell} + Cost^{tray}, \tag{34}$$

where

$$H_i = \left(\frac{N_i}{E_i^o} \cdot T_i^s \right) + D_i^s. \tag{35}$$

H_i and D_i are the height and the diameter of the column; E_i^O is the overall column efficiency set at 20%; T_i^s, D_i^s are the tray spacing and "dead" space in each column. Because the diameter is fixed by the flow rate of the gas stream, and because of the assumption for the superficial velocity (Hallale and Fraser, 1998), the expressions reduce to linear equations.

CASE A

The absorption factor defined by

$$A_F = \frac{F_L}{mG} \tag{36}$$

is set to 1.4 (King, 1980). F_L is the flow rate of the water stream, G is the flow rate of the gas stream, and m is the equilibrium constant. The design corresponding to this setting is illustrated in Figure 6.13.

The freshwater demand is 118.3 t/h and the annual freshwater cost £651,616. The cost of the columns amounts to £19,030/yr and the total annualized cost is £670,647. The development of reuse networks with and without capital cost considerations is explained next.

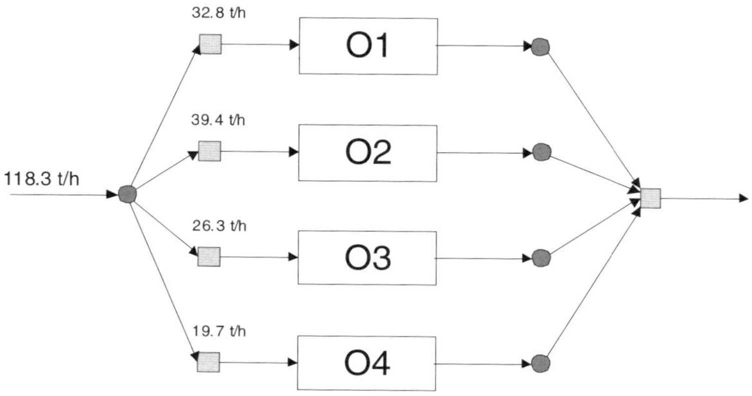

COLUMN	NTRAYS
1	4
2	3
3	7
4	2

Figure 6.13: Case A design; $A_F = 1.4$.

CASE B

The objective is the minimum freshwater cost, and Figure 6.14 illustrates the optimal design. Problem (P_1) features 173 continuous and 30 binary variables. The procedure required 4 iterations, consuming 8.2 CPU seconds in total. The freshwater demand drops by 76% (annual savings

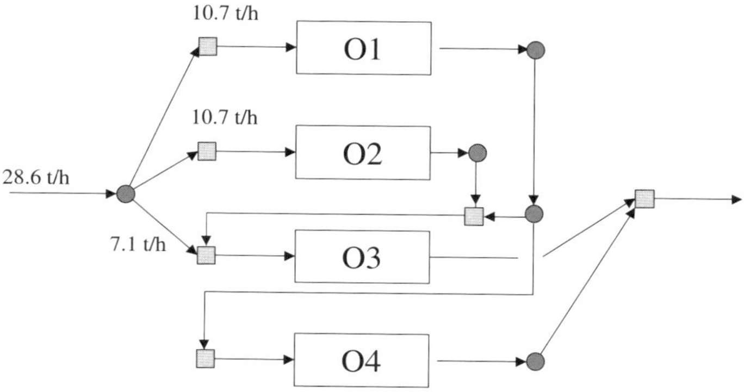

COLUMN	NTRAYS
1	26
2	22
3	581
4	8

Figure 6.14: Minimum freshwater cost.

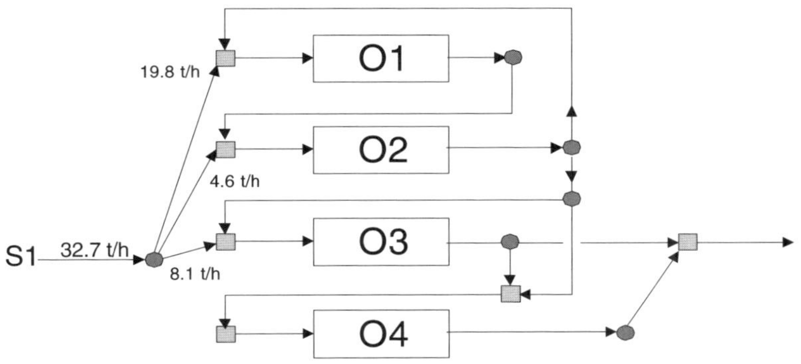

COLUMN	NTRAYS
1	6
2	15
3	36
4	6

Figure 6.15: Minimum total annual cost design. Mass exchanger costs included.

of £493,708.8/y). In column 3, however, the number of trays required to perform the separation increases dramatically. In fact, there are 581 trays needed, and the column height is about 300 m! The annualized capital cost is £297,505/y, namely over 15 times more expensive in comparison with Case A. The total annual cost is £455,063/y.

CASE C

The objective is to minimize the total annual cost (considering mass exchanger costs), and the optimal design is illustrated in Figure 6.15. The cost of the mass exchangers is reduced dramatically even though the freshwater demand is almost the same as in Case B (72 % of Case A demand). The annualized capital cost of the absorption columns is £42,530/y, a reduction of over 80 % against the previous case.

The total annualized cost is £222,495/y. A solution with similar performance has been reported by Hallale and Fraser (1998), as illustrated in Figure 6.16.

Figure 6.17 summarizes the results for the three designs. As compared with Case A, the final solution achieves total annual savings of around £448,150 (or 67 % reduction). Coupled with this overall savings, the freshwater consumption can be reduced by some 72 %; that is 85.6 t/h. This saving of freshwater is achieved without altering significantly the capital costs of the network related to the size of the mass exchangers. In these solutions the linearized approximation to the number of equilibrium stages involves five linear segments (i.e., six variables in the SOS2 set).

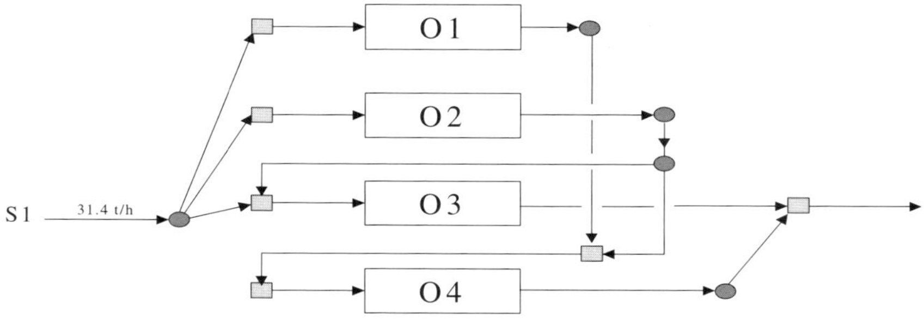

COLUMN	NTRAYS
1	13
2	11
3	58
4	5

Figure 6.16: Network design reported by Hallale and Fraser (1998).

7 CONCLUSIONS

A number of systematic methodologies are available to improve the design with a simultaneous purpose to account the environmental performance. The methods appear rather promising, but a lot of research is still required. The development of more integrated schemes, the use of advanced separation methods, and the development of alternatives for solvent substitutes are all research areas with a promising future for the environment. Computer-aided tools are expected to play a major role with their ability to automate decisions and speedily access databases with useful information for the engineer.

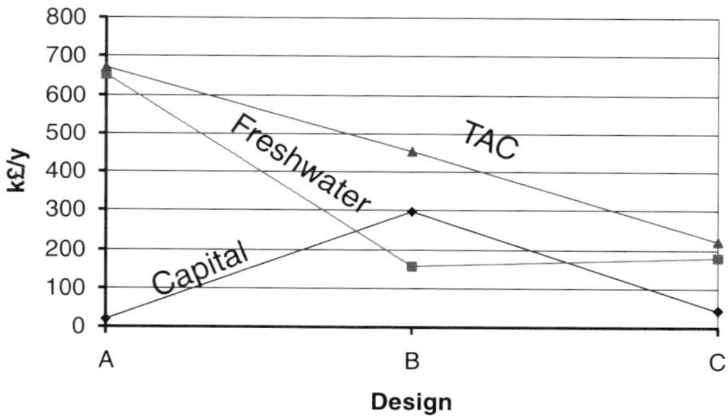

Figure 6.17: Comparison of results for Example 3.

ACKNOWLEDGMENTS

The financial support of CONACYT, ITESM Campus Morelos, EPSRC, and the Process Integration Department at UMIST is gratefully acknowledged. Gratitude is expressed to Miss D. Chow for helping in revising the manuscript.

REFERENCES

Adjiman, C. S., Androulakis, I. P., Maranas, C. D., and Floudas, C. A. (1996). A global optimisation method, αββ, for process design. *Comp. Chem. Eng.* 20(suppl.): S419–S423.

Ahmad, S., and Smith, R. (1989). Targets and design for minimum number of shells in heat exchanger networks. *Chem. Eng. Res. Des.* 67:481–494.

Al-Khayyal, F. A., and Falk, J. E. (1983). Jointly constrained biconvex programming. *Math. Oper. Res.* 8(2):273–286.

Bagajewicz, M. J., Pham, R., and Manousiouthakis, V. (1998). On the state space approach to mass/heat exchanger network design. *Chem. Eng. Sci.* 53(14):2595–2621.

Bagajewicz, M. J., and Manousiouthakis, V. (1992). Mass/heat-exchange network representation of distillation networks. *AIChE J.* 38(11):1769–1800.

Baker, T. E., and Lasdon, L. S. (1985). Succesive linear programming at Exxon. *Management Sci.* 31:264–274.

Briones, V., and Kokossis, A. C. (1999a). Hypertargets: A conceptual programming approach for the optimisation of industrial heat exchanger networks – I: Grassroots design and network complexity. *Chem. Eng. Sci.* 54:519–539.

Briones, V., and Kokossis, A. C. (1999b). Hypertargets: A conceptual programming approach for the optimisation of industrial heat exchanger networks – II: Retrofit design. *Chem. Eng. Sci.* 54:541–561.

Briones, V., and Kokossis, A. C. (1999c). Hypertargets: A conceptual programming approach for the optimisation of industrial heat exchanger networks – III: Industrial applications. *Chem. Eng. Sci.* 54:85–706.

Delaby, O., and Smith, R. (1995). Minimization of flue gas emissions. *Trans. IChemE* 73(Part B):21–32.

Douglas, J. M. (1992). Process synthesis for waste minimization. *Ind. Eng. Chem. Res.* 31:238–243.

El-Halwagi, M. M. (1997). *Pollution Prevention through Process Integration: Systematic Design Tools*. Academic Press, San Diego.

El-Halwagi, M. M., and Manousiouthakis, V. (1989a). Synthesis of mass exchanger networks. *AIChE J.* 35(8):1233–1244.

El-Halwagi, M. M., and Manousiouthakis, V. (1989b). Design and analysis of mass exchange networks with multicomponent targets. AIChE Annual Meeting, San Francisco, November.

El-Halwagi, M. M., and Spriggs, H. D. (1998). Solve design puzzles with mass integration. *Chem. Eng. Prog.* 94(August):22–44.

Floudas, C. A. (1995). *Nonlinear and Mixed-integer Optimization*, ed. 1. Oxford University Press, New York.

Gupta, A., and Manousiouthakis, V. (1994). Waste reduction through multicomponent mass exchange network synthesis. *Computers Chem. Eng.* 18(suppl.):S585–S590.

Hallale, N., and Fraser, D. M. (1998). Capital cost targets for mass exchange networks: A special case: Water minimisation. *Chem. Eng. Sci.* 53(2):293–313.

Haverly, C. A. (1977). Behaviour of LP models. *ACM Sigmap Bull.* 22:52–60.

Haverly, C. A. (1978). Studies of the behaviour of recursion for the pooling problem. *ACM Sigmap Bull.* 25:19–29.

Haverly, C. A. (1979). Behaviour of recursion model: More studies. *ACM Sigmap Bull.* 26:22–28.

Haverly, C. A. (1980). Recursion model behaviour: More studies. *ACM Sigmap Bull.* 28:39–41.

Gundersen, T., and Naess, L. (1988). The synthesis of cost optimal heat exchanger networks: An industrial review of the state of the art. *Comp. Chem. Eng.* 12:503.

King, C. J. (1980). *Separation Processes*, ed. 2. McGraw-Hill, New York.

Kuo, W. J., and Smith, R. (1997). Effluent treatment system design. *Chem. Eng. Sci.* 52(23):4273–4290

Kuo, W. J., and Smith, R. (1998). Designing for the interactions between water-use and effluent treatment. *Trans. IChemE* 76(Part A):287–301.

Kuo, W. J. (1996). A combined approach to water minimisation and effluent treatment system design. Ph.D. dissertation, Department of Process Integration, UMIST, Manchester, UK.

Linnhoff, B. (1993). Pinch analysis—A state-of-the-art overview. *Trans. IChemE* 71(part A):503–522.

Magnanti, T. L., and Wong, R. T. (1984). Network design and transportation planning: Models and algorithms. *Transport. Sci.* 18(1):1–55

Palacios-Gomez, F., Lasdon, L., and Engquist, M. (1982). Nonlinear optimization by successive linear programming. *Manage. Sci.* 28(10):1106–1120.

Papalexandri, K. P., Pistikopoulos, E. N., and Floudas, A. (1994a). Mass exchange networks for waste minimization: A simultaneous approach. *Trans. IChemE* 72 (part A):279–294.

Papalexandri, K. P., and Pistikopoulos, E. N. (1994b). A multiperiod MINLP model for the synthesis of flexible heat and mass exchange networks. *Comp. Chem. Eng.* 18(11/12):1125–1139.

Quesada, I., and Grossman, I. E. (1995). Global optimization of bilinear process networks with multicomponent flows. *Comp. Chem. Eng.* 19(12):1219–1242.

Rossiter, A. P., Spriggs, D. H., and Klee, H., Jr. (1993). Apply process integration to waste minimization. *Chem. Eng. Prog.* 89(January):30–36.

Simon, J. D., and Azma, H. M. (1983). Exxon experience with large scale linear and nonlinear programming applications. *Comp. Chem. Eng.* 7(5):605–614.

Smith, R., and Petela, E. A. (1991a). Waste minimisation in the process industries. Part 1: The problem. *Chem. Eng.* 506:24–25.

Smith, R., and Petela, E. A. (1991b). Waste minimisation in the process industries. Part 2: Reactors. *Chem. Eng.* 509/510:17–23.

Smith, R., and Petela, E. A. (1992a). Waste minimisation in the process industries. Part 3: Separation and recycle systems. *Chem. Eng.* 513:24–28.

Smith, R., and Petela, E. A. (1992b). Waste minimisation in the process industries. Part 4: Process operations. *Chem. Eng.* 517:21–23.

Smith, R., and Petela, E. A. (1992c). Waste minimisation in the process industries. Part 5: Utility waste. *Chem. Eng.* 523:32–35.

Wang, Y. P., and Smith, R. (1994a). Wastewater minimization. *Chem. Eng. Sci.* 49(7):981–1006.

Wang, Y. P., and Smith, R. (1994b). Design of distributed effluent treatment systems. *Chem. Eng. Sci.* 49(18):3127–3145.

Wang, Y. P., and Smith, R., (1995). Wastewater minimisation with flowrate constraints. *Trans. IChemE* 73 (part A):889–904.

Williams, H. P. (1987). *Model Building in Mathematical Programming*, ed. 3. John Wiley & Sons, West Sussex, UK.

Zhang, J., Kim, N. H., and Lasdon, L. S. (1985). An improved successive linear programming algorithm. *Manage. Sci.* 31(10):1312–1331.

PART FOUR

SYNTHESIS AND SCREENING OF ENVIRONMENTAL TECHNOLOGIES

CHAPTER 7

Mass Transfer Technology for Pollution Prevention

Nick Hallale

Department of Process Integration, UMIST, Manchester M60 1QD, U.K.

Abstract

This chapter discusses recent developments made in the area of mass transfer operations. Instead of focusing on individual unit operations, the chapter deals mainly with the integration of these operations into an overall process flowsheet. Systematic tools are given to ensure that optimal process designs are produced.

Keywords: mass transfer operations, mass exchange networks, pollution prevention

1 INTRODUCTION

Mass transfer operations are indispensable in the chemical process industries. They include a wide range of separations (e.g., absorption, stripping, liquid extraction, leaching, adsorption, and ion exchange), and thus there are applications in a wide range of process industries. Example applications are feed preparation, product separation, product finishing, and recovery of valuable materials. Another one, which is becoming increasingly important, is that of waste minimization. Reuse and recycling of hazardous waste streams can often be facilitated by using the appropriate mass transfer operations. These operations involve the transfer of components from a rich stream into a lean mass separating agent (MSA). Rich streams typically are process or effluent streams that contain the components of interest which could be valuable materials or undesirable contaminants. The MSAs, on the other hand, are the streams employed to do the separation (e.g., solvents, stripping agents, adsorbents, ion-exchange resins, etc.). The mass transfer is carried out by contacting the rich stream and MSA in a *mass exchanger*, which is a direct-contact, countercurrent unit.

The design of individual mass exchange units is well established and documented (e.g., Treybal, 1981); however, until recently, less attention has been given to the integration of these unit operations into an overall flow-sheet.

This chapter begins by discussing briefly the different types of mass exchangers, as well as techniques for designing them. The chapter then moves on to the synthesis of *mass exchange networks*. These networks are systems of mass exchangers, which transfer the desired components from a set of rich streams to a set of MSAs in the best way. Techniques for minimizing the MSA cost, the capital cost, and the total cost are presented.

2 MASS EXCHANGE UNITS

As mentioned earlier, there is a wide range of mass transfer operations and equipment types; however, they all have the common function of contacting a rich stream and an MSA. Figure 7.1 shows a generic mass exchanger with the streams to be contacted.

This rest of this section discusses the different types of mass exchangers and outlines the techniques for sizing them. Further details are available in El-Halwagi (1997). These techniques will be extended later on in order to determine the minimum capital cost of a mass exchange network.

When sizing the mass exchangers, the concept of equilibrium is very important. Consider a rich stream and an MSA, which are brought into contact. While the phases are in contact, there is mass transfer from the rich phase to the lean phase. With thorough mixing and sufficient contact time, the two phases will reach equilibrium. The composition of a rich stream in equilibrium with an MSA of composition x is y^*, as defined by the mass transfer equilibrium relation:

$$y^* = f(x). \tag{1}$$

Data for equilibrium relations can be obtained from the literature (Perry, 1984) or by carrying out experimental studies.

The difference between the actual composition (rich stream or MSA) and the equilibrium composition is a measure of the mass transfer driving force and is used to size the exchangers. This can be represented on a set of y-x axes, as shown in Figure 7.2. This is a plot of the composition in the rich stream against that in the MSA. The actual exchanger operation is plotted

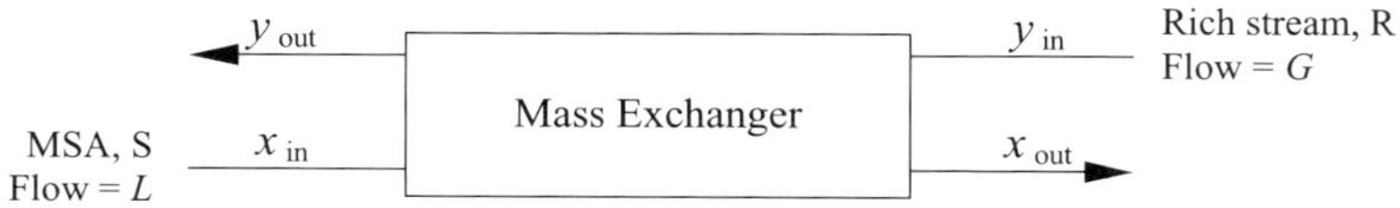

Figure 7.1: A mass exchanger.

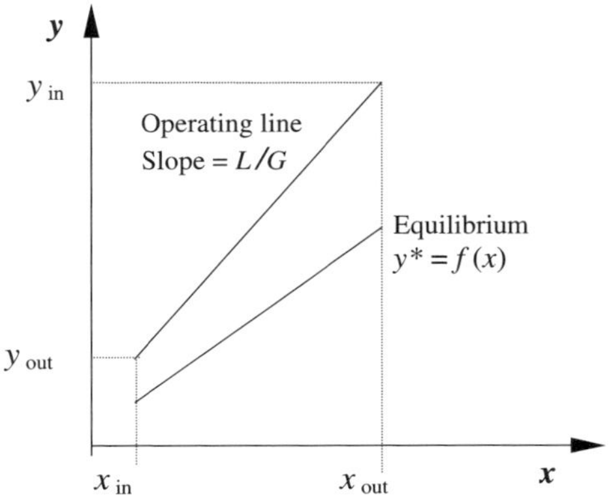

Figure 7.2: A mass exchanger represented on *y*-*x* axes.

as an operating line. As shown, the slope of this line is equal to the ratio of MSA flowrate to rich stream flowrate in the exchanger.

Notice that the equilibrium relation is drawn on the same set of axes. For simplicity, it has been shown here as a linear function (i.e., $y^* = mx + b$). This representation is useful, as the separation between the operating line and the equilibrium line is clearly shown. Mass exchangers can be classed into two main categories: *stagewise exchangers* and *continuous-contact exchangers*.

2.1 STAGEWISE EXCHANGERS

A stagewise exchanger consists of a series of stages in which the rich and lean phases are repeatedly contacted and then separated. If the phases leaving the stages are assumed to be in equilibrium with one another, the stage is referred to as an *ideal* or *equilibrium stage*.

The most common types of stagewise exchangers are tray or plate columns (Fig. 7.3) and systems of staged vessels, such as mixer-settlers or agitated tanks.

The most important piece of information required for sizing stagewise exchangers is the number of equilibrium stages. This can be determined graphically or analytically.

As shown in Figure 7.4, the number of equilibrium stages, N_{stages}, can be determined by "stepping off" between the two ends of the exchanger (Treybal, 1981). Note that the distance between the operating line and the equilibrium line affects the number of stages. The closer together these lines are, the more stages will be needed. If these lines touch or cross one another, an infinite number of stages will be required (i.e., infeasible operation).

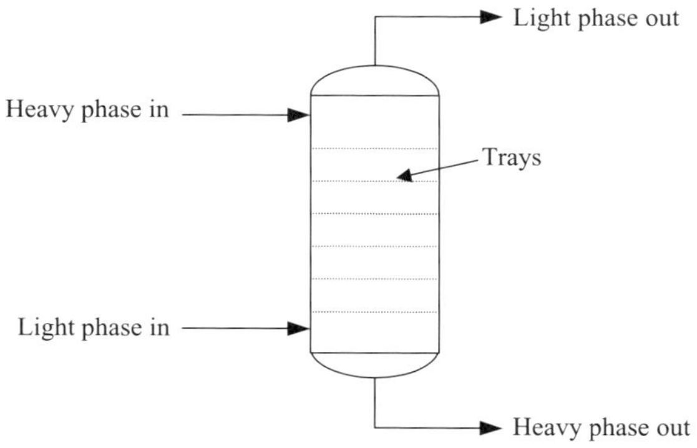

Figure 7.3: A tray column.

If the operating and equilibrium line are both straight, the Kremser equation (Treybal, 1981) can be used to calculate N_{stages} analytically:

$$\text{If } A \neq 1, \qquad N_{\text{stages}} = \frac{\ln\left[\left(\frac{y_{\text{in}} - mx_{\text{in}} - b}{y_{\text{out}} - mx_{\text{in}} - b}\right)\left(1 - \frac{1}{A}\right) + \frac{1}{A}\right]}{\ln A}; \tag{2}$$

$$\text{If } A = 1, \qquad N_{\text{stages}} = \frac{y_{\text{in}} - y_{\text{out}}}{y_{\text{out}} - mx_{\text{in}} - b}, \tag{3}$$

where

$$A = \frac{L}{mG}. \tag{4}$$

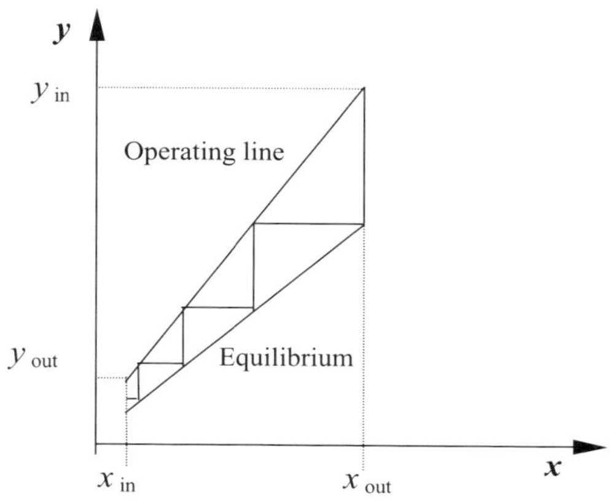

Figure 7.4: Graphical method for determining the number of equilibrium stages.

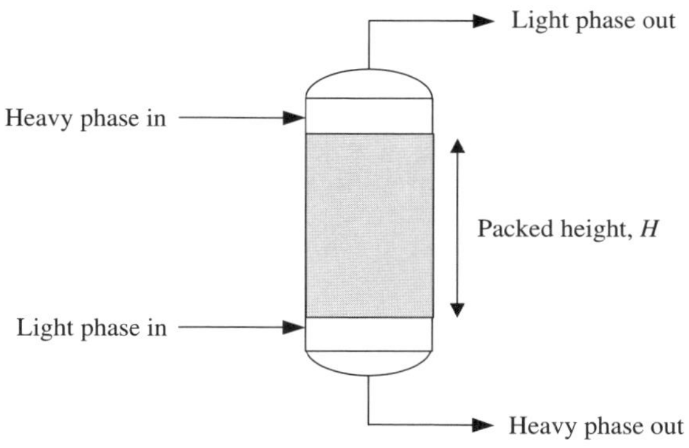

Figure 7.5: A packed column.

A is termed the *removal factor* and is seen to be the ratio of the slope of the operating line to the slope of the equilibrium line.

In practice, the number of *real* stages may not be the same as the number of equilibrium stages. This is because equilibrium may not be reached in each stage. This can be accounted for using overall or stage efficiencies. Details are given by El-Halwagi (1997).

Other important parameters are the column diameter and tray spacing (tray columns) and vessel volume (mixer-settlers or agitated tanks). Further details are available (Treybal, 1981; Coulson et al., 1993). Hallale (1998) summarizes some useful shortcut procedures and rules of thumb to estimate these parameters for costing purposes.

2.2 CONTINUOUS-CONTACT EXCHANGERS

In this type of exchanger, the rich and lean phases are in continuous contact throughout with no intermediate phase separation and recontacting. The most common exchanger type belonging to this category is the packed column (Fig. 7.5).

Other types include spray columns and mechanically agitated columns. The characteristic size of a continuous-contact exchanger is the height, which can be estimated using the *transfer unit method*. The following equation (El-Halwagi, 1997) gives the exchanger height, H:

$$H = HTU_y NTU_y, \tag{5}$$

where HTU_y and NTU_y are the overall height of a transfer unit and the number of transfer units, respectively. This equation is based on the rich phase (hence the subscript y), but could also be based on the lean phase.

The overall transfer unit heights are calculated by

$$HTU_y = \frac{G_i'}{K_y a}, \tag{6}$$

where G is the superficial flowrate of the rich phase (flowrate per unit cross-sectional area) and $K_y a$ is the overall rich phase mass transfer coefficient.

Values for $K_y a$ may be provided by the equipment manufacturer, estimated from experience with similar systems or else calculated using correlations. El-Halwagi (1997) provides a list of sources for correlations for estimating mass transfer coefficients.

For straight operating and equilibrium lines, the number of transfer units can be calculated as

$$NTU_y = \frac{y_{\text{in}} - y_{\text{out}}}{\Delta y_{\text{lm}}}, \tag{7}$$

where

$$\Delta y_{\text{lm}} = \frac{\Delta y_{\text{in}} - \Delta y_{\text{out}}}{\ln\left(\frac{\Delta y_{\text{in}}}{\Delta y_{\text{out}}}\right)}. \tag{8}$$

The composition differences, Δy, are shown in Figure 7.6. Notice that these are the differences between the actual (operating) compositions and the equilibrium compositions.

El-Halwagi (1997) states that the exchanger diameter normally is determined by selecting a superficial velocity for one of the phases. This velocity is intended to ensure proper contact while avoiding hydrodynamic problems, such as flooding, weeping, or entrainment. Correlations for estimating superficial velocities are available in the literature (Coulson et al., 1993).

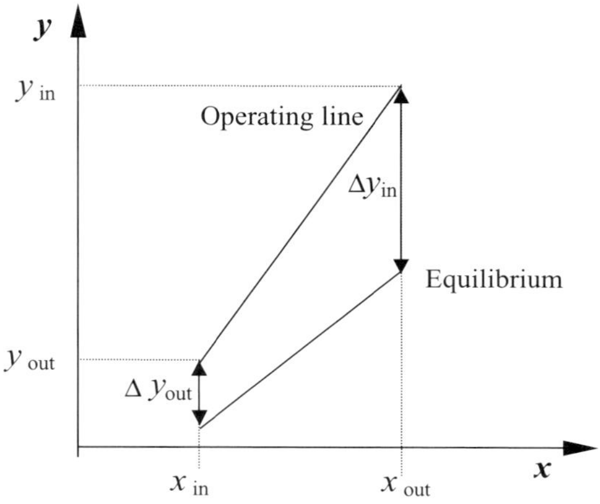

Figure 7.6: Composition differences in a continuous-contact exchanger.

Example 1: Mass Exchanger Design

A copper-etching plant produces a waste water stream with a flowrate of 0.1 kg/s. This stream contains copper at a concentration of 0.06 kg copper/kg water. To recycle this stream, the copper concentration needs to be reduced to 0.015 kg copper/kg water. The required treatment is to be carried out via solvent extraction.

The solvent to be used is P_1 (an aromatic β-hydroxyoxime), which is available at a flowrate of 0.025 kg/s. This solvent already contains some copper at a composition of 0.001 kg copper/kg solvent. The equilibrium relation governing copper extraction can be approximated by the following linear relation (El-Halwagi and Manousiouthakis, 1990a):

$$y^* = 0.148x + 0.013.$$

The mass exchanger is a sieve-tray column.

Estimate the number of stages.

Solution

The Kremser equation for calculating the number of stages is demonstrated below.

The removal factor, A, is given by equation (4):

$$A = \frac{0.025}{(0.148)(0.1)} = 1.69.$$

This is not equal to 1, and so the number of stages is given by equation (2):

$$N_{\text{stages}} = \frac{\ln\left[\left(\dfrac{0.06 - 0.148(0.001) - 0.013}{0.015 - 0.148(0.001) - 0.013}\right)\left(1 - \dfrac{1}{1.69}\right) + \dfrac{1}{1.69}\right]}{\ln 1.69}$$

$$= 4.56 \text{ equilibrium stages.}$$

Rounding up to the next whole number gives a value of 5 equilibrium stages.

3 SYNTHESIS OF MASS EXCHANGE NETWORKS

The previous section of this chapter discussed how individual mass exchange units are designed. As mentioned earlier, this is a well-established topic in chemical engineering. This section discusses technology for synthesizing the best *systems* of mass exchange units to deliver a specified separation. This is a far more challenging problem.

In Example 1 we saw that it was fairly straightforward to design an exchanger when we knew what the rich stream and MSA were and also when we knew their flowrates and inlet and outlet compositions; however, what would happen if these were not specified? Consider, for example, what would happen if the problem had consisted of two streams requiring treatment and if there were not one, but two different solvents available. As soon as the number of rich streams or MSAs becomes more than one, the problem is no longer that of unit design, but that of network or system design.

Before we could think about exchanger design, we would have to decide which MSAs to employ, what flowrates to use, which MSA should be used to treat which rich stream, and what the flowsheet configuration should be. These decisions are all interrelated and therefore need to be considered simultaneously. The multitude of possibilities means that an exhaustive or trial-and-error approach is extremely unlikely to yield the best possible solution. This is true, even with the availability of large and powerful computers today.

This section presents *systematic* techniques to approach this problem. First, the formal mass exchange network problem statement is given. Then, techniques for targeting the minimum MSA cost of a mass exchange network are presented. This is followed by methods for targeting the minimum network capital cost. We then examine how these costs can be traded off in order to minimize the total network cost. Finally, design methods that enable the targets to be met are presented.

3.1 PROBLEM STATEMENT

The mass exchange network synthesis task has been introduced by El-Halwagi and Manousiouthakis (1989a) and can be stated as follows:

> Given a number N_R of rich streams and a number N_S of MSAs (lean streams), it is desired to synthesize a cost-effective network of mass exchangers that can preferentially transfer certain species from the rich streams to the MSAs. Given also are the flowrate of each rich stream, G_i, its supply (inlet) composition, y_i^s, and its desired (outlet) composition, y_i^t, where $i = 1, 2, \ldots, N_R$. In addition, the supply and desired compositions, x_j^s and x_j^t, are given for each MSA, where $j = 1, 2, \ldots, N_S$. The mass transfer equilibrium relations are also given for each MSA. The flowrate of each MSA is unknown and is to be determined as part of the synthesis task.
>
> The candidate MSAs (lean streams) can be classified into N_{SP} process MSAs and N_{SE} external MSAs (where $N_{SP} + N_{SE} = N_S$). The process MSAs already exist on the plant site and can be used for a low cost (often virtually free). The flowrate of each process MSA, L_j, is bounded by its availability in the plant and may not exceed a value of L_j^c. On the other hand, the external MSAs can be purchased from the market and their flowrates are to be determined by economic considerations.

The desired composition of the transferred species in each MSA is determined by the specific circumstances of the application. The nature of such circumstances may be physical (e.g., maximum solubility), technical, (e.g., to avoid excessive corrosion, viscosity or fouling), environmental (e.g., to comply with environmental regulations), safety (e.g., to stay away from flammability limits), or economic (e.g., to optimize the regeneration cost).

3.2 WHAT IS A TARGET?

Before proceeding, it is worth examining the concept of a target. We know that to design a mass exchange network there are a huge number of alternatives to consider in order to find the best solution. Without some reference point, it is very difficult to know whether we have found the best solution or whether we could be doing better.

It would be very useful if we could predict, before making any design decisions, what the best performance (e.g., minimum cost) would be. This would provide a great incentive to find improvements and would also give confidence that a particular design was truly the best one. Of course, the amount of time spent searching for the best solution would be drastically reduced as well.

There are now systematic methods for predicting the minimum MSA cost, minimum capital cost, and minimum total cost ahead of any design. These will now be presented.

3.3 MINIMUM MSA TARGETS

To minimize the cost of MSAs, it is necessary to make maximum use of the process MSAs before using external MSAs. The applicability of the process MSAs is limited by thermodynamics, and this is accounted for in the targeting method. The method shown here was first developed by El-Halwagi and Manousiouthakis (1989a) and later adapted slightly by Hallale (1998). To perform the targeting, we need to specify the *minimum composition difference* allowable in the network, which we designate as $\Delta y_{\min}$.

Recall that for any mass exchanger to be feasible, the operating line always needs to be above the equilibrium line (see Fig. 7.2). If these lines intersect, there will be a zero driving force, and hence an infinitely large exchanger will be required. By specifying a minimum composition difference, we are specifying that no mass exchanger in the network will have its operating line and equilibrium line separated vertically by less than $\Delta y_{\min}$.

The next step is to draw all the rich streams as lines on a plot of composition versus mass load transferred. This is shown in Figure 7.7a.

Notice how the inlet and outlet compositions of the rich streams form *composition intervals*. The slope of each line is equal to the inverse of the stream flowrate. The next step is to combine the rich streams within these intervals by drawing lines with slope equal to the inverse of the total stream flowrate in each interval (Fig. 7.7b). This gives the *rich composite curve*, which shows the cumulative mass to be removed at each rich stream composition.

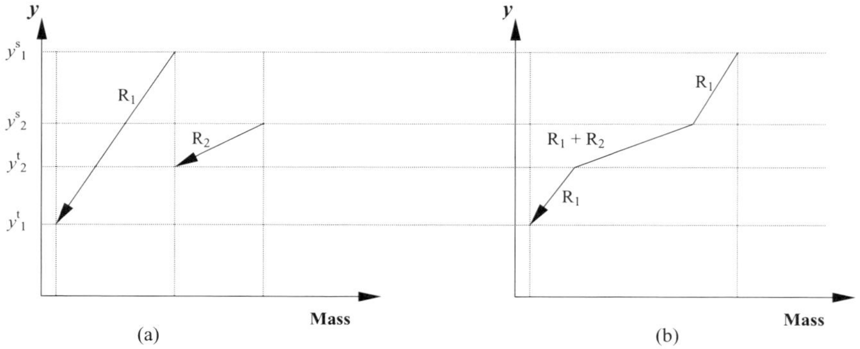

Figure 7.7: Construction of the rich composite curve.

Next, the same is done for the MSAs. Plotting the MSAs on composition-mass load axes is not as straightforward as for the rich streams because MSA compositions are not equivalent. This is because the MSAs have different equilibrium relations and cannot be compared directly. One approach to dealing with this is to use multiple composition axes, whose scales correspond via the equilibrium relations. This has been explained in full detail (El-Halwagi and Manousiouthakis, 1989a; El-Halwagi, 1997).

Another approach has recently been developed (Hallale, 1998) and will be used here. Each MSA composition, x, has a corresponding equilibrium rich stream composition, y^*, as defined by equation (1). Referring to MSA compositions as equilibrium-rich stream compositions accounts for the different equilibrium relations and allows all MSAs to be compared on a common basis.

Figure 7.8 demonstrates this for a problem with linear equilibrium relations (i.e., $y^* = m_j x_j + b_j$). The individual MSA streams are plotted in

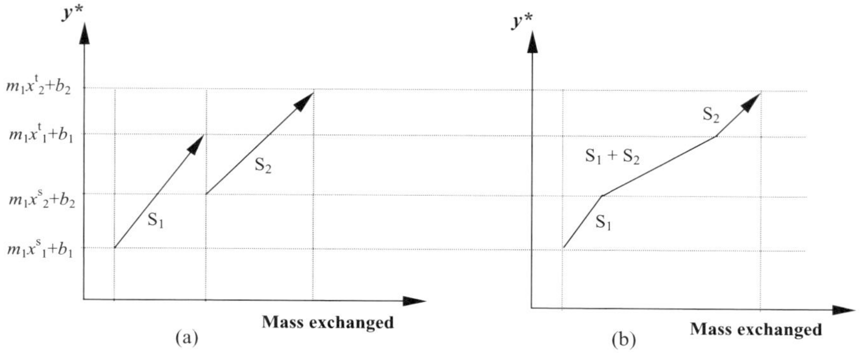

Figure 7.8: Construction of the lean composite curve (note y^* is plotted against mass exchanged).

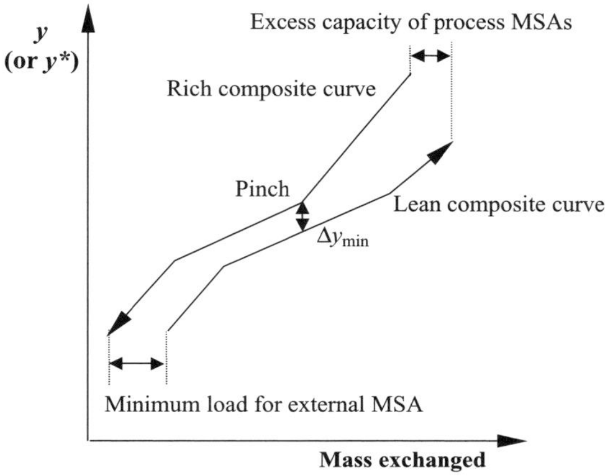

Figure 7.9: Plotting both composite curves on the same axes shows the pinch and determines the MSA targets.

Figure 7.8a. Note that only the process MSAs are included and that they are taken to be at their maximum flowrates.

It will be noticed that the slopes of the lines are not equal to the inverse of the MSA flowrate, but now are equal to the ratio of equilibrium constant, m_j, to MSA flowrate, L_j. The reason for this is because we are using y^* and not actual MSA compositions.

Similarly to before, the MSAs are combined within composition intervals in order to give a *composite lean stream*. This shows the cumulative mass that can be removed at a given y^* value. Next, both composite curves are plotted on the same set of axes. The lean composite curve is shifted horizontally until the two composites are separated vertically by Δy_{min} (Fig. 7.9). The point where this occurs is termed the *mass transfer pinch* and is the thermodynamic bottleneck to mass transfer between process streams.

Figure 7.9 gives some very useful information regarding the problem. First, the vertical overlap between the composite curves shows the extent to which mass can be transferred to the process MSAs (i.e., the maximum recovery possible).

The distance by which the lean composite curve extends past the rich composite curve shows the excess capacity of the process MSAs to remove mass. This excess can be eliminated by lowering the flowrate and/or outlet composition of one or more process MSAs.

The distance by which the rich composite curve extends past the lean composite curve shows the *minimum* mass that needs to be removed by an external MSA. This information can then be used to give the minimum cost of MSAs.

The pinch divides the problem into two separate regions (that above the pinch and that below it). We will see later that to achieve the minimum MSA targets, no mass may be transferred across the pinch.

TABLE 7.1
Stream data for dephenolization problem (Examples 2–6)

Rich streams	G *(kg/s)*	y^s *(mass fraction)*	y^t *(mass fraction)*		
R1	2	0.050	0.010		
R2	1	0.030	0.006		
MSAs	L^c *(kg/s)*	x^s *(mass fraction)*	x^t *(mass fraction)*	m	b
S1	5	0.005	0.015	2	0
S2	3	0.01	0.030	1.53	0
S3	–	0.0013	0.015	0.71	0.001

Example 2: MSA Targeting for a Dephenolization Mass Exchange Network

This example is adapted from one presented by El-Halwagi (1997). It involves the removal of phenol from two wastewater streams, R_1 and R_2, by solvent extraction. Two process MSAs are available: these are gas oil (S_1) and lube oil (S_2), and they are available for free. The addition of phenol to these oil streams is actually beneficial for them, and it is specified that the entire gas oil stream should be used. The external MSA is a light oil (S_3) with a cost of \$0.01/kg. Table 7.1 gives stream data for the problem.

Find the minimum MSA cost for a $\Delta y_{\min}$ of 0.001.

Solution

Figure 7.10 shows the composite curves plotted on the same axes and separated by a $\Delta y_{\min}$ of 0.001. The pinch point is at $y = 0.0163$ ($y^* = 0.0153$). The excess capacity of the process MSAs is read off as 0.01565 kg/s phenol. Similarly, the load for the external MSA is read off as 0.00965 kg/s phenol.

The excess capacity means that one of the process MSAs can have its flow reduced. Because it has been specified that all of S_1 must be used, S_2 can have its flow decreased. The new flow of S_2 is given by

$$\text{Flow } S_2 = 3 - \frac{0.01565}{(0.03 - 0.01)} = 2.22 \text{ kg/s}.$$

The flowrate of the external MSA, S_3, is determined as

$$\text{Flow } S_3 = \frac{0.00965}{(0.015 - 0.0013)} = 0.704 \text{ kg/s}.$$

Because the cost of S_3 is \$0.01/kg, the MSA cost target is \$0.00704/s. Assuming an operating time of 8600 hours per year gives a cost of \$217,960/yr.

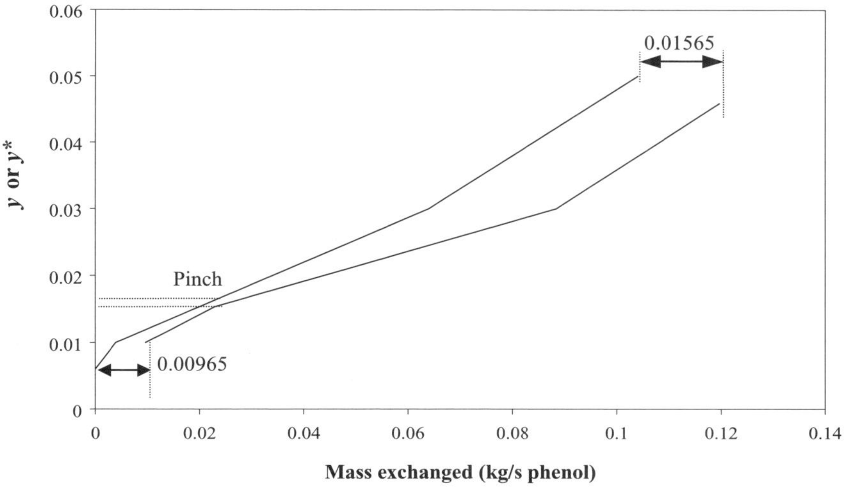

Figure 7.10: Mass transfer composite curves for Example 1.

3.4 MINIMUM CAPITAL COST TARGETS

This section discusses two important targets that are used to give an indication of the minimum capital cost of a network. The first target is the minimum number of mass exchange units; the second one is the minimum total exchanger size.

3.4.1 Minimum Units Target

This target is very well established (El-Halwagi and Manousiouthakis, 1989a). The minimum number of mass exchange units is simply one less than the total number of streams (rich streams plus process MSAs plus external MSAs). Because the pinch divides the problem into two regions, unit targets should be calculated for each region and then added to give the overall target.

Example 3: Minimum Units Targeting for Dephenolization Mass Exchange Network

What is the minimum number of units for the dephenolization problem of Example 2?

Solution

Figure 7.11 shows the streams for the network as a *grid diagram*. This representation was introduced by El-Halwagi and Manousiouthakis (1989a). It may appear unusual at first but is much more convenient than a

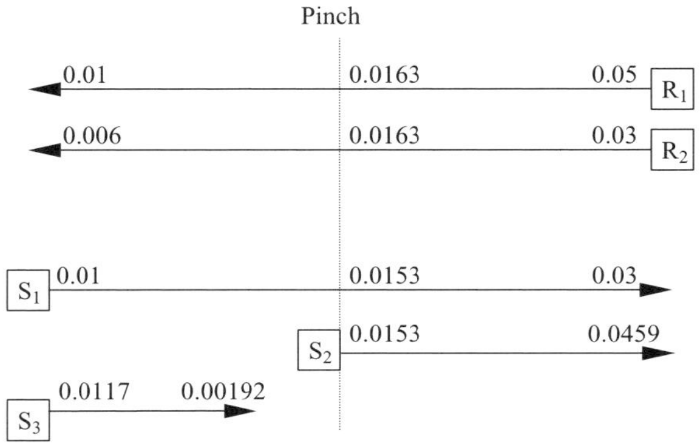

Figure 7.11: Grid diagram for Example 3.

conventional flowsheet. Rich streams are drawn running from right to left and lean streams the opposite way. Compositions are given above each stream. Notice that the lean stream compositions are represented as y^* values and not as their actual values.

Above the pinch there are four streams (R_1, R_2, S_1, and S_2), and so the minimum number of units for this region is 3. Below the pinch there are also four streams (R_1, R_2, S_1, and S_3), which again gives a units target of 3. The total target is therefore 6 units.

3.4.2 Minimum Exchanger Size Targets

Although knowing the minimum number of units is very useful, it alone is not sufficient to indicate the minimum capital cost. First, it does not give an actual monetary value. Second, the exchanger sizes are also important. There often are several alternative network designs that will achieve the minimum number of units; however, the exchanger sizes could be very different in different designs, leading to differing capital costs.

Recently, targets have been developed to predict the minimum total exchanger size—i.e., total number of stages or continuous-contact height—for a given Δy_{min} (Hallale and Fraser, 1998a; Hallale, 1998). The method is based on a newer graphical tool called the *y-y* composite curve plot*. As the name implies, this is a plot of rich stream composition against MSA composition and is similar to the *y-x* diagram (see Fig. 7.2). The new plot consists of a *composite operating line* as well as an equilibrium line.

Figure 7.12 shows how the composite operating line is constructed. The first step is to redraw the mass transfer composite curves, but with the MSAs set at their target flowrates. The external MSAs are now included (Fig. 7.12a). Notice that the curves are now in perfect mass balance.

The composite operating line is then drawn by plotting the y values versus the y^* values, as shown in Figure 7.12b. The equilibrium line is now simply the line $y = y^*$ (i.e., a line with a slope of 1).

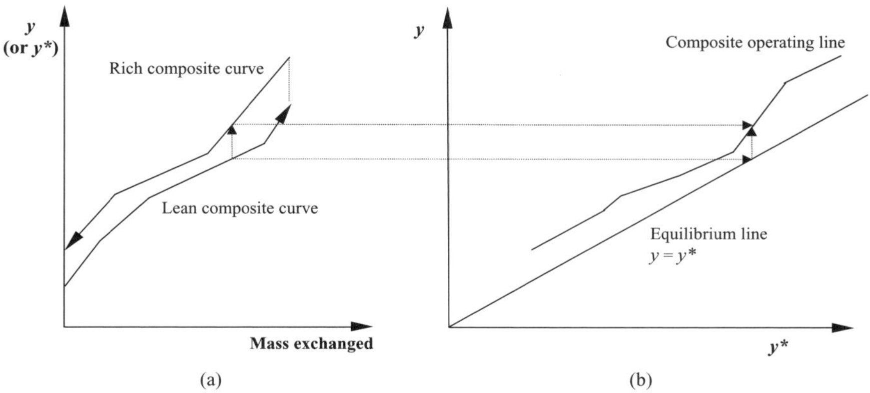

Figure 7.12: Construction of the y-y^* composite curve plot. (a) The balanced mass transfer composite curves; (b) the y-y^* composite curve plot.

Notice that the composition differences along the mass transfer composite curves are the same as those between the composite operating line and the equilibrium line.

Now, the problem is made up of composition intervals, which are the linear segments of the composite operating line. Each interval can be treated as an imaginary mass exchange unit and can have a number of equilibrium stages or transfer units assigned to it, as discussed in Section 2 of this chapter.

For stagewise systems, the target for the total number of stages can be estimated as

$$N_{\text{stages, min}} = \sum_{k}^{\text{Intervals}} N_{\text{stages}, k} N_{\text{R}, k}, \tag{9}$$

where $N_{\text{R},k}$ is the number of rich streams in interval k. This equation has been shown to give consistently good predictions of the total number of stages, provided the number of MSAs is not more than the number of rich streams. This usually is the case. More complex methods exist to deal with situations where this is not so, but will not be discussed here; see Hallale (1998) for details. Techniques are also given for dealing with nonequilibrium stages.

For continuous-contact systems, the minimum total exchanger height can be targeted as

$$H_{\text{min}} = \sum_{k}^{\text{Intervals}} NTU_{y,k} \left(\sum_{i}^{\substack{\text{Rich}\\ \text{streams}}} HTU_{y,i} \right)_k . \tag{10}$$

This equation gives the true minimum height if all the exchangers have the same cross-sectional surface area and overall mass transfer coefficient, $K_y a$; however, provided these parameters do not vary by more than an order of magnitude, the target has been shown to be within a few percent of the true minimum (Hallale, 1998).

3.4.3 Capital Cost Estimation

In this chapter it is assumed that the capital cost of a mass exchanger is a function of the number of equilibrium stages or the continuous-contact height; i.e.,

$$\text{Cost} = a + bN_{\text{stages}^c} \tag{11a}$$

or

$$\text{Cost} = a + bH^c, \tag{11b}$$

where a, b, and c are constants in the cost law. To estimate a capital cost, it is necessary to estimate the distribution of the minimum number of stages or height among the minimum number of units, N. The simplest way is to assume that the distribution takes place evenly. Thus,

$$\text{Capital cost target} = N\left[a + b\left(\frac{N_{\text{stages, min}}}{N}\right)^c\right] \tag{12a}$$

or

$$\text{Capital cost target} = N\left[a + b\left(\frac{H_{\text{min}}}{N}\right)^c\right]. \tag{12b}$$

Obviously, this can lead to some inaccuracy, but the effects have been found not to be significant by Hallale (1998), who presents more detailed costing methods, such as those based on exchanger mass or volume. Techniques are also given for problems with nonuniform exchangers.

Example 4: Targeting Capital Cost for Dephenolization Mass Exchange Network

Target the minimum capital cost for the dephenolization problem of Example 2 with Δy_{min} set as 0.001. The mass exchangers for this application are all sieve-tray extraction columns with a capital cost of \$22,760 N_{stages}, annualized over 5 years.

Solution

The first step is to target the minimum number of stages using the method described previously. Figure 7.13 shows the balanced mass transfer composite curves which use the MSA targets found in Example 2.

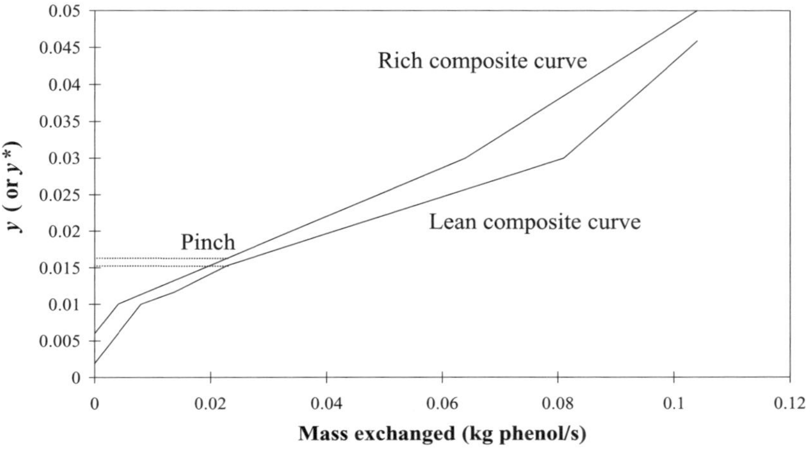

Figure 7.13: Balanced mass transfer composite curves for Example 4.

This is then used to give the y-y^* composite curve plot shown in Figure 7.14. As shown, there are seven composition intervals in the network. Each interval can have a number of equilibrium stages calculated for it, as demonstrated for Interval 1.

Figure 7.15 is a grid diagram for the problem, showing clearly all the composition intervals and their corresponding numbers of stages. It also shows the streams in each interval.

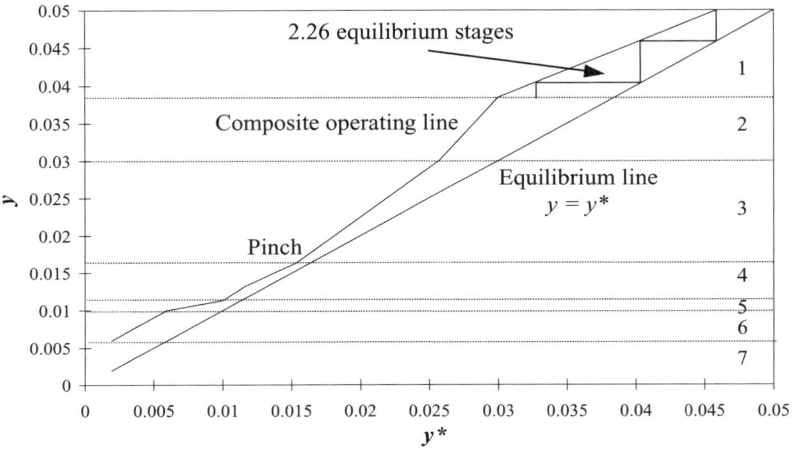

Figure 7.14: The y-y^* composite curve plot for Example 4.

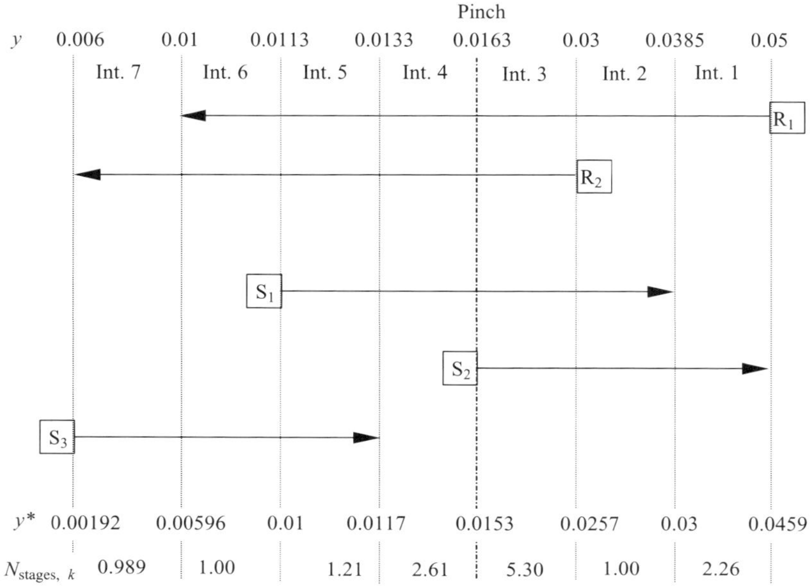

Figure 7.15: Grid diagram for Example 4.

Using equation (9), the number of stages target is

$$\begin{aligned} N_{\text{stages, min}} &= 2.26(1) + 1.00(1) + 5.30(2) + 2.61(2) + 1.21(2) \\ &\quad + 1.00(2) + 0.989(1) \\ &= 24.5 \text{ stages (round up to 25).} \end{aligned}$$

Using the given cost correlation, the minimum capital cost is estimated as \$569,000. With the annualization, this becomes \$113,800/yr. Note that in this problem, the capital cost is a linear function of the number of stages, and thus the number of units target did not affect the capital cost.

3.5 TOTAL COST TARGETS

Example 2 showed that the MSA cost target for the dephenolization problem is \$217,960/yr, and Example 3 showed the capital cost target is \$113,800/yr. Does this mean that the minimum total cost is the sum of these two (i.e., \$331,760/yr)? Not necessarily. Recall that all the targets we have estimated were for a Δy_{min} of 0.001; however, is this the best value to use? Using a larger Δy_{min} tends to increase the MSA cost targets but decreases the capital cost targets (Hallale, 1998). Thus there is a trade-off between these two costs, and the optimum Δy_{min} would be where the total cost was minimized. Without the benefit of targeting, there would be no way to locate this optimum without doing many repeated designs at different Δy_{min}

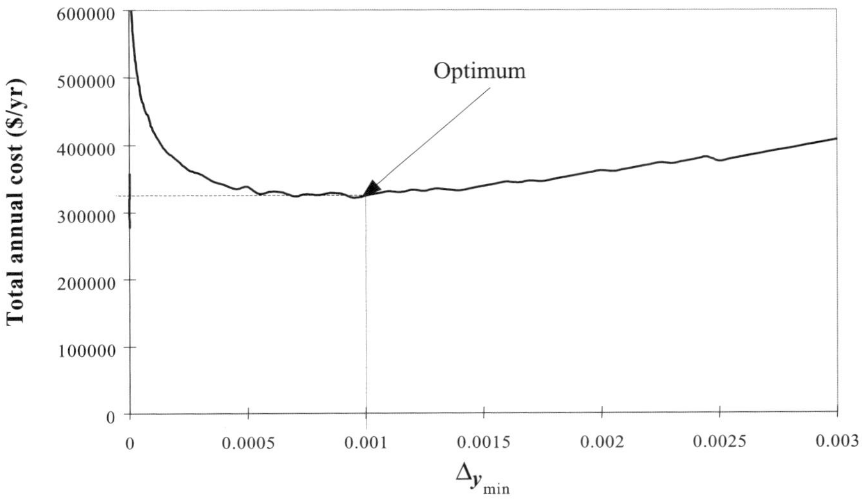

Figure 7.16: Total cost target profile for Example 5.

values (El-Halwagi and Manousiouthakis, 1990a). Fortunately, this can now be achieved without any design work, but simply by calculating *targets* over a range of $\Delta y_{\min}$ values.

Example 5: Targeting Total Cost for Dephenolization Mass Exchange Network

What is the minimum total cost of the dephenolization network of Example 2? What is the best $\Delta y_{\min}$ to use? Use the same capital cost data as in Example 4.

Solution

Repeating the procedures used in Examples 2 and 4 gives the total cost targets for a range of $\Delta y_{\min}$ values. These are plotted in Figure 7.16.

As shown, the optimum $\Delta y_{\min}$ for this example is actually 0.001. The minimum total annual cost target is therefore $331,760/yr.

3.6 MASS EXCHANGE NETWORK DESIGN

So far, we have seen how to target the minimum MSA cost, capital cost, and total cost for a mass exchange network; however, we have not yet explored how realistic these targets are. Obviously they are not of much use if it is not possible to meet them, or at least approach them, in an actual design.

This section presents some systematic design rules which ensure that the targets can be achieved. Design is carried out using the grid diagram introduced earlier.

3.6.1 Design for Minimum MSA

The minimum MSA design method was developed by El-Halwagi and Manousiouthakis (1989a). To meet the minimum MSA target, the regions above and below the pinch must be designed separately with no mass being transferred across the pinch. The pinch is the most constrained part of the network, so design should start at the pinch and move away from it (El-Halwagi and Manousiouthakis, 1989a).

There are two feasibility criteria for matching streams at the pinch (El-Halwagi and Manousiouthakis, 1989a):

1. *Stream population.* Immediately above the pinch, the number of rich streams must be greater than or equal to the number of lean streams, and immediately below the pinch, the opposite must be true. Splitting of rich or lean streams at the pinch may be needed to meet these requirements.
2. *Operating line versus equilibrium line.* When selecting matches at the pinch, it is important to recall that it is here where the minimum composition difference, Δy_{min}, will be experienced. If this is to be the minimum driving force, the operating line and equilibrium line must diverge away from the pinch. Thus, immediately above the pinch, the slope of the operating line should be greater than or equal to the slope of the equilibrium line, or

$$\left(\frac{L_j}{m_j}\right)_{\text{above pinch}} \leq G_{i,\text{above pinch}}. \tag{13a}$$

Immediately below the pinch, the opposite must be true:

$$\left(\frac{L_j}{m_j}\right)_{\text{below pinch}} \geq G_{i,\text{below pinch}} \tag{13b}$$

These rules are very useful for selecting between alternate matches. Once again, some stream splitting may be required to meet these conditions. See El-Halwagi and Manousiouthakis (1989a) and El-Halwagi (1997) for more details. Away from the pinch there usually is more freedom in choosing matches.

There usually are several possible designs that achieve the minimum MSA targets, but their capital costs can vary significantly. More guidelines are required in order to approach the minimum capital cost.

3.6.2 Design for Minimum Capital Cost

The minimum units target can be met simply by having each match exactly satisfy or tick-off the mass load of one stream (El-Halwagi and Manousiouthakis, 1989a). Recall, however, that achieving the minimum number of units is not enough to guarantee a minimum capital cost.

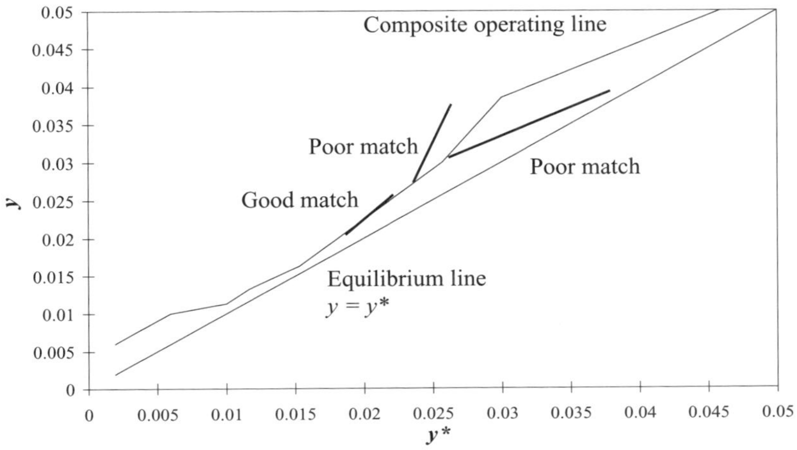

Figure 7.17: Match evaluation using the y-y^* composite curves.

A method for approaching the minimum number of stages or height targets has recently been developed (Hallale and Fraser, 1998a; Hallale, 1998). In this method, the operating line of any proposed match is superimposed on the y-y^* composite curve plot, as demonstrated in Figure 7.17.

Good matches will fit the ideal profile relatively closely (ideal ones will fit exactly), whereas poor ones will deviate from it. The general design philosophy is to use a low (minimum or near-minimum) number of units, each of which should provide a reasonable fit to the composite operating line. If the ideal profile cannot be reasonably approximated with the minimum number of units, additional units may be required.

By using this method, the designer can analyze the goodness of each individual match as it is placed *without* having to complete the network. If the designer is faced with having to decide between several matches, this method allows poor alternatives to be readily identified and discarded.

Example 6: Dephenolization Network Design

Design a mass exchange network to meet the total cost target set in Example 5.

Solution

As determined earlier, the value of Δy_{min} to be used is 0.001 and the minimum total cost target is \$331,760/yr. Figure 7.18 shows a possible network designed using the guidelines presented above. A mass exchanger is represented as a pair of circles joining the matched streams, with the mass load and number of stages shown below it. The total cost of the network is \$345,400/yr, which is only 4% above the target. This agreement is certainly good enough for preliminary design purposes. Notice that

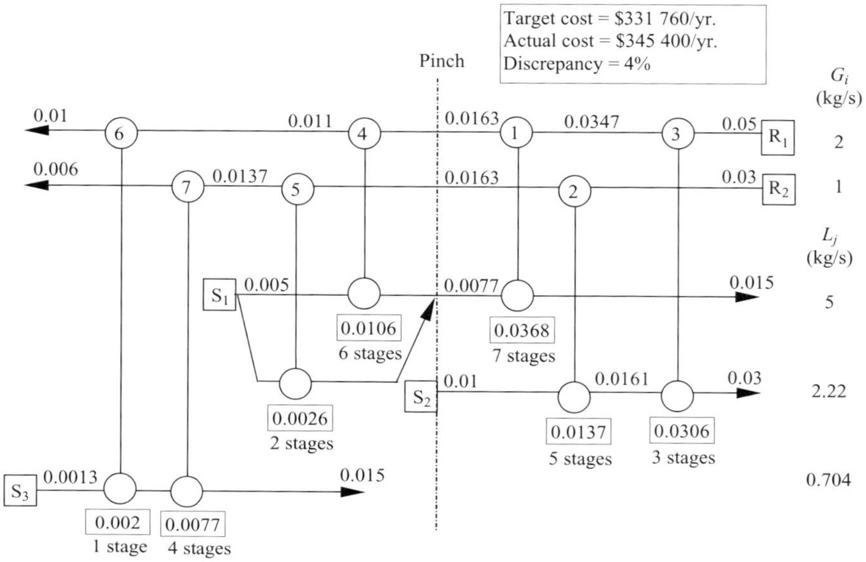

Figure 7.18: Network design for Example 6.

one more unit than the minimum had to be used. Also, note that actual MSA compositions and flows are now shown. For more detail, see Hallale (1998).

Some readers may be wondering whether it is worth having such a complex design. Wouldn't it be a lot simpler to use only the external MSA? The answer is yes, but at a price. Figure 7.19 shows the simplest possible design, in which only S_3 is used. The flow of S_3 is now 7.59 kg/s, which is far greater than if the process MSAs were used efficiently. The total cost of this design is $2,368,500, which is almost 7 times as great as that of Figure 7.18.

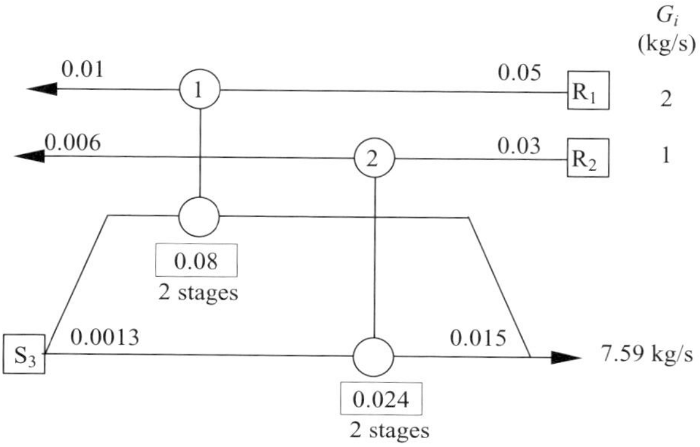

Figure 7.19: A simple, yet expensive, alternative.

4 CONCLUSIONS AND FURTHER READING

This chapter has shown how the technology for designing individual mass transfer operations has been extended to the design of process systems. The chapter presented systematic tools for tackling the complex problems associated with process and flowsheet design. The two-stage approach was presented for the design of mass exchange networks. First, performance targets are set based on thermodynamic and economic analysis. Then, the system is designed to approach the targets. This approach gives confidence about the quality of a design while reducing the amount of design effort required.

It should be mentioned that the chapter covered a fairly simple type of problem. For example, only single-component transfer has been considered, and linear equilibrium relations are assumed. Also, MSA regeneration has not been considered. The original concepts of El-Halwagi and Manousiouthakis (1989a) have been extended to deal with a much wider range of problems. A comprehensive review of these developments is given by El-Halwagi (1997), but they include Simultaneous Synthesis of Mass Exchange and Regeneration Networks (El-Halwagi and Manousiouthakis, 1990b), Reactive Mass Exchange Networks and Nonlinear Equilibrium (El-Halwagi and Srinivas, 1992; Srinivas and El-Halwagi, 1994), Simultaneous Heat and Mass Exchange (Srinivas and El-Halwagi, 1994b), Multicomponent Problems (El-Halwagi and Manousiouthakis, 1989b; Gupta and Manousiouthakis, 1994; Wang and Smith, 1994), and Synthesis of Waste-interception Networks (El-Halwagi et al., 1996). All of these works have aimed primarily at minimizing MSA (operating) costs and the number of units; however, they can now be combined with methods for minimizing capital and total cost that were presented in this chapter. The results are process designs that often are significantly cheaper than those generated by other methods. Several successful applications are discussed in Hallale (1998) and Hallale and Fraser (1998b, 1999).

REFERENCES

Coulson, J. M., Richardson, J. F., and Sinnott, R. K. (1993). *Chemical Engineering*, Vol. 6, ed. 2. Pergamon Press, Oxford.

El-Halwagi, M. M. (1997). *Pollution Prevention through Process Integration: Systematic Design Tools*. Academic Press, San Diego.

El-Halwagi, M. M., Hamad, A. A., and Garrison, G. W. (1996). Synthesis of waste interception and allocation networks. *AIChE J.* 42:3087–3101.

El-Halwagi, M. M., and Manousiouthakis, V. (1989a). Synthesis of mass exchange networks. *AIChE J.* 35(8):1233–1244.

El-Halwagi, M. M., and Manousiouthakis, V. (1989b). Design and analysis of mass exchange networks with multicomponent targets. AIChE Annual Meeting, San Francisco.

El-Halwagi, M. M., and Manousiouthakis, V. (1990a). Automatic synthesis of mass exchange networks with single-component targets. *Chem. Eng. Sci.* 45(9):2813–2831.

El-Halwagi, M. M., and Manousiouthakis, V. (1990b). Simultaneous synthesis of mass exchange and regeneration networks. *AIChE J.* 36(8):1209–1219.

El-Halwagi M. M., and Srinivas, B. K. (1992). Synthesis of reactive mass exchange networks. *Chem. Eng. Sci.* 47(8):2113–2119.

Gupta, A., and Manousiouthakis, V. (1994). Waste reduction through multicomponent mass exchange network synthesis. *Computers Chem. Eng.* 18(S):S585.

Hallale, N. (1998). Capital Cost Targeting for the Optimal Synthesis of Mass Exchange Networks. Ph.D. thesis, University of Cape Town, South Africa.

Hallale, N., and Fraser, D. M. (1998a). Capital cost targets for mass exchange networks—a special case: Water minimisation. *Chem. Eng. Sci.* 53(2):293–313.

Hallale, N., and Fraser, D. M. (1998b). Synthesis of a cost-optimum gas treating process using pinch analysis. *Adv. Environ. Res.* 2(2):167–178.

Hallale, N., and Fraser, D. M. (1999). Optimum design of mass exchange networks using pinch technology. *Computers Chem. Eng.* 23(S):S66–168.

Perry, R. H. (1984). *Perry's Chemical Engineers' Handbook*, ed. 6. McGraw-Hill, New York.

Srinivas, B. K., and El-Halwagi, M. M. (1994a). Synthesis of reactive mass exchange networks with general nonlinear equilibrium functions. *AIChE J.* 40(3):463–472.

Srinivas, B. K., and El-Halwagi, M. M. (1994b). Synthesis of combined heat and reactive mass exchange networks. *Chem. Eng. Sci.* 49(13):2059–2074.

Treybal, R. E. (1981). *Mass Transfer Operations*, ed. 3. McGraw-Hill, Singapore.

Wang, Y. P., and Smith, R. (1994). Wastewater minimisation. *Chem. Eng. Sci.* 49:981–1006.

CHAPTER 8

Screening of Recovery and Treatment Options for Waste Minimization

Andreas A. Linninger and Aninda Chakraborty
Laboratory for Product and Process Design, Department of Chemical Engineering, University of Illinois, Chicago, Chicago, IL 60607

George Stephanopoulos
Laboratory for Intelligent Systems in Process Engineering, Department of Chemical Engineering, Massachusetts Institute of Technology, Cambridge, MA 02139

Abstract

In pharmaceutical and specialty chemical production, stringent product purity and quality requirements tend to cause high waste-to-product ratios. Therefore, a large number of different types and levels of waste are leaving pharmaceutical production campaigns and need material recovery and/or adequate treatment. For a given waste stream, numerous alternatives, including physical separation techniques for the recovery of valuable resources, as well as destructive treatment options, may be available. This chapter will demonstrate a hybrid methodology that can synthesize optimal plant-wide waste management policies. In phase one, a superstructure of all technically feasible flowsheets is generated using heuristic search techniques. Informed reasoning based on a set of relaxed rules ensuring technical limitations and effectiveness deduces a short list of feasible operations. Short-cut models predict the composition of residuals and the associated operating cost. Repeated application of the inference mechanism combined with simulation gives rise to a superstructure of all feasible treatment alternatives for all waste streams. In phase two, this superstructure is rigorously optimized to construct the most competitive plant-wide treatment policy subject to capacity, emission, and logistic constraints.

Keywords: waste treatment, process synthesis, pollution prevention

INTRODUCTION

Traditionally, chemical engineers develop the operating procedures for the manufacturing of a desired product with a given laboratory recipe. Today, pollution prevention commands consideration of the process in its entirety, including environmental effects caused by the choices of reactants, auxiliary materials, such as mass separating agents, and its energy consumption. Pollution prevention is most effective in the early design phases, when many structural decisions are not finalized yet (Dechema, 1990). In a new design approach, many different chemical routes using distinct reaction pathways, particular choices of solvents, and operations have to be considered. Discrimination between candidate flowsheets from a pollution prevention point of view would require measures for the environmental impact caused by effluent streams as well as consideration of recycle, reuse, and recovery options of valuable resources. A systematic approach of this new design paradigm could be composed of two phases. First, what are the alternative chemical routes for the production of the desired molecule? This first problem is a synthesis problem for identifying the best choice of raw materials and reactions for the production of the target molecule. Second, what are the amount and composition of unavoidable byproducts, effluent streams, and expected final emissions for each of the candidate recipes? The second task involves the refinement of the flowsheet developed in phase one through the integration of material recovery and/or necessary destructive waste treatment steps.

Although both questions are related synthesis problems, the first question is much more challenging. Some research articles report progress in true synthesis of chemical recipes (Ali et al., 1998; Jaksland et al., 1995; Friedler et al., 1994; Lakshmanan and Biegler, 1994; Siirola and Rudd, 1971). Unfortunately, theoretical approaches for computer-aided process synthesis are still immature. This chapter only addresses the less difficult second problem. The objective is to synthesize operating procedures with optimal exploitation of material recovery and treatment options for all byproduct streams of the base recipe. This will facilitate the comparison of different technological approaches produced in phase one. Nevertheless, the findings of the second problem are expected to provide valuable insights for the ultimate goal of systematic process synthesis for pollution prevention.

For automated treatment selection, two basic questions need to be addressed: (i) What operations for recovery and treatment are feasible; (ii) what is the best combination of steps in order to reach desired impact levels at minimum cost at a plant-wide level? Technological and economic criteria for each byproduct stream determine the first question. The optimal answer to the second question requires a trade-off between competing alternatives, their associated cost, and availability of treatment technologies at specific sites.

OUTLINE OF THE CHAPTER

In this chapter we focus on recipes from batch pharmaceutical productions. In the discussion a computer-aided methodology for the computer-aided synthesis of waste treatment polices will evolve. A review of prior work for waste minimization is given in Section 1. Section 2 will examine the required knowledge for property-based treatment selection. Section 3 will present an approach to generate a superstructure of all feasible treatment policies on a plant-wide level. We will demonstrate in Section 4 how combinatorial optimization of the superstructure leads to plant-wide optimal waste policies that maximize customized cost functions subject to site-specific constraints. A case study will be the theme of Section 5. We will close the chapter with conclusions and an outlook to future research directions.

1 CURRENT PRACTICE AND PRIOR RESEARCH: TECHNOLOGY SELECTION

Batch production of fine chemicals and pharmaceuticals is characterized by complex organic reaction chemistry. Due to stringent purity and operating constraints, stage intermediates and products are preferably obtained as solid crystals. These operations generate spent liquids or slurries containing organic solvents, quench materials, diluents, or other chemical substances extensively used for dilution of raw materials, as liquid catalysts to facilitate or control reactions, or as mass separating agents. Other waste streams include filter cakes or fowled catalysts. In effect, large amounts of organic wastes containing solvents are produced in each batch. Uncertainties associated with the effects of trace species on the reaction mechanisms often prohibit reuse of recovered substances for the same campaign. The presence of multicomponent mixtures containing hazardous and toxic species and the complicated process integration make management of byproduct streams from pharmaceutical manufacturing an attractive problem. Bottlenecks in solvent recovery and treatment facilities at many multipurpose batch production sites underscore its practical relevance.

1.1 HEURISTICS

Responsibility for handling of byproducts in pharmaceutical manufacturing is often delegated to plant managers. Process reintegration and treatment allocation decisions are frequently taken in an ad-hoc fashion using heuristic selection criteria. Recently, computerized collections of treatment knowledge, as well as cost data, are available commercially (AEA, 1996). Environmentally concerned companies maintain compilations of treatment standards. These procedures commend decision trees whose nodes delineate characteristic property ranges of applicability for a certain treatment option (Fig. 8.1). In addition, expected benefits or cost can be inferred.

1.2 EXPERT SYSTEM APPROACHES

Expert systems have been conceived for medical diagnosis (e.g., MYCIN, Shortliffe, 1976). The basic idea is to capture "expert" knowledge through quantitative rules. Rules imply statements relating the if-part, or antecedent,

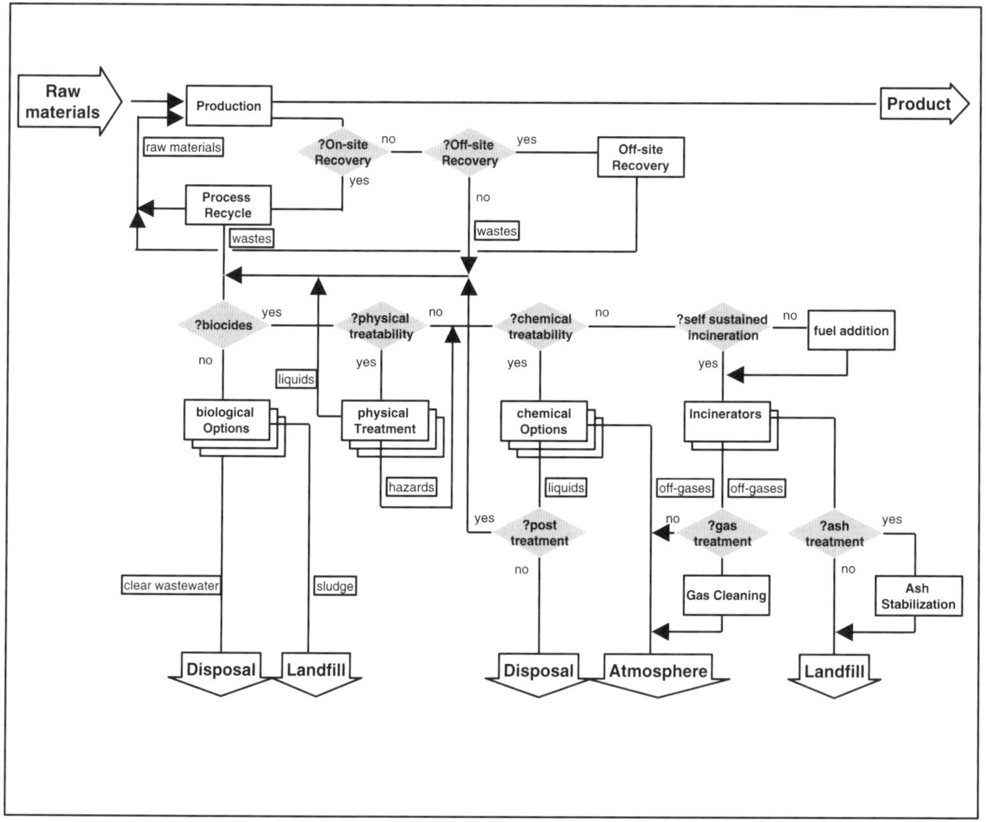

Figure 8.1: Decision tree for material recovery and treatment options (Linninger, 1996).

with the then-part, the consequent. Facts capture the problem's initial state and its evolution. Execution of rules whose antecedents agree with the given facts produces new consequential facts, which gradually approach a desired solution.

A significant contribution to synthesis of waste treatment policies is Merck's EASY system (Venkataramani et al., 1990). Technology selection is decided using a rule-base of duplets <chemical substance–preferred treatment>. Entries like <benzene-distillation> allowed inference of appropriate treatment for chemicals found in the effluents. The final results consisted of a treatment sequence including possible pre- and posttreatment. Solutions of the EASY system propose the "ideal" sequence for each waste stream without generating alternatives or considering wastes from the same stage or campaign. Several expert system approaches address the physical separation selection and design (Barnicki and Fair, 1990; Barnicki and Fair, 1992; Wahnschafft et al., 1991).

Rule-based systems can be classified as search techniques for environments in which the problem can be represented by a state space. Candidate solutions are easily found even in large search spaces. Detecting the optimum among multiple solutions is difficult. For optimality considerations, mathematical programming is a superior choice.

1.3 PROGRAMMING METHODS

The process synthesis problem has been tackled by mathematical programming (Grossman, 1996, 1985). Flowsheet optimization seeks the best values for discrete as well as continuous design variables (i.e., design decisions and operating conditions, respectively) (Floudas, 1995; Floudas and Pardalos, 1999; Viswanathan and Grossman, 1990). Therefore, programming involves binary as well as continuous design variables (Garfinkel and Nemhauser, 1972; Parker and Radin, 1988). El-Halwagi and Manousiothakis (1989, 1990) developed a comprehensive theory for the optimal design of separation systems. Their Mass-Exchanger-Network methodology was applied and extended advantageously in many academic and industrial problems (El-Halwagi, 1997; Dunn et al., 1994; Spriggs, 1994; Wang and Smith, 1994; Bagajewicz and Manousiouthakis, 1992). Furthermore, the more general concept of mass integration has been introduced to provide a holistic approach to the allocation, generation, and separation of streams and species (El-Halwagi, 1997; El-Halwagi and Spriggs, 1998). Solvent Recovery Targeting and Design Targets for Reaction/Distillation Networks have been addressed by Ahmad and Barton (1999).

Mathematical models of chemical equilibrium and kinetic effects, constitutive equations in separation techniques, or polynomial cost correlations involve nonlinear functions. These nonlinearities, in combination with structural decisions, lead to *mixed integer nonlinear programs* (MINLP). Finding a globally optimal solution in MINLPs of more than hundreds of binary decision variables is difficult. Larger systems evade rigorous optimization altogether. In addition, programming methods depart from a predefined *superstructure*, which contains all possible structural candidate solutions. Only solutions accommodated in the superstructure can be "found" by MINLP; however, for selecting plant-wide recovery and treatment options, it is not obvious how to postulate or guess such a "complete" superstructure.

1.4 PLANNING THEORY FOR WASTE TREATMENT

A hybrid approach for the synthesis of plant-wide optimal material recycle and waste management policies was developed by Linninger and Chakraborty (1999) and Chakraborty (1998). Their methodology combines the advantages of knowledge-based search with combinatorial optimization techniques. A schematic of the two-phase hybrid approach is depicted in Figure 8.2:

- Phase 1: Generate a set of treatment alternatives—superstructure generation.
- Phase 2: From among this superstructure of feasible treatment options for all waste streams, select rigorously the plant-wide policy that optimizes a particular performance function subject to site-specific constraints—superstructure optimization.

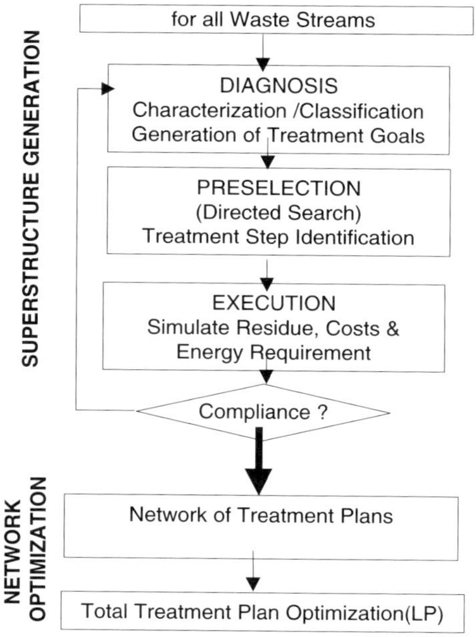

Figure 8.2: Overview of the methodology for waste treatment synthesis.

Before we can do computer-aided treatment selection, a state space representation of treatment technologies, regulatory constraints, and quantitative technology selection criteria is necessary.

2 TREATMENT KNOWLEDGE

This section derives separation and treatment targets from the physicochemical properties of the byproducts. For the prototype program, expertise was extracted from the study of applied thermodynamics and engineering economics, specialized literature (EPA, Technology Profiles, 1994), and professional publications (LaGrega et al., 1994; Breton, et al., 1988), which reflect state-of-the-art material separation and waste management practices. The information found in these textual sources was not suitable for direct use in a computer-based methodology. Consequently, the extracted knowledge was organized into categories, converted into atomic rules and procedures (Chakraborty and Linninger, 1999a, 1999b).

2.1 OVERVIEW OF TREATMENT OPTIONS

The available treatment technologies fall into five major categories: (i) physical separation, (ii) chemical, (iii) biological, (iv) final treatment, and (v) incineration.

Physical Separations. Physical separations are the predominant recourse for recycling and recovery of valuable resources. They do not chemically alter species, but they change the state of waste mixture by separating its subcomponents due to physical property differences such as distillation (i.e., relative volatilities, extraction-activities, crystallization-liquidus points). Usually recovered materials will not be used for subsequent batches in production of high-valued drugs. A less sensitive use is equipment washing, or bulk application can occur in practice.

Chemical Treatment. Destruction of toxic compounds in low concentrations often calls for special operations. Chemical treatment encompasses chemically reactive methods that destroy organic species through wet air oxidation, oxidation with hydrogen peroxide, and emerging technologies, such as supercritical fluid oxidation. For highest qualities and purities, chemical chlorination can even reach drinking water quality, provided there are no indestructible inorganics.

Biological Treatment. In biological treatment, organic species are reduced by living organisms called biomass. Chemically this means that the biomass oxidizes higher organic molecules and lives on the energy surplus. The metabolic end products are carbon dioxide and water (as well as benign slurries). Biological treatment is frequently the last step in wastewater treatment. Unfortunately not all species are biodegradable.

Changing waste compositions and loads are detrimental to the destruction performance of the biological treatment methods.

Incineration. Incineration is divided into thermal and catalytic incineration. It is common to all methods to destroy organic and certain inorganic species by oxidation with oxygen. Incineration achieves high volume reductions. Due to its robustness, it is the default method for all organic wastes. Unfortunately, incineration is one of the most expensive treatment methods.

Final Treatments. Final treatments describe the ultimate destination of all compliant streams. Technically these methods are nonreactive, although some species transformation continues in the sewer, the atmosphere, and at landfills.

2.2 DEDUCTION OF FEASIBILITY

Table 8.1 gives a partial view of quantitative rules for inference of feasibility for Wet Air Oxidation. From Table 8.1 it becomes evident that all selection rules fall into two principal categories: (i) limitations and (ii) effectiveness. Limitations describe necessary conditions for a treatment to apply. Effectiveness is a second and sufficient condition to ensure that a treatment also resolves a certain task. Formally we can conclude that a treatment is *feasible* if the mixture properties obey all limitation rules and satisfy at least one effectiveness rule.

In addition, procedural knowledge is used for the calculation of material separation and destruction efficiencies, prediction of the residue

TABLE 8.1
Technology selection rules for wet air oxidation

Wet Air Oxidation (WAO)	(1) Flowrate: 6 < flow < 3,000 gal/min (2) COD = 15,000–20,000 mg/L (3) If COD < 15,000 mg/L, then WAO is used with auxiliary energy (max 500–600 BTU/gal). (4) C_water > 85% (5) Organics < 10% and C_solids ~ 5%–6%	(1) Destruction of cyanides, ethylene, olefins, acrylnitrile, and aliphatic compounds of low molecular weight such as alcohols, aldehydes, ketones, carboxylic acids, and aromatic hydrocarbons like toluene and pyrene. (2) Halogenated-condensed rings like pesticides, halogenated aromatic compounds like PCB and chlorobenzene, and low molecular weight compounds are not treatable by WAO.

composition, and computation of the associated treatment costs. Details on each separation and treatment technology can be found elsewhere (Chakraborty, 1998).

3 SUPERSTRUCTURE GENERATION

The objective of superstructure optimization is to identify a set of feasible recovery and treatment operations. The feasible alternatives shall be derived from the properties of the effluent. This material-based paradigm drives the search in the linear planning algorithm developed by Linninger and Chakraborty. In-depth discussion of the planning theoretical aspects exceeds the scope of this presentation, but is given in Linninger and Chakraborty (1999). Their approach can automatically synthesize the superstructure using the following assumptions:

- **State Description.** The state of waste streams and property model for the species is known. This information can be obtained from plant data or results from conceptual design consideration using a computer-aided design environment such as BDK (Linninger and Stephanopoulos, 1998; Linninger et al., 1998).
- **Regulations.** There exists a database $\mathfrak{R}$, which contains quantitative rules for all relevant species. These rules quantify recovery targets and regulate permissible emission thresholds for the final disposal of waste into the atmosphere, the sewer, or land deposit.
- **Dual State Representation of Treatment Steps.** A model for each treatment t is available, describing the application limitations, L_t, as well as effectiveness rules, E_t. Furthermore, mathematical models for state transitions predict the composition of the residual(s) from the application of a feasible treatment step, t, onto the waste mixture, w.

3.1 LINEAR PLANNING

The planner iterates through three phases as depicted in Figure 8.2, top. Equations (3.1) to (3.3) express the three phases in first-order logic.

Diagnosis:

$$\forall c(c \in w)[\exists\, p,r(p = property(c)) \wedge (r \in \mathfrak{R}) \wedge (violates(p,r))] \tag{3.1}$$

$$\Rightarrow Goal(c);$$

Preselection:

$$\forall t,w,c(c \in w)(t \in T): [\,!\,\exists l,\, p(l \in \mathrm{L}_t) \wedge (p = property(c)) \wedge violates(p, l)]$$
$$\wedge [\exists e, g : (e \in \mathrm{E}_t) \wedge (g = Goal(c)) \wedge (satisfies(g, e))] \tag{3.2}$$

$$\Rightarrow treatment\_option(t, w);$$

Execution:

$$\underline{r} \leftarrow execute\,(t,w)\;\forall\, t,w\;treatment\_option\,(t,w). \qquad (3.3)$$

Diagnosis generates consistent initial and goal states. The presence of recycle targets or regulatory violations defines the initial state. For each compound property that violates a regulation, a treatment goal is generated. The fact, *Goal(c)*, is inserted into the knowledge base. This goal, *Goal(c)*, needs to be addressed through recovery or treatment. The final state is in full compliance (i.e., absence of any goals). The targets are deduced by examining possible adverse properties or chemical substances in the waste streams against the set of environmental goals and the regulation database, $\mathfrak{R}$. Table 8.2 exemplifies regulations concerning liquid organic wastes (Belhateche, 1995). More details on the assessment of regulatory offences, the required physical property databases and their property prediction models, and the generation of a rule-base for it can be found (Chakraborty, 1998; Linninger et al., 1994, 1995, 1996).

Preselection is the core of the planner that finds all feasible treatment options for all goals by examining the limitation and effectiveness rules of t.

Limitations. The limitations describe ranges of state parameters that disqualify an operation. Determining exact cut-off values is hard. That is why relaxed upper limits were specified beyond which no economical operation can be expected; compare Rule A-3. Very often, the subsequent optimization phase will render feasible operation close to their threshold value

TABLE 8.2
Industrial wastewater disposal regulation

Parameter	*Concentration, mg/L*
COD	300–2000
BOD	100–300
Oil and Grease/TPH	15–55
TSS	15–45
pH	6.0–9.0
Temperature	Less than 40°C
Color	2 Color units
NH_3/NO_3	1.0–10
Phosphates	0.2
Heavy metals	0.1–5.0
Surfactants (total)	0.5–1.0 total
Sulfides	0.01–0.1
Phenol	0.1–1.0
Toxic organics (total)	1.0 total
Cyanide	0.1

uneconomical. These may emanate from (i) property constraints (e.g., *do not distill mixtures whose boiling point is above flammable limit*) or (ii) economical constraints:

> *Rule A-3: if organic concentration exceeds 8%, adsorption is not economically feasible* (*Rationale: too much activated carbon required*).

Typically, a short list of characteristic mixture properties appears in limitation rules. These *key properties*, such as organic content, concentration of particulate solids, and volatile organic compounds (VOC), delineate boundaries between the treatment alternatives. Clearly the boundaries for different options overlap, resulting in multiplicity of applicability. If all limitation rules hold, then the system proceeds to assess the treatment's effectiveness.

Effectiveness. These criteria determine which species can effectively be handled by the particular treatment technology. As an example, consider the removal of organics from wastewater. Satisfaction of the concentration limitations is insufficient if the major contaminant is too light for adsorption. Whenever any goal is satisfied by any effectiveness rule, the system produces a fact stating that the treatment is *a-treatment-option* for a waste w_i, or *treatment_option* (t, w_i).

Execution enforces the mass and energy balances that govern the separation or treatment step. Step execution of the treatment t on the mixture, w, produces one or more residua and predicts their new state (e.g., amount, temperature, pressure, composition). Residual calculation relies on short-cut calculation for mass and energy balances using an operation-centered modeling paradigm as described in the BDK project (Linninger and Stephanopoulos, 1998). Frequently these newly produced effluents are not compliant and need further treatment. The necessity for further treatment steps depends again on the material properties of the residuum, which in turn is a function of amount and composition.

As an example, take thermal incineration of combustible organic wastes. Often incineration produces two residuals: an offgas and ashes. The total number of carbon, hydrogen, nitrogen, and halogen atoms of feed and simple stoichiometric oxidation influences the composition of the residual off gases. In addition, contaminants like hydrochloric acid and nitric oxides may be found in the flue gas. Inorganic salts and oxides build ashes and slags (i.e., secondary residual). Other models include simple vapor-liquid equilibria for operations like evaporation or condensation.

3.2 CONTROL STRUCTURE

Data-driven search space pruning. Exhaustive evaluation of limitation and efficiency rules of all treatments may lead to an unacceptably large number of rule execution and physical property queries. This number may be prohibitive for large-scale industrial applications. Preselection can be expedited

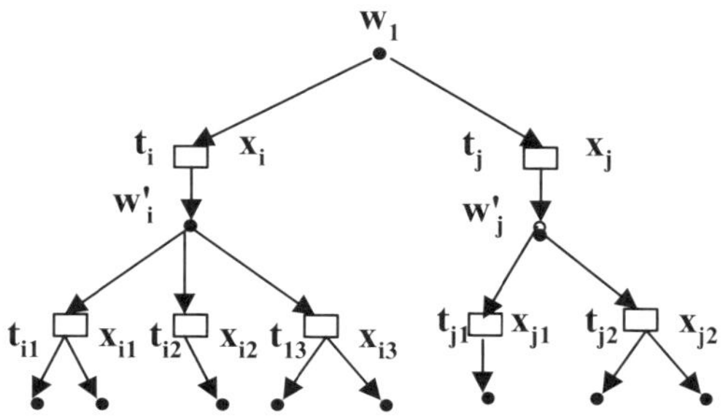

Figure 8.3: State task network.

using "informed search" (Russel and Norvig, 1995; Anada et al., 1986). The limitation rules are chained following an *AND-logic*. Violation of the first rule produces the fact *not-treatable-by* (*wi*, *t*) and triggers the continuation of the search for other options without firing any other rule in L. All treatment goals g are then opportunistically checked against the effectiveness rules. The effectiveness rules are *OR*-chained, leading to a positive treatment identification with the first affirmative effectiveness rule firing. This property-based elimination of possible steps drastically reduces the edge branching factor. This expedites the speed of the planning algorithm.

3.3 CAPTURING THE COMBINATORIAL COMPLEXITY OF THE RESULTS

After the completion of *diagnosis, preselection*, and *execution*, each noncompliant residuum is submitted to another reasoning cycle. Repeated application leads to a tree of all treatment alternatives for each waste stream. This treatment tree τ was encoded using a State Task Network (STN) representation (see Fig. 8.3), where wastes w_i stand for the nodes, whereas the treatments t constitute the arcs (Kondili et al., 1993). A subtree containing ordered sequences of treatment options that treat a waste w_{i0} and all its residuals to compliance is called a treatment path p. This subtree p may branch, as some treatments produce more than a single residual. The union of all treatment paths p for a given waste w_{i0} makes up the treatment tree τ. Two treatment paths are called *disjunct* if they do not belong to the same tree τ. The superstructure S is the union of all treatment trees τ for all wastes W.

4 SUPERSTRUCTURE OPTIMIZATION

The superstructure generated in phase one implies all combinations of operations for waste reduction through recycling as well as treatment steps. The next subsection will describe how to find optimal waste management

TABLE 8.3
Overview of the performance of policy optimization

Simulation experiment	*No. of stages*	*No. of streams*	*No. of binary variables*	*No. of policies*	*No. of iterations*	*Execution time (sec)*
A	1	7	88	10,368	37	0.12
B	2	9	146	823,200	95	0.23
C	3	16	228	10^{10}	108	0.27
D	6	32	456	10^{20}	530	0.68
E	12	64	912	10^{40}	1665	1.20

policies using combinatorial optimization techniques (see Fig. 8.2, bottom). The objective is to construct policies that maximize a subjective performance function and satisfy site-specific constraints.

4.1 PROBLEM FORMULATION

A treatment policy in S is a set of *disjunct* treatment paths, p, with exactly one path per waste. The number of embedded policies is equal to the cardinality of the cross product of all trees. Therefore, explicit problem formulation or solution using mathematical programming techniques alone would be prohibitive. There are more than 10,000 different combinations of operations for the first stage alone (Table 8.3). In the next subsection we will describe how the logic of the superstructure in its STN encoding can be easily transformed into the equivalent mathematical programming formulation. Equations (4.1) to (4.5) describe superstructure optimization as a constrained linear integer program.

$$\min z = \sum_{\forall i} x_i \text{cost}(x_i) + \sum x_i \psi(x_i), \qquad \textit{Performance Function}, \quad (4.1)$$

$$\sum_{\forall \text{ children of } p} x_i = x_p \quad \forall p \in S, \qquad \textit{Path Constraints}, \quad (4.2)$$

$$\sum_{x_i \subset Incineration} x_i Mass(x_i) \leq Incineration_{\max}, \qquad \textit{Capacity Constraint}, \quad (4.3)$$

$$\sum_{\forall \text{ leaves}} x_i BOD(x_i) \leq BOD_{\max}, \qquad \textit{Emission Constraint}, \quad (4.4)$$

$$\sum_{\forall \text{ leaves}} x_i CO_2(x_i) \leq CO_{2_{\max}}, \qquad \textit{Emission Constraint}, \quad (4.5)$$

Path Constraints and Logical Consistency. Binary design variables x_i are assigned to each treatment task t. A value of one represents the fact that treatment t_i is part of a treatment strategy. A value of zero means that the step t_i is not included in the final policy. Path consistency can be expressed as a chain of conditional statements. Figure 8.3 shows a situation in which two treatments, t_i and t_j, are applicable for waste/residue w_i. The binary variables x_i and x_j correspond to steps t_i and t_j. In the solution, either t_i or t_j has to be chosen in order to maintain path consistency. This is mathematically represented by the equality constraint $x_i + x_j = 1$. Similarly, there are three options, t_{i1}, t_{i2}, t_{i3}, for the residue w'_i of waste w_i. For any of these options, t_{i1}, t_{i2}, t_{i3}, to enter a valid treatment policy, the prior step t_i has to be picked too. This dependency of the succeeding operations can be expressed through an equation $x_{i1} + x_{i2} + x_{i3} = x_i$. In general form, path constraints are formulated for each node as prescribed by equation (4.2). A similar argument gives rise to an expression for the alternative branch involving treatment t_j.

Capacity Constraints. In modern production plants, available solvent recovery and treatment facilities are shared for several production lines. Frequently, limited capacities and availability of these processes may hinder further production expansion. That is why it is important to include these site-specific limitations for a plant-wide optimum design. Using the design variables, capacity constraints can be expressed by simple summation over the nodes of a particular treatment category. Equation (4.3) signifies that the total mass of all wastes sent to incineration, $\sum x_i$ Mass (x_i), does not exceed the plants incinerator capacity, $\text{Incineration}_{\text{max}}$.

Environmental Constraints. Environmental constraints express maximum levels of emissions per unit time. Even though the superstructure only contains compliant terminal nodes, the total amount of emissions may be restricted by (i) governmental regulations on the federal, local, or municipal level, or (ii) be subject to voluntary corporate environmental standards. Environmental constraints can be considered by adding the masses of the respective emissions associated with the final treatment steps. Equations (4.4) and (4.5) enforce a site-specific maximum emission level for BOD in the wastewater and a maximum total CO_2 emission, respectively. The left-hand side of equations (4.4) and (4.5) may also be used to construct environmental targets Ψ in the objective function.

Treatment Cost. The total annualized cost (TAC) for a treatment step consists of the annualized capital cost (ACC) (e.g., equipment and capital charge) and the annual operating cost (AOC) (e.g., operating and maintenance cost). Only variable costs are reflected in the proposed methodology. Consideration of annualized capital cost is straightforward, but is typically of lesser importance due to long payback periods and lifetime of the facilities.

Cost estimation models can be classified broadly into three categories: (i) composition dependent, (ii) composition independent, and (iii) marginal cost. Short-cut estimation formulas are assumed for the composition-dependent cost models. Equations (4.6a) and (4.6b) give the cost of incinerators as a function of the fuel value of the waste streams:

$$\textit{Cost of Incineration}\,[\$/\text{yr}] = F \times (\text{price of fuel}), \tag{4.6a}$$

$$\text{where } F = \begin{cases} \left(6000 H_v^{\text{waste}}\right) & \text{if } H_v^{\text{waste}} < 5000\ \text{Btu/lb}, \\ 0 & \text{otherwise.} \end{cases} \tag{4.6b}$$

F is the amount of fuel required [kg/yr], and H_v^{waste} is the net heating value of the waste mixture. The cost of neutralization is exemplified as follows:

$$\textit{Cost of Neutralization}\,[\$/\text{yr}] = N \times (\text{price of neutralizing agent}). \tag{4.7}$$

N is the amount of neutralization agent required for the acid-base neutralization reaction. Similar short-cut methods were developed for treatments like the evaporator and the condenser. The cost for reuse, recycle, and recovery is given as follows:

$$\textit{Reuse or Recovery Cost}\,[\$/\text{yr}] = \eta \times (\text{virgin price of material}), \tag{4.8}$$

where

$$\eta = 0.9 \quad \{\text{Reuse}\},$$
$$\eta = 0.8 \quad \{\text{Onsite Recovery}\},$$
$$\eta = 0.7 \quad \{\text{Offsite Recovery}\}.$$

Composition-independent estimation uses treatment-specific intensive cost models as depicted in Figure 8.4. The price of the different utilities is listed in Table 8.4. The cost estimates make no claim for up-to-date precision. The purpose of this compilation is to capture cost dependencies and their relative expenditure. The search for an absolute scale appears futile, as treatment cost may differ from plant to plant.

Performance Function. Typically, cost correlations appear as the objective function in the combinatorial optimization problem [compare equation (1)]. For true pollution prevention, environmental criteria Ψ must be

TABLE 8.4
Cost of utilities required for different treatments

Utility	*Price*
Fuel	\$4 per million BTU
Cooling water	\$0.25 per 1000 gal
Steam at 100 °C	\$3.60 per 1000 lb
Water at 100 °F	\$0.05 per 1000 gal
Hydrochloric acid	\$0.427 per kg
Calcium hydroxide	\$0.135 per kg

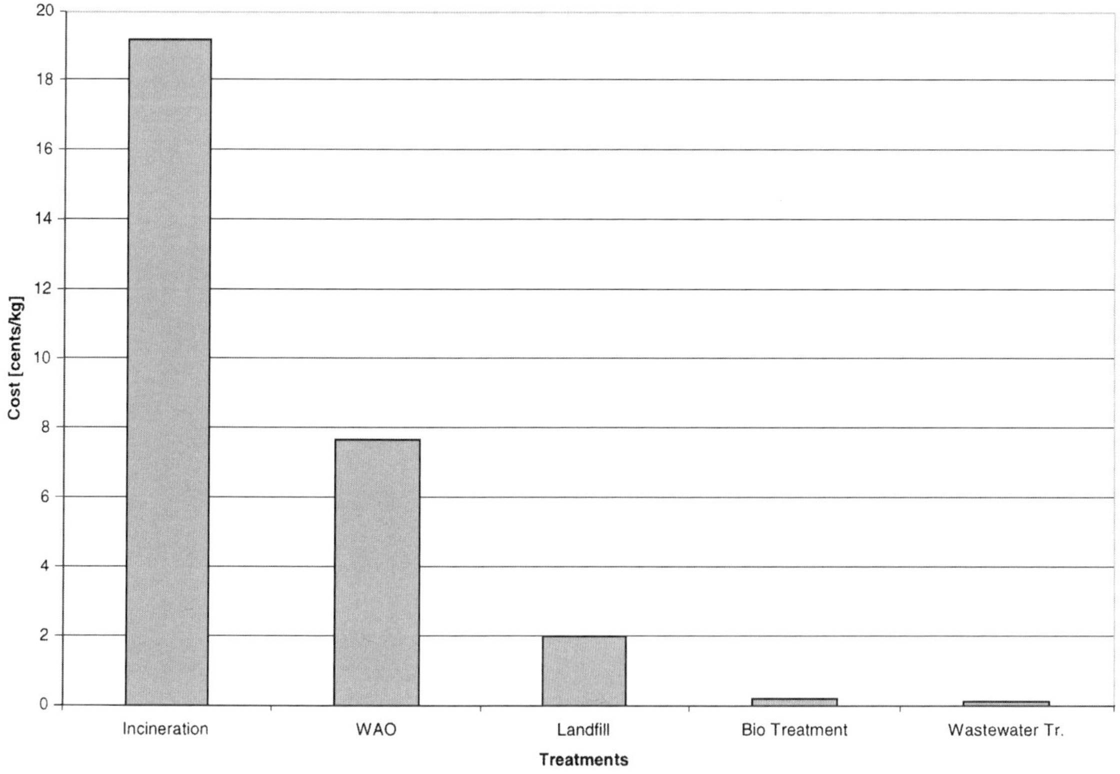

Figure 8.4: Comparative cost of different treatment options.

considered equally. Thus, subjective performance function may also incorporate

- desired level of material recovery
- minimal levels of specific emissions
- combination of the above (i.e., multiobjective optimization problem)

Alternatively, performance functions may be used that balance multiple and conflicting objectives, such as cost versus minimal emissions. The formulation of equations (4.1) to (4.5) is directly amenable to multiobjective optimization methods, such as goal programming (Ignizio, 1976, 1982).

4.2 SOLUTION OF THE OPTIMIZATION PROBLEM

The symbolic features of the proposed environment automatically generate mathematical expressions for the cost function, emission, capacity, and path constraints. The resulting mathematical problem is a constrained linear integer program. This is due to the fact that the nonlinear cost and residue calculations (i.e., physical, chemical, environmental property predictions, residue composition, cost estimation) have been executed in the superstructure synthesis step. The performance function and constraints are all linear in the binary decision variables x_i with constant technology coefficients. Despite its simplicity, it implicitly embraces all combinations of treatment policies on a plant-wide level. The proposed policy screening can be more precisely qualified as a multibin knapsack problem. Although this problem is still NP-hard, the compact linear formulation allows solution of the synthesis problem for a large number of wastes with commercial algorithms.

Solvability and Performance. Branch-and-Bound (BB)–type algorithms are adequate for this integer program (IP) expressed by equations (4.1) through (4.5). The BB repeatedly solves a relaxed linear problem in terms of continuous variables y_i. These variables y_i are the continuous relaxations of the discrete variables x_i. The solution of the relaxed problem, LP, is mainly determined by the number of constraints. Because the number of constraints is relatively small [i.e., p equations of type (4.2), as well as equations (4.3) to (4.5)], the relaxed subproblem is easily solved, even for extremely large design spaces. The LP solution gives bounds for the original integer program. By comparing the bounds of the LP, inferior branches can be discarded. If the relaxed LP solution only contains integer variables, the LP is also the solution to the IP. Branching on the binary variable associated with the relaxed solution iteratively leads to the optimal integer solution. This feature of BB is called implicit enumeration. This rationale explains the excellent performance of the superstructure optimization step in the simulation experiments. It is a direct consequence of the particular problem formulation of the proposed methodology.

5 RESULTS: CASE STUDY

This section discusses results from a case study for three stages of the production of a maleate salt, a pharmaceutical intermediate product. The chemical recipe is also discussed in Linninger et al. (1994).The production level of the first stage is set at an annual rate of 352,000 kg of intermediate and 3,175,000 kg of 7 distinct waste streams. The compositions of the streams WS1 through WS7 are listed in Table 8.5. Four streams can be categorized as organic liquids; one stream is aqueous. The last two effluents constitute a solid filter cake and slurry.

Automatic Synthesis of Treatment Tree for Waste Stream WS3. The streams are submitted to automatic treatment selection. Figure 8.5 shows the treatment options for the waste stream generated by the reasoning mechanism for waste stream WS3. In this tree, T_i's stand for the different recycle and treatment steps, and S_i's are the residual streams arising from each of these different treatment steps. This liquid waste stream, WS3, is mainly composed of dichloromethane and ammonium hydroxide. Table 8.6 summarizes the distinct paths embedded in the treatment tree τ. Dichloromethane is a valuable byproduct, and hence it can be reused as shown by the path WS3-A (see Table 8.6). The waste stream can be neutralized (T2), and dichloromethane can be recovered from the neutralized waste stream in an onsite recovery plant (Path WS3-B). The residue from the onsite recovery plant can be dried (T4) and the compliant residue from the drier can be sent to landfill.

Alternatively, the neutralized waste stream can be submitted to ion-exchange, as shown in path WS3-E, WS3-H, and WS3-I, or incinerated, as

TABLE 8.5
Composition of the different streams in a single stage of the case study

	Composition (kg)						
Compounds	*WS1*	*WS2*	*WS3*	*WS4*	*WS5*	*WS6*	*WS7*
Ammonium acetate	39847	970	6	0	0	453600	0
Hydroxylamine	1173	2624	1	14	24727	0	10510
Ammonium hydroxide	16287	19062	14136	0	0	0	0
Dichloromethane	422	1028	439455	362152	0	0	0
Acetonitrile	63831	156	0	0	0	0	0
Sodium chloride	30211	425638	0	0	4013	0	1729
Water	224375	2197	0	0	0	0	0
Ammonium dichloroacetate	77436	1885	0	0	0	0	0
Cyclohexane	0	0	0	91432	422478	0	326873

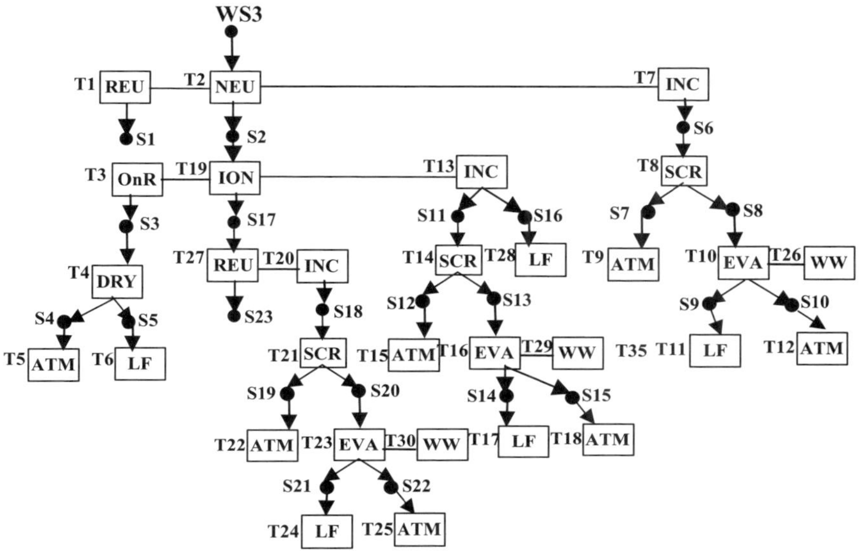

Figure 8.5: Treatment tree for waste stream three, WS3. (WW – Waste Water. SCR – Scrubbing. DRY – Drying. LF – Landfill. REU – Reuse. ION – Ion Exchange. INC – Incineration. EVA – Evaporation. NEU – Neutralization. ATM – Atmosphere)

shown in paths WS3-F and WS3-G. Another possible option is to incinerate the original waste (e.g., paths WS3-C and WS3-D). In total, there are nine distinct treatment paths for this waste stream. Table 8.6 also compares path cost. Negative cost signifies a benefit from material recovery. The least expensive is path WS3-A, which proposes the reuse of the original waste stream, WS3.

TABLE 8.6
Total cost estimation of all the treatment paths for WS

Path name	*Treatment paths*	*Total cost ($)*
WS3-A	T1	–50.60
WS3-B	T2-T3-T4-T5-T6	–31.90
WS3-C	T7-T8-T9-T10-T11-T12	29.54
WS3-D	T7-T8-T9-T26	25.70
WS3-E	T2-T19-T27	–48.89
WS3-F	T2-T13-T14-T15-T16-T17-T18-T28	31.70
WS3-G	T2-T13-T14-T15-T29	27.75
WS3-H	T2-T19-T20-T21-T22-T23-T24-T25	31.26
WS3-I	T2-T19-T20-T21-T22-T30	27.42

TABLE 8.7
Total cost of different policies and associated relative cost for pollution prevention

Waste treatment policy	*Total cost ($/yr)**	*Cost of pollution Prevention effort ($/yr)***
Unconstrained (base case)	−1349087	—
Most expensive	1526979	2876066
No onsite recycle	−633898	715188
80% reduced reuse capacity	1451706	2800794
25% reduced incineration capacity	−1171066	178020

* Negative sign reflects benefits for reuse of raw materials.
** Total cost–cost of base case.

5.1 EVOLUTION OF OPTIONS FOR STAGE 1 AND THE PLANT-WIDE SOLUTION

Table 8.3 illustrates the combinatorial features for the first stage of the case study. The seven waste streams from stage 1 lead to a total of 32 treatment paths of 88 individual treatment steps. The superstructure for stage 1 alone gives rise to a total of 10,368 distinct waste treatment policies. The superstructure for three stages has a cardinality of 228 individual treatments and plans involving only 2 to 15 alternative paths. In its totality, it embeds over 10^{10} alternative treatment policies, as illustrated in Table 8.3. Despite the size of the search space, optimal plant-wide policies were found below 1 CPU second on a 300 MHz Pentium II PC using the CONOPT algorithm of (GAMS, 1996), as shown in Table 8.3. Table 8.7 depicts the optimal policies emerging from different scenarios and the cost for pollution prevention associated with each of them. Negative cost implies benefits due to recycle, recovery, or reuse. The unconstrained optimal solution is considered the base case. Comparison with the base case allows assessment of the effort associated with a different waste reduction strategies. The cost for pollution prevention is the difference between the cost of a given policy and the cost for the base case. Figure 8.6 demonstrates the high degree of flexibility that can be obtained by computer-aided synthesis of waste management. It compares the policy with highest and lowest demand on select treatment technologies. For the same set of wastes, the required incineration capacity may vary by 70% between the highest and the lowest scenario. The policy that uses incineration extensively produces approximately 90% more CO_2 emissions than the one, which avoids combustion.

5.2 MULTIOBJECTIVE OPTIMIZATION

Design with ecological considerations is a multiobjective problem whose solution is given by the "optimal" trade-off among competing goals. Pollution prevention design balances noncommensurate objectives, such as process

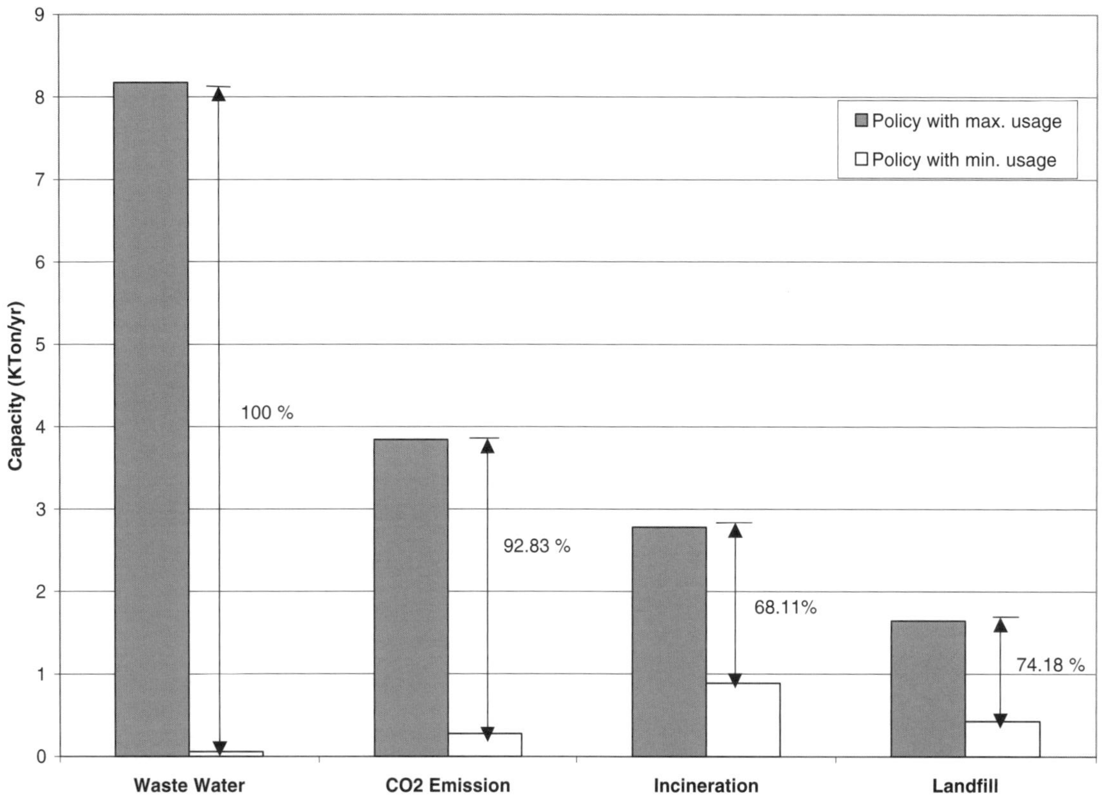

Figure 8.6: Extremal resource utilization of different treatment policies.

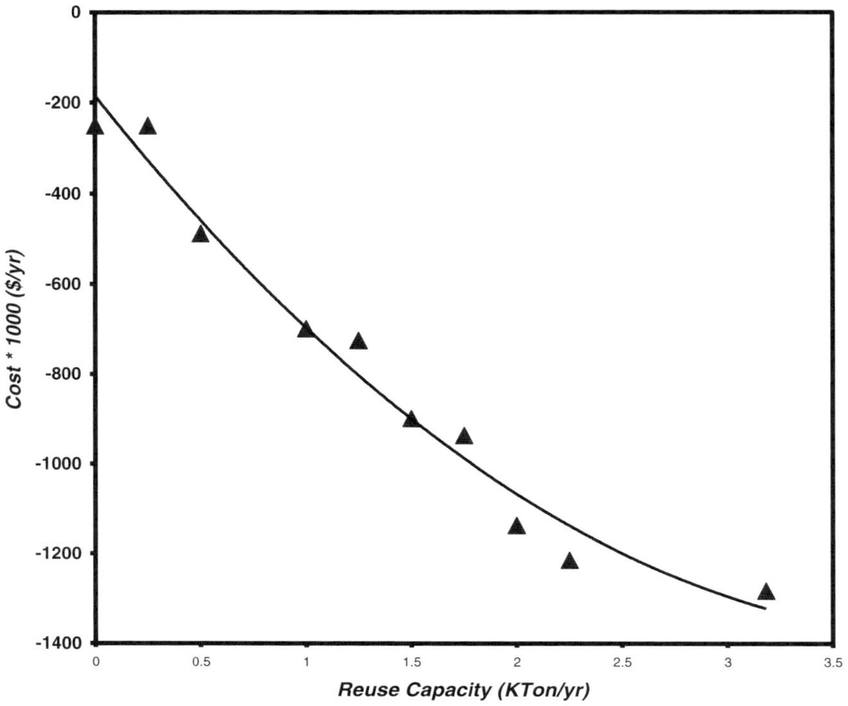

Figure 8.7: Pareto-optimal policies for different reuse capacities.

economics alongside environmental impact. A solution, π, to a multiobjective optimization problem is *Pareto-optimal* if no other feasible solution is at least as good as π with respect to every objective, and strictly better than π with respect to at least one goal.

The parabolic-shaped efficiency frontier or trade-off curve of Figure 8.7 includes all *Pareto-optimal* policies with respect to the goals treatment cost versus reuse capacity. Above it lies the space of inferior designs. Under the curve no feasible solution exists. The globally least expensive policy* recovers 3,250 tons of solvents per year. Reduction of the recovery capacity would lead to a shift of the dominating solution toward the left of the abscissa. Hence, smaller recovery benefits or higher operating cost would be the effect.

Figure 8.8 depicts the trade-off among the alternative goals—recycling versus CO_2 emissions. The locus of *Pareto-optima* delineates the policies with smallest CO_2 emissions for a specific level of recovery per year. Clearly, a reduction in the availability of solvent recovery capacities would lead to higher atmospheric emissions.

*The globally unconstrained cost minimum.

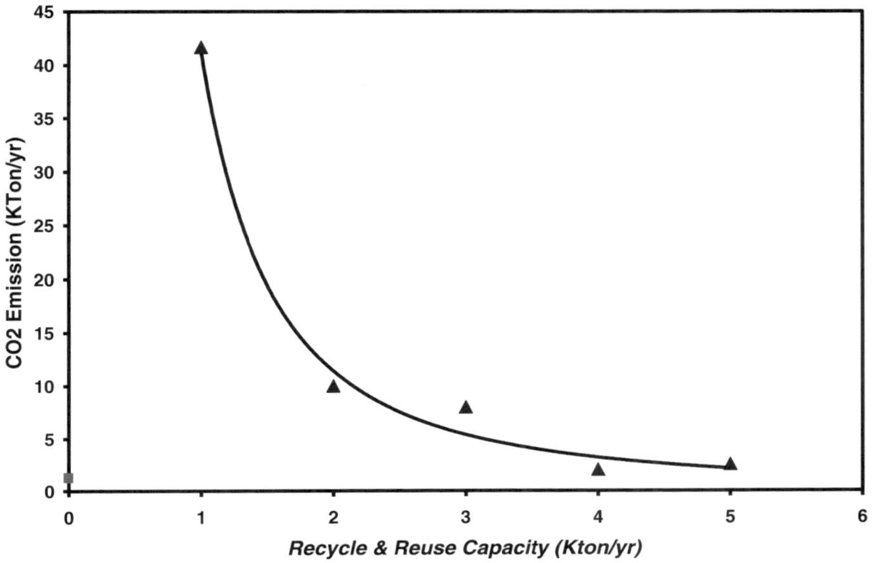

Figure 8.8: Pareto-optimal curve of material recovery vs. CO_2 emissions.

6 CONCLUSIONS AND SIGNIFICANCE

A hybrid methodology for the computer-aided synthesis and optimization of plant-wide optimal policies was presented. The selection of recovery and treatment options was accomplished through a material-based and goal-driven search through relaxed technology selection rules. A linear planning algorithm synthesized a superstructure containing all feasible treatment alternatives. Consistency of the evolving flowsheets is maintained by means of short-cut models for prediction of residuals amount and composition. The number of binary decision variables scales linearly despite exponential growth of embedded treatment policies. The superstructure optimization step leads to a linear integer programming formulation for finding the optimal treatment policy subject to environmental, logistic, and path constraints. A combinatorial explosion is effectively averted by the equivalent mathematical representation of the superstructure as a linear integer program. Through the separation of knowledge-based superstructure generation from the optimization problem, the combinatorial complexity of waste management policies could be reduced and solved in reasonable computer effort in little CPU time. The methodology may serve a computer-aided tool for the assessment of variable cost and unavoidable emissions from recycle and waste treatment strategies. It can be used for the evaluation of different base case production routes in the conceptual design phase. Comparison of different base case batch recipes and their associated monetary and environmental impact provides a quantitative measure for the degree of pollution prevention. Pareto-optimal solutions reflecting the best trade-off between conflicting objectives, such as CO_2 emissions, waste water, and treatment cost, are found for any desired level of impact or treatment capacity. This feature enables debottlenecking capacity overloads in batch pharmaceutical manufacturing sites. This approach promotes ecologically balanced processes design by considering pollution prevention as one of the process objectives rather than just a constraint.

The procedure can also be applied for targeting of plant-wide harmonized treatment policies or for the automatic synthesis of treatment flowsheets. Subsequently only a subset of interesting candidate solutions could be submitted to a second, more refined optimization. Then it is warranted to simultaneously consider operating conditions and mild number of structural variations (e.g., operating conditions and column sequences).

Although the presentation illustrated waste treatment selection, the methodology may lead to new approaches for computer-aided process synthesis.

REFERENCES

AEA Technology (1996). EpSelon Database & EpSelon Expert.

Ali, S., Linninger, A. A., and Stephanopoulos, G. (1998). Synthesis of Batch Processing Schemes as Synthesis of Operating Procedures: A Means-Ends Analysis and Non-Monotonic Planning Approach. Paper 216c, AICHE Annual Meeting, Miami, FL.

Ahmad, B. S., and Barton, P. I. (1999). Solvent recovery targeting. *AIChE J.* 45:335–349.

Anada, A. L., Foo, S. M., and Gunasingham, H. (1986). Knowledge representation using an augmented planning network: Application to an expert system for planning HPLC separations. *J. Chem. Inf. Comput. Sci.* 28:82.

Bagajewicz, M. J., and Manousiouthakis, V. (1990). Mass/heat-exchange network representation of distillation networks. *AIChE J.* 38(11):1769.

Barnicki, S. D., and Fair, J. R. (1990). Separation system synthesis: A knowledge-based approach. 1: Liquid mixture separation. *Ind. Eng. Chem. Res.* 29:421–432.

Barnicki, S. D., and Fair, J. R. (1992). Separation system synthesis: A knowledge-based approach. 2: Gas vapor mixture. *Ind. Eng. Chem. Res.* 31:1679–1694.

Belhateche, D. H. (1995). Choose appropriate wastewater treatment technologies. *Chem. Eng. Prog.* 32:48.

Breton, M., Frillici, P., Palmer, S., Spears, C., Arienti, M., Kravett, M., Shayer, A., and Suprenant, N. (1988). *Treatment Technologies for Solvent Containing Wastes*. Noyes Publications, New Jersey.

Chakraborty, A. (1998). Synthesis of Waste Treatment Alternatives for Batch Pharmaceutical Wastes: A Case Study. UIC-LPPD Report, Chicago, IL.

Chakraborty, A., and Linninger, A. A. (1999a). Systematic Assessment of Treatment Policies for Liquid Wastes from Pharmaceutical Plants: A Case Study. Paper 98c, AICHE Spring Meeting, Houston, TX.

Chakraborty, A., and Linninger, A. A. (1999b). Synthesis of Treatment Flowsheets for Pharmaceutical Wastes, Paper 24c, AICHE Spring Meeting, Houston, TX.

Dechema (1990). Produktionsintegrierter Umweltschutz in der Chemischen Industrie. ISBN 3–926959, Frankfurt.

Dunn, R.F., Zhu, M., Srinivas, B. K., and El-Halwagi, M. M. (1994). Optimal design of energy-induced separation networks for VOC recovery. *AIChE Symp. Ser.* 90: 74–81.

El-Halwagi, M., and Manousiouthakis, V. (1989). Synthesis of mass exchange networks. *AIChE J.* 35(8):1233.

El-Halwagi, M., and Manousiouthakis, V. (1990). Simultaneous synthesis of mass-exchange and regeneration networks. *AIChE J.* 36(8):1209.

El-Halwagi, M., and Spriggs, H. D. (1998). Solve design puzzles with mass integration. *Chem. Eng. Prog.* 94:22–44.

EPA (1974). Superfund innovative technology evaluation program. *Technology Profiles* ed. 7. Cincinnati, US Environmental Protection Agency.

Floudas, C. A. (1995). *Non-Linear and Mixed-Integer Optimization.* Oxford University Press, New York.

Floudas, C. A., and Pardalos, P. A. (1999). Recent Developments in Global Optimization and Their Relevance to Process Design. Paper I09, FOCAPD 99, CO.

Friedler, F., Varga, J.B., and Fan L.T. (1994). Algorithmic approach to integration of total flowsheet synthesis and waste minimization. *Pollution Prevention via Process and Product Modifications. AIChE Symp. Ser.* 90:86.

GAMS Release 2.25 (1996). *A Users Guide.* GAMS Development Corporation.

Garfinkel, R., and Nemhauser, G. (1972). *Integer Programming.* Wiley, New York.

Grossman, I. E. (1996). Mixed-integer optimization techniques for algorithmic process synthesis. in *Process Synthesis* (J. Anderson). Academic Press, New York.

Grossman, I. E. (1985). Mixed integer programming approach for the synthesis of integrated process flowsheets. *Comp. Chem. Eng.* 9:463.

Ignizio, J. P. (1976). *Goal Programming and Extensions.* Heath (Lexingtons Series), Lexington, MA.

Ignizio, J. P. (1982). *Linear Programming in Single and Multiple Objective Systems.* Prentice-Hall, Englewood Cliffs, NJ.

Jaksland, C., Gani, R., and Lien, K. (1995). Separation process design and synthesis based on thermodynamic insights. *Chem. Eng. Sci.* 50(3):511.

Kondili, E., Pantelides, C. C., and Sargent, R. W. H. (1993). A general algorithm for short-term scheduling of batch operations. *Comp. Chem. Eng.* 17:21.

Lakshmanan, A., and Biegler, L. (1994). Reaction network targeting for waste minimization. *AIChE Symp. Ser.* 90(303):128.

LaGrega, M. D., Buckingham, P. L., and Evans, J.C. (1994). *Hazardous Waste Management.* McGraw-Hill, New York.

Linninger, A. A., Ali, S. A., Stephanopoulos, E., Han, C., and Stephanopoulos, G. (1994). Synthesis and assessment of batch processes for pollution prevention. *Pollution Prevention via Process and Product Modifications. AIChE Symp. Ser.* 90(303):46–58.

Linninger, A. A., Ali, S. A., Stephanopoulos, E., Han, C., and Stephanopoulos, G. (1995). Generation and assessment of batch processes with ecological considerations. *Comp. Chem. Eng.* 19:S7–S13.

Linninger, A. A. Ali, S. A., and Stephanopoulos, G. (1996). Knowledge-based validation and waste management of batch pharmaceutical process designs. *Comp. Chem. Eng.* 20:S1431–1436.

Linninger, A. A., Salomone, E., Ali, S. A., Stephanopoulos, E., and Stephanopoulos, G. (1998). Pollution prevention for production systems of energetic materials. *Waste Manage. J.* 17:165–173.

Linninger, A. A., and Stephanopoulos, G. (1998). A Natural Language Approach for the Design of Batch Operating Procedures. *Informatica, Special Issue on Natural Language Processing and Multi-Agent Systems* 22(4):5550–5596.

Linninger, A. A., and Chakraborty, A. (1999). Synthesis of optimization of waste treatment flowsheets. *Comp. Chem. Eng.* 23:1415–1425.

Parker, G., and Radin, R. (1988). *Discrete Optimization.* Academic Press, San Diego.

Russel, S., and Norvig, P. (1995). *Artificial Intelligence – A Modern Approach.* Prentice-Hall, Englewood Cliffs, NJ.

Siirola, J. J., and Rudd, D. F. (1971). Computer-Aided synthesis of chemical process designs. *Ind. Eng. Chem. Fund.* 10:353.

Shortliffe, E. F. (1976). *Computer-based Medical Consultations: MYCIN.* Elsevier/North Holland, Amsterdam, London, New York.

Spriggs, H. D (1994). Design for pollution prevention. *AIChE Symp. Ser.* 90:1.

Venkataramani, E. S., House, M. J., and Bacher, S. (1990). *An Expert System Based Environmental Assessment System (EASY).* Merck & Co.

Viswanathan, J., and Grossman, I. E. (1990). A combined penalty function and outer approximation for MINLP approximation. *Comp. Chem. Eng.* 14:769.

Wahnschafft, O. M., Jurain, T. P., and Westerberg, A. W. (1991). Split – A separation process designer. *Comp. Chem. Eng.* 15(8):565–581.

Wang, Y. P., and Smith, R. (1994). Wastewater minimization. *Chem. Eng. Sci.* 49(7): 981–1006.

CHAPTER 9

Pollution Prevention with a Middle Vessel Column

Weiyang Cheong and Paul I. Barton

Department of Chemical Engineering and Energy Laboratory, Massachusetts Institute of Technology, Cambridge, MA 02139

Abstract

A novel separation process termed the middle vessel (batch distillation) column (MVC) is discussed as an alternative to traditional batch distillation configurations (rectifier and stripper). The MVC is shown to be more environmentally friendly than corresponding combinations of batch rectifiers and strippers that achieve the same separation. It also results in possible savings in time and cost, savings which may or may not be substantial, depending on the nature of the separation. The flexibility of the MVC is best epitomized by its ability to separate an azeotropic mixture into its corresponding pure components at high purity. Finally, the MVC is also shown to be a time-analog of traditional continuous distillation sequences. The MVC thus handles, in a batch multipurpose mode, separations that would otherwise require a dedicated continuous distillation sequence. Separations, possible with a continuous column, but not carried out due to a low or periodic demand, can now be conducted periodically when required on a batchwise basis using an MVC.

Keywords: azeotropic batch distillation, solvent recovery, flexible manufacturing

INTRODUCTION

Batch distillation, with its lower energy efficiency and higher labor requirements, continues to be an important technology due to its greater operational flexibility, compared with continuous distillation. This flexibility is especially attractive to the pharmaceutical and specialty chemical industries, where products are typically required in small volumes and subject to short product cycles and fluctuating demand. Also associated with these industries are waste streams that are typically more toxic and concentrated than commodity chemical industry waste due to specialized reactions which require specific organic solvents and possibly toxic reagents. This is accentuated by the fact that batch processes usually generate additional wastes associated with transfer spillage and washing of vessels.

A batch separation system that exhibits the same flexibility as batch rectifiers or strippers and reduces the transfers required between vessels would thus reduce the amount of pollution and waste associated with a batch process. The MVC is one such process, as it offers the same flexibility as either a batch rectifier, a batch stripper, or some combination of the two while reducing the number of transfers required due to the presence of only one vessel that can perform any required distillation-based separation in a batch manufacturing system.

First proposed by Robinson and Gilliland (1950), the MVC was subsequently "rediscovered" and actively studied during the 1990s by various researchers (Davidyan et al., 1994; Meski and Morari, 1995; Safrit et al., 1995, 1997a, 1997b, 1997c; Barolo et al., 1996a, 1996b, 1997, 1998; Hasebe et al., 1996, 1997; Cheong and Barton, 1999a, 1999b, 1999c). In the MVC, feed from the main holdup vessel is introduced in the middle of the column such that there exists both a stripping and a rectifying section. This is in contrast to the batch stripper, where feed from the holdup vessel is introduced at the top of the column, and the batch rectifier, where feed from the vessel is introduced at the bottom of the column. An analysis of the characteristic dynamics of the MVC will be presented, and a short discussion on planning separations with the MVC will also be provided, derived from understanding its characteristic dynamics and the appropriate use of bifurcation graphs characterizing the MVC batch distillation regions.

Finally, it is shown that a single MVC is able to mimic the operation of an entire continuous distillation sequence while maintaining its batch character. The MVC is thus able to conduct separations that are more environmentally friendly but previously infeasible because they would have entailed continuous distillation sequences in the primarily batchwise pharmaceutical and specialty chemical industries. Other than being extremely attractive to batchwise chemical-manufacturing industries, the MVC can also be employed as a general solvent-recovery technology for any flexible manufacturing system that produces waste solvent streams with fluctuating concentration and/or fluctuating waste species (e.g., metal finishing, printing, and automobile painting). A continuous distillation sequence optimized for a specific set of species/concentration may be unsuitable for use with other waste species/concentrations.

The MVC thus provides a flexible batchwise separation process with which waste solvent streams of any composition and/or concentration may easily be treated by varying the operational characteristics of the MVC between batches. The flexibility of the MVC and its ability to reproduce many separations within the same physical column thus make it much more environmentally friendly than constructing many sequences of tiny distillation columns, each optimized for a particular separation of chemicals/solvent waste with a well-defined composition and concentration.

THE MIDDLE VESSEL COLUMN

Previous studies include the use of continuous entrainers in the MVCs to split azeotropes (Safrit et al., 1995, 1997a; Lelkes et al., 1998), the use of the MVC in the optimal sequencing of batch separations so as to maximize efficiency (Safrit and Westerberg, 1997b, 1997c), and experimental characterizations of the MVC using a pilot plant modified from an existing continuous column (Barolo et al., 1996a, 1996b, 1998). Although the above may be interesting, the emphasis here is to characterize the core qualitative dynamics of a normal MVC based on mathematical models that best characterize the operation of the MVC (Davidyan et al., 1994; Meski and Morari, 1995; Cheong and Barton, 1999a, 1999b, 1999c).

PHYSICAL STRUCTURE OF THE MVC

The schematic structure of a typical MVC is represented in Figure 9.1. The middle vessel is physically located at ground level to avoid unnecessary structural difficulties in supporting a heavy vessel in mid-air.

As illustrated in Figure 9.1, there are *ND* trays and a total condenser in the rectifying section of the MVC, and *NB* trays and a total reboiler in the stripping section of the MVC. Although the use of a total reboiler is not common in industry, we assume its use to maintain symmetry of our model. For the purposes of industrial usage, the use of a partial reboiler is equivalent to a total reboiler and one additional stage in our model.

The mixture to be separated has *NC* components, with a total molar holdup in the middle vessel of M. Two different vapor and liquid flow rates exist in the rectifying and stripping sections of the MVC. In the rectifying section, the vapor and liquid flow rates are V_d and L_d, respectively, whereas in the stripping section, the vapor and liquid flow rates are V_b and L_b, respectively. There is heat removed at the condenser on top of the rectifying section, as denoted by Q_C, and heat introduced at the total reboiler at the bottom of the stripping section, as denoted by Q_R. Further, there is an optional heat exchanger at the middle holdup vessel which either removes or adds heat, depending on the separation needs of the MVC, and is denoted by Q_{MVC}. This heat exchanger in the middle holdup vessel allows for differing vapor flow rates (V_d and V_b) in the two sections of the MVC. Finally, distillate is drawn from the top of the rectifying column at a rate of D, and bottoms product is drawn from the bottom of the stripping column at a rate of B.

It should be noted that Davidyan et al. (1994), Meski and Morari (1995), and Barolo et al. (1996a, 1996b, 1997, 1998) all studied structures similar to Figure 9.1, with one exception. In maintaining the symmetry of the MVC, we specify that the vapor from the stripping section be bubbled into the holdup vessel, where it is equilibrated with the liquid in the holdup vessel, before it is introduced into the rectifying section of the column. At the same time, we introduce the possibility of adding or removing heat from the middle holdup vessel, which allows us to achieve a Variable Vapor Flow (VVF)-MVC, where

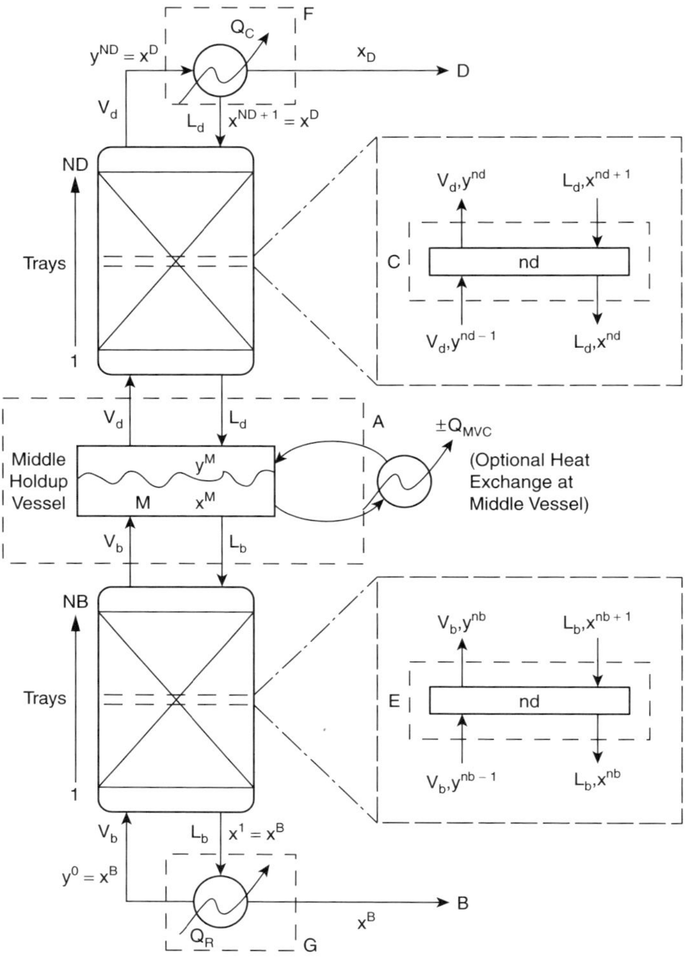

Figure 9.1: Schematic structure of typical MVC.

the vapor flow rate in the rectifying section need not be similar to the vapor flow rate in the stripping section of the MVC.

A different configuration was proposed by Hasebe et al. (1996), where they used two existing rectifiers connected with the appropriate piping and modified one of the rectifiers to take on the role of the stripping section; however, they too proposed the direct introduction of the vapor from the stripping section into the rectifying section, thereby forgoing the ability to vary the vapor flow rates between the two sections arbitrarily (i.e., the vapor rate can be increased, but not decreased). This ability to vary the vapor flow rate between the two sections provides an extra degree of freedom in the control of the column operation that proves to be crucial in attempts to use the MVC to mimic separations previously possible only with a spatial sequence of continuous distillation columns (Cheong and Barton, 1999b).

MATHEMATICAL MODEL

The MVC is modeled based on a differential model of the rate of composition change in the middle vessel, assuming negligible holdup on the trays and constant molar overflow (CMO). For formulating the MVC model, a "middle vessel parameter" λ is defined as

$$\lambda(t) = \frac{D(t)}{D(t) + B(t)}, \tag{1}$$

where λ may vary with time, as the product withdrawal flow rates, D and B, are variable in time, defined by the user to achieve the required separation.

Considering the overall and component mass balances around the whole column, and defining a dimensionless warp time ξ (Bernot et al., 1990),

$$d\xi = \frac{D + B}{M} dt. \tag{2}$$

The dynamics of the composition for each of the components in the middle holdup vessel is characterized for $i = (1, \ldots, NC)$ by

$$\frac{dx_i^M}{d\xi} = x_i^M - \lambda x_i^D - (1 - \lambda)x_i^B, \tag{3}$$

or in vector notation, where $\boldsymbol{x}$ s are the NC vectors of composition,

$$\frac{d\boldsymbol{x}^M}{d\xi} = \boldsymbol{x}^M - \lambda \boldsymbol{x}^D - (1 - \lambda)\boldsymbol{x}^B. \tag{4}$$

Superscripts indicate location of the composition, M for middle vessel, D for distillate product, and B for bottoms product. Detailed derivations are presented in Cheong and Barton (1999a). Equations (3) and (4) thus characterize completely the dynamic behavior of the middle holdup vessel composition.

It should be noted that the model described above serves as a generalization of traditional batch distillation models, as it encompasses both the batch stripper (at $\lambda = 0$, which implies that the distillate flow rate is zero) and the batch rectifier (at $\lambda = 1$, which implies that the bottoms flow rate is zero) as special cases.

Note also that composition of the distillate product $\boldsymbol{x}^D$ and bottom product $\boldsymbol{x}^B$ can be obtained via tray-to-tray calculations and adopting an appropriate nonideal VLE model, such as the NRTL or Wilson local-composition activity coefficient models, which are able to account for azeotropes in the mixture to be separated (Cheong and Barton, 1999a). Details of the appropriate modeling and design equations for the MVC are summarized in Appendix A.

Alternatively, assuming limiting conditions (infinite reflux, infinite number of trays), $\boldsymbol{x}^D$ and $\boldsymbol{x}^B$ can also be obtained theoretically via construction of the alpha and omega limit sets (Ahmad and Barton, 1996) of the middle vessel composition. Briefly explained, the alpha and omega limit sets of a given composition are the limiting sets of the residue curve which pass through that given composition, with definitions as follows:

$$\alpha(\boldsymbol{x}) = \lim_{\xi \to -\infty} \varphi(\xi, \boldsymbol{x}) \tag{5}$$

and

$$\omega(\boldsymbol{x}) = \lim_{\xi \to +\infty} \varphi(\xi, \boldsymbol{x}), \tag{6}$$

where $\varphi(\xi, \boldsymbol{x})$ is the solution to

$$\frac{d\varphi}{d\xi} = \varphi - \boldsymbol{y}(\varphi), \qquad \varphi(0) = \boldsymbol{x}; \tag{7}$$

i.e., the equation for simple distillation with an appropriate warped time (ξ) applied. A short treatise on residue curves and the usefulness of residue curve maps is presented in Appendix B.

Under limiting conditions (LC), the column composition profile of the MVC will be equivalent to the column composition profile of a continuous column operating at the same reflux and reboil ratios (infinite) and with a feed tray composition equivalent to the middle vessel composition (which, in general, will be different from the feed composition to the continuous column). Given that the infinite-reflux continuous column composition profile (i.e., distillation lines) can be approximated by the appropriate residue curve passing through the middle vessel composition ($\boldsymbol{x}^M$) (Wigdado and Seider, 1996), the column composition profile of the LC-MVC will also be approximated by the residue curve passing through the middle vessel composition, [i.e., $\varphi(\boldsymbol{x}^M)$]. The resulting distillate product composition, ($\boldsymbol{x}^D$), which equals the composition at the top of the LC-MVC, is thus given by the alpha limit set of the middle vessel composition [$\alpha(\boldsymbol{x}^M)$]. Correspondingly, the resulting bottoms product composition, ($\boldsymbol{x}^B$), which equals the composition at the bottom of the LC-MVC, is given by the omega limit set of the middle vessel composition [$\omega(\boldsymbol{x}^M)$] (Cheong and Barton, 1999a).

By extension (Cheong and Barton, 1999a), even in the presence of azeotropes that bound the trajectory of the residue curves due to the presence of separatrices or bundles of curves containing separatrices within the $NC - 1$ dimensional composition space, it remains true that

$$\boldsymbol{x}_{LC}^{D} = \alpha(\boldsymbol{x}^M) \tag{8}$$

and

$$\boldsymbol{x}_{LC}^{B} = \omega(\boldsymbol{x}^M). \tag{9}$$

Thus, with a given $\boldsymbol{x}^M$, it is possible to predict the distillate and bottoms product composition, provided a knowledge of the residue curve map of the system to be separated is available.

Furthermore, even as the middle vessel composition varies with time, as long as the alpha and/or omega limit sets of the middle vessel composition remain unchanged [i.e., the middle vessel composition remains within the same rectifier distillation region (alpha limit set unchanged) and/or remains within the same stripper distillation region (omega limit set unchanged)], then $\boldsymbol{x}^D$ and/or $\boldsymbol{x}^B$ will also remain unchanged (Cheong and Barton, 1999a). This invariance of product composition over long periods of time under LC thus results in greater predictability of the characteristic dynamics of the MVC and correspondingly greater clarity in the separation possibilities for a given mixture to be separated with an MVC.

Implicit in the ease of planning a separation with an LC-MVC is the fact that under nonlimiting conditions, $\boldsymbol{x}^D$ and $\boldsymbol{x}^B$ will both be varying continuously with ξ, and as such it becomes much harder to determine the path of the middle holdup vessel composition without conducting extensive dynamic simulations; however, the separation possibilities under LC are indicative of the separation possibilities that may be possible under non-LC. In fact, in the presence of azeotropes and curved separatrices, the use of non-LC may actually increase the separation possibilities for a given mixture due to the increased curvature of the MVC column composition profile (Cheong and Barton, 1999a, 1999b).

CHARACTERISTIC DYNAMICS

Given the model describing the dynamics of the composition in the middle holdup vessel, it is now possible to derive a coherent interpretation of the dynamics of the MVC. From equation (4), we define the net product composition obtained from the MVC, $\boldsymbol{x}^P(\xi)$, as

$$\boldsymbol{x}^P(\xi) = \lambda\, \boldsymbol{x}^D(\xi) + (1-\lambda)\boldsymbol{x}^B(\xi). \tag{10}$$

The net product composition represents the composition that would be obtained if the distillate and the bottoms product were mixed together at that given point in time (ξ). Consequently, it also represents the point that the middle vessel composition would move away from, in a straight line, if the middle vessel composition and the net product were both plotted in the mixture composition space, as seen by substituting equation (10) back into equation (4):

$$\frac{d\boldsymbol{x}^M(\xi)}{d\xi} = \boldsymbol{x}^M(\xi) - \boldsymbol{x}^P(\xi). \tag{11}$$

This is demonstrated graphically in Figure 9.2a. Thus, a knowledge of $\boldsymbol{x}^D(\xi)$, $\boldsymbol{x}^B(\xi)$, and $\lambda(\xi)$ would mean that the path of middle vessel composition motion can easily be obtained. Conversely, it implies that for a given path of middle vessel composition that we require, we could plan some combination of $[\boldsymbol{x}^D(\xi), \boldsymbol{x}^B(\xi), \lambda(\xi)]$ that would satisfy our requisite path.

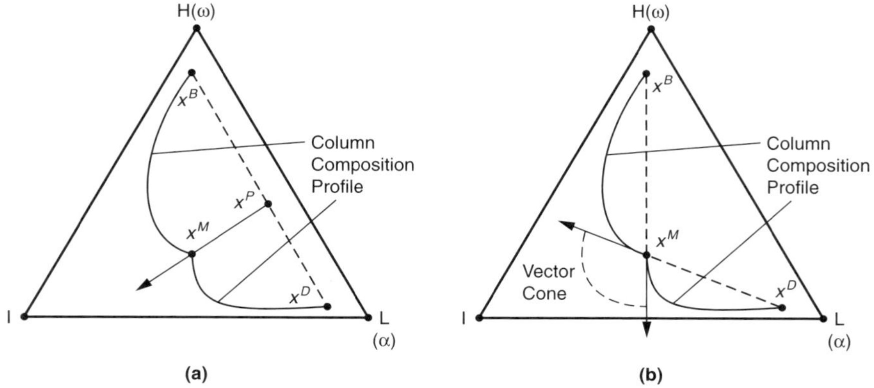

Figure 9.2: Graphical representation of middle vessel composition motion.

Alternatively, equation (4) can also be rearranged to form

$$\frac{d\boldsymbol{x}^M(\xi)}{d\xi} = \lambda[\boldsymbol{x}^M(\xi) - \boldsymbol{x}^D(\xi)] + (1 - \lambda)[\boldsymbol{x}^M(\xi) - \boldsymbol{x}^B(\xi)], \tag{12}$$

which implies that

> The middle vessel composition in the MVC moves in a direction that lies within the vector cone swept out by two vectors: one connecting the distillate composition to the middle vessel composition ($\boldsymbol{x}^M - \mathbf{x}^D$) and one connecting the bottoms composition to the middle vessel composition ($\boldsymbol{x}^M - \boldsymbol{x}^B$).

The vector cone spanned by direction vectors ($\boldsymbol{x}^M - \boldsymbol{x}^D$) and ($\boldsymbol{x}^M - \boldsymbol{x}^B$) thus defines the possible range of motions that the middle vessel composition can undergo from its current composition, with the actual direction of motion dependent on the value of $\lambda(\xi)$. This concept is illustrated graphically in Figure 9.2b, for which arbitrary values are assigned for $\boldsymbol{x}^M$, $\boldsymbol{x}^D$, and $\boldsymbol{x}^B$.

Equations (11) and (12) can thus be used to describe the characteristic dynamics of the MVC, depending on the nature of the problem formulation. The vector-cone model, as given by equation (12), is useful in preliminary design stages, where a broad overview of the plausible middle vessel composition paths is required. This conceptualization of the vector cone allows the designer to visualize the limits of separation easily, thereby allowing for feasibility analyses that are relatively straightforward. Even if the exact values of $\boldsymbol{x}^M$, $\boldsymbol{x}^D$, and $\boldsymbol{x}^B$ are not known (i.e., limiting conditions are not assumed), a knowledge of their possible values would allow for the formation of estimates for the possible vector cones that can be formed. Although not 100% accurate, it allows the designer to eliminate obviously infeasible separations.

The net-product model [as given by equation (11)], on the other hand, is useful for determining the exact path of the middle vessel composition, given the conditions of operation [i.e., $\boldsymbol{x}^D(\xi)$, $\boldsymbol{x}^B(\xi)$, and $\lambda(\xi)$]. Further, the net-product model is also useful in helping the designer pinpoint values of λ at which bifurcation behavior occurs for given points in the composition space.

Finally, these representations of the MVC characteristic dynamics apply to systems of a higher dimension (i.e., with more components in the mixture to be separated) and to azeotropic systems with separatrices. Note, however, that as long as only two products are drawn from the MVC (i.e., distillate and bottoms), the vector cone remains two-dimensional, in a plane defined by $\boldsymbol{x}^M$, $\boldsymbol{x}^D$, and $\boldsymbol{x}^B$, even if we apply the MVC to a quaternary (or higher dimension) mixture for which the composition space is three-dimensional (or higher). It is possible, however, to use multivessel columns (i.e., columns with many holdup vessels distributed along their length, each with substantial holdup and a characteristic composition) (Hasebe et al., 1997; Wittgens et al., 1996; Skogestad et al., 1997) to draw more than two products in a controlled fashion by modeling the multivessel column with an appropriate warped-time model similar to that of the MVC (Cheong and Barton, 1999a). Using this warped-time model, it is possible to design separations with more degrees of freedom in operating the column due to a higher-dimensional vector cone of possible motion.

The presence of azeotropes and separatrices, on the other hand, only serves to restrict the composition space within which separation is possible, as bundles of residue curve trajectories containing these separatrices form pot composition boundaries for the MVC (Ahmad and Barton, 1996), which restrict the middle vessel composition to lie within or on these boundaries. The product compositions under limiting conditions, $\boldsymbol{x}^D_{LC}$ and $\boldsymbol{x}^B_{LC}$, also tend to undergo discrete changes more often, as the restricted composition space implies that the boundaries demarcating the batch-rectifying and/or batch-stripping regions are more easily encountered, such that the alpha and omega limit sets corresponding to $\boldsymbol{x}^M$ are changed with higher frequency. Furthermore, azeotropic compositions may be the alpha and/or omega limit sets in an azeotropic distillation, as opposed to a system without azeotropes, where the alpha and omega limit sets are only the constitutive pure components of the mixture to be separated.

SEPARATING AZEOTROPES

To illustrate the planning of an azeotropic separation with the MVC, the complete separation of a mixture of acetone, benzene, and chloroform (denoted as ***A***, ***B***, and ***C***, respectively) will be examined as an example.

STEP ONE: OBTAIN THE RESIDUE CURVES MAP FOR ABC

Figure 9.3 shows the residue curves map for the mixture of ***ABC***. As illustrated, there is one maximum boiling azeotrope ***AC*** in the system, and a very curved separatrix connecting ***AC*** to the simplex vertex at ***B***. Curvature

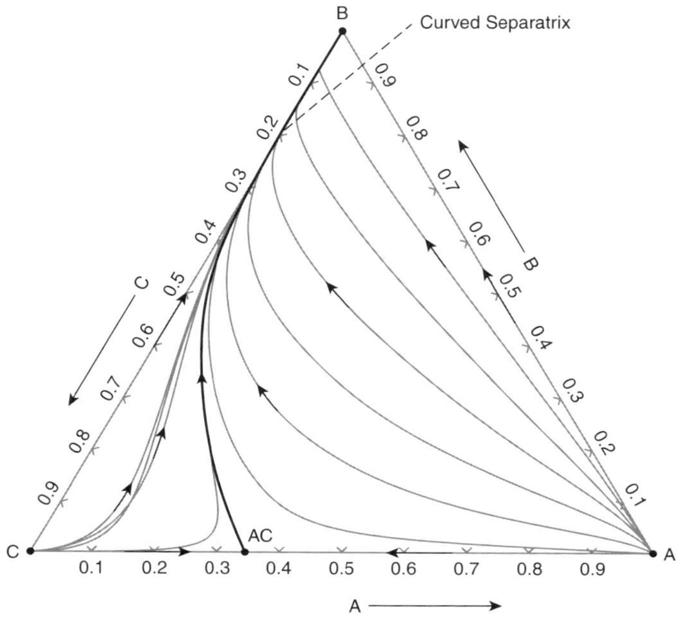

Figure 9.3: Residue curves map for the mixture of acetone, benzene, and chloroform (*ABC*).

in separatrices is expected (Reinders and De Minjer, 1940) and has been shown to allow continuous column separations believed to be impossible in the presence of straight separatrices (Wahnschafft et al., 1992, 1993).

From Figure 9.3, the alpha and omega limit sets for each section of the composition simplex were determined, and are summarized in Table 9.1.

STEP TWO: OBTAIN THE MVC BATCH DISTILLATION REGIONS FOR $\lambda = 0, \lambda = 1/2$, and $\lambda = 1$

Using the net-product model, the MVC batch distillation regions were determined for $\lambda = 0$ (Fig. 9.4a), for $\lambda = 1$ (Fig. 9.4b), and for the intermediate value of $\lambda = 1/2$ (Fig. 9.4c).

The curved separatrix is a region of interest because its curvature results in a net-product that lies outside of the separatrix. This causes the middle vessel composition to leave the separatrix as it draws this net-product that lies outside the separatrix (Fig. 9.5a), however, exiting the separatrix, the MVC middle vessel composition would enter interior region ***C-AC-B***, resulting in a change in the alpha limit set to pure ***C***. The new net-product composed of ***C*** (alpha limit set) and ***B*** (omega limit set), which would necessarily push the middle vessel composition back onto the separatrix (Fig. 9.5b), with

TABLE 9.1
Alpha and omega limit sets of sections in the ***ABC*** composition simplex

Section in composition simplex	*Alpha limit set*	*Omega limit set*
Simplex vertex ***A***	Simplex vertex ***A***	Simplex vertex ***A***
Simplex vertex ***B***	Simplex vertex ***B***	Simplex vertex ***B***
Simplex vertex ***C***	Simplex vertex ***C***	Simplex vertex ***C***
Azeotrope ***AC***	Azeotrope ***AC***	Azeotrope ***AC***
Simplex edge ***A-B***	Simplex vertex ***A***	Simplex vertex ***B***
Simplex edge ***B-C***	Simplex vertex ***C***	Simplex vertex ***B***
Simplex edge ***A-AC***	Simplex vertex ***A***	Azeotrope ***AC***
Simplex edge ***C-AC***	Simplex vertex ***C***	Azeotrope ***AC***
Curved separatrix ***AC-B***	Azeotrope ***AC***	Simplex vertex ***B***
Interior region ***A-AC-B***	Simplex vertex ***A***	Simplex vertex ***B***
Interior region ***C-AC-B***	Simplex vertex ***C***	Simplex vertex ***B***

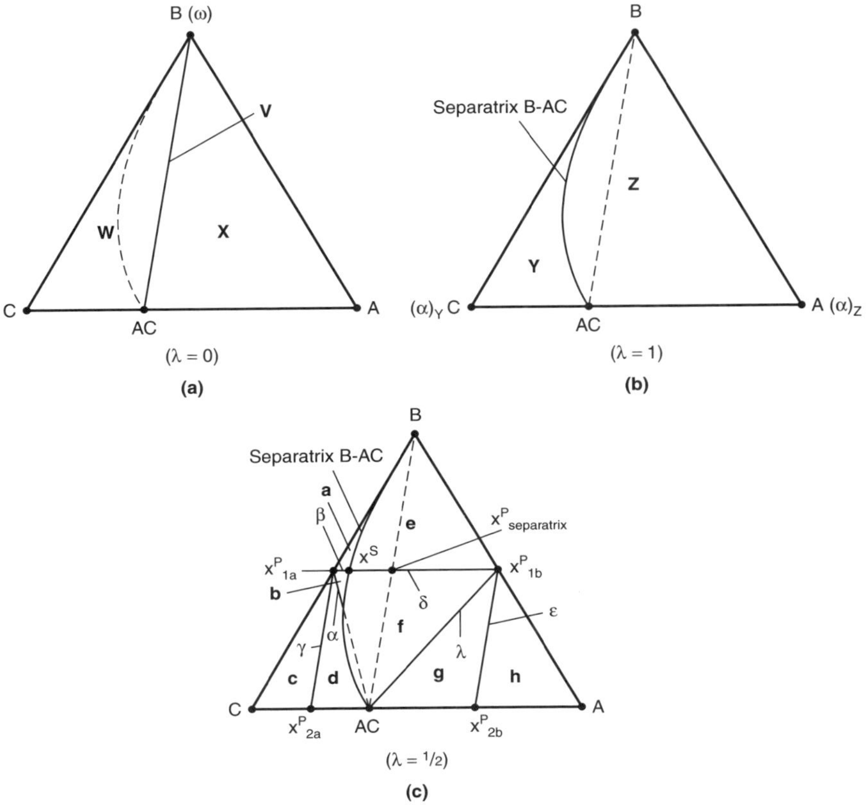

Figure 9.4: MVC batch distillation regions for the ***ABC*** mixture.

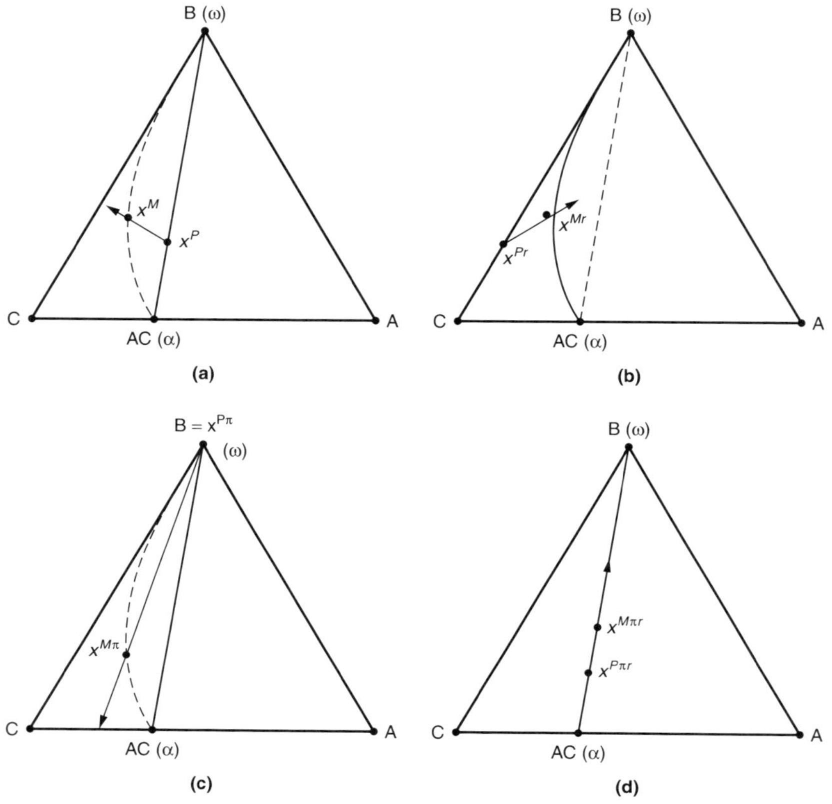

Figure 9.5: Effect of curved separatrix on the nature of products drawn for the ***ABC*** mixture.

its corresponding change in the alpha limit set back to ***AC*** and a repeat of the process. Thus, once the middle vessel composition encounters the separatrix, this seesawing change in the alpha limit set ensures that the middle vessel composition traces out the separatrix throughout the duration of the operation. The only exception occurs at $\lambda = 0$, where a change in the alpha limit set has no effect on the net-product (the multiplier on the alpha limit set in the net-product is zero). A composition that lies on the separatrix, in an MVC operated at $\lambda = 0$, moves away from the net-product (pure ***B***) into the interior region ***C-AC-B*** and remains in it until it encounters simplex edge ***C-AC*** (Fig. 9.5c). Note that if the separatrix was straight, then a still pot composition that lies on the separatrix would necessarily yield a net-product that is on the separatrix itself, and the analysis would be straightforward (Fig. 9.5d).

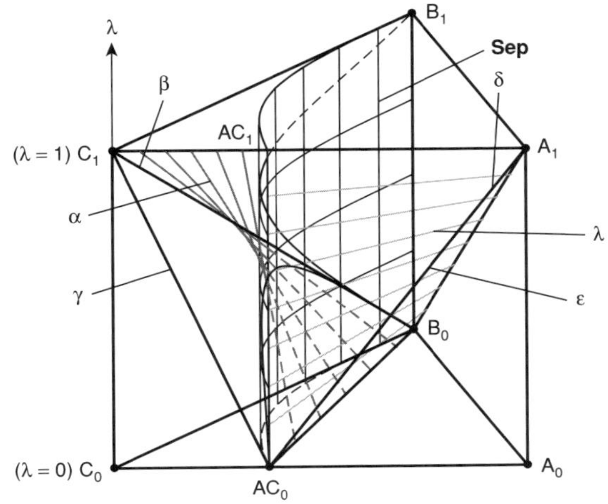

Sep - separatrix curved surface demarcated by AC_1, B_1, AC_0 and B_0 as denoted by lines

α - curved surface demarcated by C_1, AC_1, AC_0 and B_0 as denoted by lines

β - planar surface demarcated by C_1, AC_1 and B_0

γ - planar surface demarcated by C_1, AC_0 and B_0

λ - curved surface demarcated by AC_1, A_1, AC_0 and B_0 as denoted by lines

δ - planar surface demarcated by A_1, AC_1 and B_0

ε - triangular planar surface demarcated by A_1, AC_0 and B_0

a - region bounded by surfaces **Sep**, $B_0B_1C_1$, $B_1C_1AC_1$ and β

b - region bounded by surfaces **Sep**, α and β

c - region bounded by surfaces **Sep**, $C_1AC_0AC_1$, α and γ

d - region bounded by surfaces **Sep**, $B_0C_0C_1$, $C_0C_1AC_0$, $B_0C_0AC_0$ and γ

e - region bounded by surfaces **Sep**, $A_1B_1AC_1$, $A_1B_0B_1$ and δ

f - region bounded by surfaces **Sep**, δ and λ

g - region bounded by surfaces $A_1AC_0AC_1$, λ and ε

h - region bounded by surfaces $A_0B_0AC_0$, $A_0A_1AC_0$, $A_0A_1B_0$ and ε

Figure 9.6: Three-dimensional bifurcation graph for the *ABC* mixture.

STEP THREE: OBTAIN THE BIFURCATION GRAPH FOR ABC

Varying λ between 0^+ and 1^-, the bifurcation graph for the mixture of acetone, benzene, and chloroform ***(ABC)*** is obtained in Figure 9.6. Transformations of each of the regions labeled in Figure 9.4c, as $\lambda \rightarrow 0$ and $\lambda \rightarrow 1$, are listed in Table 9.2. It should be noted that there exists a discontinuity in the bifurcation graph due to curvature of the separatrix. The region demarcated by the separatrix at $\lambda = 0^+$ disappears when $\lambda = 0$. All other transformations are continuous.

TABLE 9.2
Transformation of MVC batch distillation regions for the ***ABC*** mixture

PCB/Region	$\lambda = 0$	$0 < \lambda < 1$	$\lambda = 1$
α	Boundary ***V***	Line segment $x^P{}_{1a}AC$	Simplex edge ***AC-C***
β	Simplex vertex ***B***	Line segment $x^P{}_{1a}x^S$	Simplex edge ***AC-C***
γ	Boundary ***V***	Line segment $x^P{}_{1a}x^P_{2a}$	Simplex vertex ***C***
a	Simplex vertex ***B***	*a*	Simplex interior ***Y***
b	Boundary ***V***	*b*	Simplex edge ***AC-C***
c	Boundary ***V***	*c*	Simplex edge ***AC-C***
d	Simplex interior ***W***	*d*	Simplex vertex ***C***
χ	Boundary ***V***	Line segment $x^P{}_{1b}AC$	Simplex edge ***AC-A***
δ	Simplex vertex ***B***	Line segment $x^P{}_{1b}x^S$	Simplex edge ***AC-A***
ε	Boundary ***V***	Line segment $x^P{}_{1b}x^P_{2b}$	Simplex vertex ***A***
e	Simplex vertex ***B***	*e*	Simplex interior ***Z***
f	Boundary ***V***	*f*	Simplex edge ***AC-A***
g	Boundary ***V***	*g*	Simplex edge ***AC-A***
h	Simplex interior ***X***	*h*	Simplex vertex ***A***

STEP FOUR: DETERMINE THE SEPARATION CRITERIA

Armed with the bifurcation graph of ***ABC***, it is now possible to design an operating procedure based on the separation criteria. In our example, we attempt to separate a mixture of ***ABC*** completely into its constituent pure components. This can be achieved either with the use of an azeotropic waste cut (***AC***) that can be appropriately recycled to the next batch or with the introduction of additional quantities of ***A***, ***B***, or ***C*** as entrainers as long as these entrainers added can be completely retrieved during the separation. The usual argument against the addition of one of the component species to aid in the separation of an azeotropic mixture is that the amount of that component species recovered will be less than the additional amount added; however, if we are able to recover more than we add, then the addition of this component would help in achieving the desired separation.

STEP FIVE: DECIDE AN OPERATING PROCEDURE FOR λ

Observing the bifurcation graph, we find that the key to our separation lies not in the variation of λ, but in the alteration of the middle vessel initial composition, which can be achieved by the appropriate addition of one of the constituent components as an entrainer or via the addition of a recycled azeotropic cut. In fact, observation of the bifurcation graph shows that it is the discontinuity in the bifurcation graph between $\lambda = 0$ and 0^+ that allows us to separate a mixture of ***ABC*** completely into pure ***A***, pure ***B***, and pure ***C***.

From $\lambda = 0^+$ to $\lambda = 1$, the stable separatrix is a pot composition barrier, but due to the discontinuity in the bifurcation graph between $\lambda = 0^+$ and $\lambda = 0$, the separatrix ceases to be a pot composition barrier at $\lambda = 0$ (i.e., we are able to cross the separatrix at $\lambda = 0$). In general, all stable separatrices are MVC pot composition barriers for all λ, except $\lambda = 0$, and all unstable separatrices are MVC pot composition barriers for all λ, except for $\lambda = 1$. Furthermore, it is observed from the MVC batch distillation regions for $\lambda = 0^+$ to $\lambda = 1$ that the azeotropic off-cut is minimized when the distance between the simplex vertex ***A*** and the point where the middle vessel composition trajectory intersects the separatrix (or separatrix intersection point) is maximized, as it results in the middle vessel composition encountering the ***C-AC*** edge at a position furthest from ***AC***, hence minimizing the amount of azeotrope that will be leftover at the end of the separation. This "crossing of the separatrix" allows the middle vessel composition to move into the interior region ***C-AC-B*** from interior region ***A-AC-B***, thereby drawing pure ***A*** and ***B*** (in interior region ***A-AC-B***) and pure ***C*** (in interior region ***C-AC-B***).

Thus, an operating schedule with no addition of batch-wise entrainers but with the recycle of an azeotropic off-cut (mode A), valid only for interior region ***A-AC-B***, can be formulated (Fig. 9.7):

1. The azeotropic off-cut from the previous separation (***AC***) is mixed with the initial composition (*F*), resulting in a middle vessel composition (***M***) from which we begin our operation of the MVC.

2. Operate the MVC at $\lambda_1 = 1$ (i.e., equivalent to a rectifier), drawing only the alpha limit set (pure ***A***) as the distillate product. Drawing ***B*** as a bottoms product (omega limit set) is undesirable, as it displaces the trajectory of the middle vessel composition away from ***B*** toward ***A***, thereby decreasing the distance that we should maximize (between ***A*** and the separatrix intersection point *D*).

3. When the middle vessel composition reaches the separatrix intersection point *D*, the middle vessel composition is induced to cross the separatrix by operating the MVC at $\lambda_2 = 0$ (i.e., equivalent to a stripper) such that the separatrix is no longer a pot composition barrier. Pure ***B*** is drawn as the MVC bottoms product, and the middle vessel composition moves away from ***B*** until it encounters the ***C-AC*** simplex edge at point *E*.

4. Finally, with the middle vessel composition on the ***C-AC*** simplex edge, we could operate the MVC either in a quasi-static state by drawing proportional amounts of ***C*** (alpha limit set) and ***AC*** (omega limit set) ($\lambda = \lambda_3$) such that the middle vessel composition remains unchanged until the MVC runs dry (i.e., $\xi \rightarrow \infty$), or operate the MVC as a rectifier (i.e., with $\lambda = 1$) such

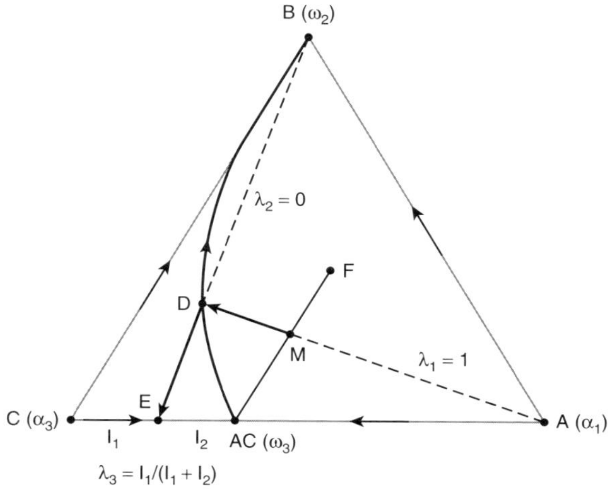

Figure 9.7: Seperation for the azeotropic mixture of ***ABC*** with azeotropic recycle (model A).

that pure ***C*** is obtained as the distillate product and the azeotropic off-cut is collected in the middle vessel.

The above schedule results in the separation of a mixture of ***ABC*** that lies in the interior region of ***A-AC-B***, into its constituent components, with the constant recycle of an azeotropic waste cut (***AC***); however, it precludes the separation of an initial composition that lies within the interior region of ***C-AC-B*** and generates an azeotropic off-cut that results in waste generation (from the last batch).

An operating schedule may also be formulated for the entire composition simplex in which pure ***B*** is added as a batch-wise entrainer, and azeotropic off-cuts are not formed (mode B) (Fig. 9.8):

1. Pure ***B*** is mixed as a batch-wise entrainer with the representative initial compositions ($\delta_1, \delta_2, \delta_3$) to obtain the middle vessel compositions ($\gamma_1, \gamma_2, \gamma_3$) which all lie in the region σ, deemed to be the region within which the middle vessel composition must begin its operation in order for the initial mixture to be separated *completely* into its pure constituent components with no azeotropic recycle.
2. Operate the MVC at $\lambda_1 = 1$, drawing only the alpha limit set (pure ***A***) as the distillate product. As before, we avoid the withdrawal of pure ***B*** (omega limit set) as the bottoms.
3. When the middle vessel composition reaches the separatrix intersection point *G*, the MVC is operated at $\lambda = 0$ for a short period of time, such that the separatrix is crossed (at $\lambda = 0$, the separatrix is not a pot composition barrier).

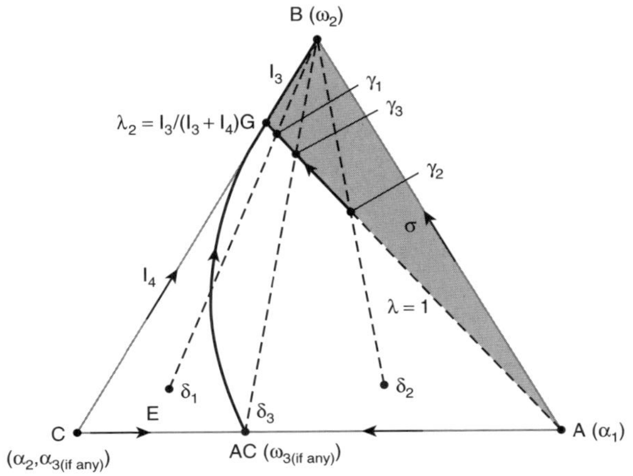

Figure 9.8: Separation for the azeotropic mixture of ***ABC*** with batch entrainer (model B).

4. After the separatrix is crossed, with the middle vessel composition lying almost on the ***C-B*** simplex edge, we could operate the MVC either in a quasi-static state by drawing proportional amounts of ***C*** (alpha limit set) and ***B*** (omega limit set) with ($\lambda = \lambda_2$) such that the middle vessel composition remains unchanged at G until the MVC runs dry (i.e., $\xi \to \infty$), or continue to operate the MVC as a stripper with $\lambda = 0$ such that pure ***B*** is obtained as the bottoms product and pure ***C*** is collected in the middle vessel (or the MVC can be operated as a rectifier with $\lambda = 1$, so as to separate the pure ***C*** from the still pot residue that would exist in the middle vessel).

In summary, it usually is presumed that an azeotropic mixture cannot be separated completely into its constituent components, and some azeotropic off-cut is expected; however, as shown in the example above, a careful observation of the bifurcation graph to identify exploitable curvatures in pot composition boundaries, and the formulation of a clever operating procedure, results in the complete separation of an azeotrope into its constituent pure components, with the help of a batch-wise entrainer and without the recycle of an azeotropic cut. A detailed explanation of the derivation for the two MVC operating schedules (mode A and mode B) is provided in Cheong and Barton (1999b).

ANALOGS TO SEQUENCES OF CONTINUOUS COLUMNS

With a configuration inclusive of both stripping and rectifying sections, MVCs are similar to continuous columns. In fact Barolo et al. (1996a, 1996b, 1998) constructed their pilot plant of an MVC from an existing continuous

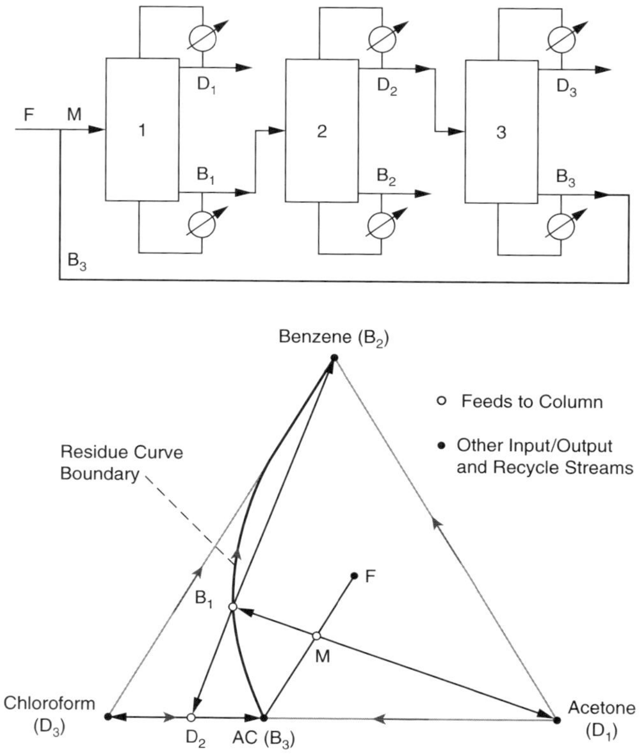

Figure 9.9: Continuous column sequence analog to model A of ***ABC*** separation.

column. Furthermore, the MVC has been shown to mimic separations achievable in a single continuous column or sequences of continuous columns (Cheong and Barton, 1999b).

The MVC is able to handle, in a batch multipurpose mode, separations that will otherwise require a dedicated sequence of continuous distillation columns. For example, the two modes of separating the azeotropic mixture of acetone, benzene, and chloroform using the MVC (mode A and mode B) are almost exactly analogous to two separation schemes proposed using a sequence of continuous columns for separating the ***ABC*** mixture (Wahnschafft et al., 1992, 1993). These sequences of continuous distillation columns are illustrated in Figure 9.9 for the mode A equivalent, and in Figure 9.10 for the mode B equivalent. Note the similarity in the schematic representations of Figures 9.9 and 9.10 with those in Figures 9.7 (mode A) and 9.8 (mode B), respectively. A detailed comparison of the two sets of analogous operations showed that this is due to the congruency of principles underlying the designs (Cheong and Barton, 1999b). The only difference is that in the continuous column sequence, separation was achieved using many columns

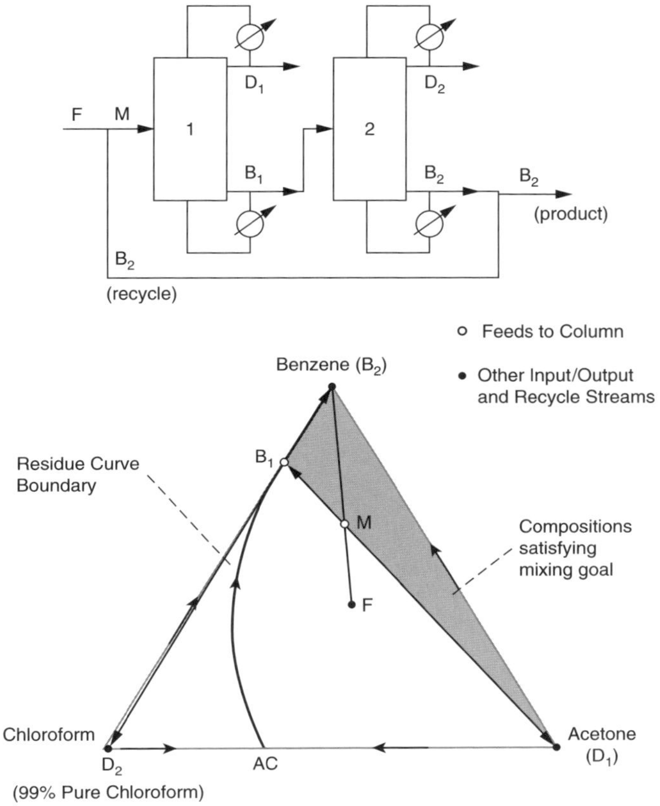

Figure 9.10: Continuous column sequence analog to model B of ***ABC*** separation.

at the same point in time, whereas for the MVC, separation was achieved using one column but over many operating phases. Thus, the MVC achieved in time domain, the same separation as that achieved by the continuous column sequence in the space domain.

Note that the MVC does not replace the continuous column, as the continuous column continues to be optimal as a continuous separation process; however, it is a useful flexible multipurpose separation process if a continuous production needs to be converted to a batch-wise production.

MVC VERSUS COMBINED STRIPPERS AND RECTIFIERS

The environmental friendliness of the MVC is due to its ability to combine both the batch stripper and the batch rectifier in one unified separating column, capable of reproducing any separation usually processed by

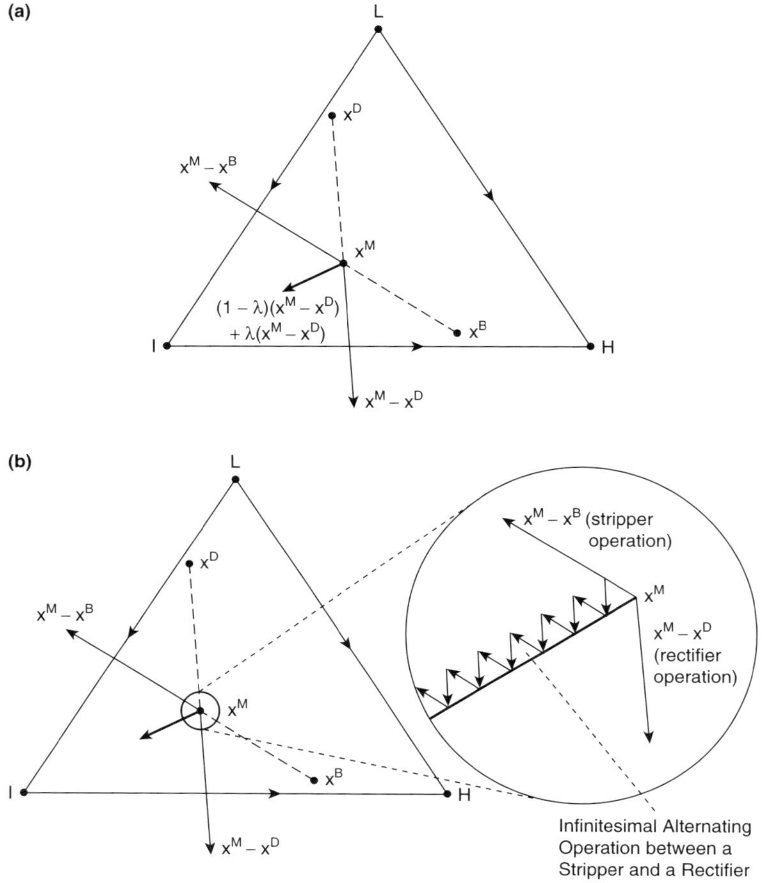

Figure 9.11: Equivalence of MVC to infinitesimal stripper and rectifier operations.

combinations of batch strippers and/or rectifiers while reducing the number of transfers necessary between vessels in a batch process. Accordingly, the MVC operated at $0 < \lambda < 1$ is equivalent to a middle vessel holdup continuously transferred between a batch rectifier and a batch stripper, operated for infinitesimally short periods of time in each of the rectifiers and strippers (as illustrated in Fig. 9.11). This is because for $0 < \lambda < 1$, the MVC holdup vessel is concurrently separated with both a rectifying and a stripping column (Cheong and Barton, 1999a).

While the combined operation of rectifiers and strippers could reproduce some separation results from applying the MVC, using the MVC also results in (1) halving of the energy costs of production, (2) savings in campaign time, and (3) ease in navigating constraints in the composition simplex (Cheong and Barton, 1999a). Briefly explained, the MVC draws both the distillate and bottoms product from the same column (i.e., with only one associated condenser and reboiler load). In comparison, the combined rectifier/stripper yields a condenser and reboiler load for both the rectifier and the stripper, each operated over the same length of time as the equivalent MVC. Loosely speaking, because the reflux/reboil ratios in the MVC will be equivalent to the reflux ratio in the rectifier and reboil ratio in the stripper, the associated energy loads per column will be similar; however, because two columns are used in the rectifier/stripper combination [or twice the amount of time is required (Meski and Morari, 1995)], as compared to the single MVC, energy consumption will be approximately halved if the MVC is used in preference over the rectifier/stripper combination.

If the rectifier and stripper columns are operated sequentially, the rectifier/stripper combination would double the campaign time, as compared to the MVC. Although it is possible to reduce the campaign time savings of the MVC with an appropriate staggering of the operation for the rectifier/stripper combination (Cheong and Barton, 1999a), there will still be overhead times associated with transfers between the rectifier/stripper that will not be present in the MVC. Thus, there are unambiguous savings in the campaign time when the MVC is used preferentially. The ability to set λ at any value between 0 and 1 also implies that there is fine control in the MVC still pot composition path, as compared to the rectifier/stripper combination, where the still pot composition must unambiguously move away from the distillate (rectifier) and bottoms product (stripper). This fine control in the MVC thus allows for ease of navigating the still pot composition within the constraints that may be present in the composition simplex.

If a new batch processing plant is designed from scratch, use of the MVC is preferred, as capital costs are comparable and the added flexibility of the MVC makes it attractive to batch plants, which have to be flexible and multipurpose in design. Furthermore, the MVC can be operated as a stripper or rectifier as required (i.e., with $\lambda = 0$ or $\lambda = 1$) by shutting off one of the product flows. The MVC, however, does require a more complex control system, especially if λ is to be varied.

Finally, use of a single MVC reduces the transfers between units, which minimize the risk of spillage and/or contamination. It also reduces the amount of waste generated from residue left in the still pots at the end of an operation, as only one still pot is required for the entire separation process. Thus, product quality, safety, and environmental friendliness all indicate that the MVC is preferred to a stripper-rectifier combination.

CONCLUSION

The MVC was discussed as a multipurpose separation tool that is capable of reproducing the separations presently possible with batch rectifiers, strippers, and even sequences of continuous distillation columns. There are huge environmental advantages for flexible manufacturing afforded by the use of a single multipurpose equipment that enables almost any continuous distillation process to be reproduced as sequential operations over time in the batch-wise MVC. The versatility of MVCs also results in the ability to separate azeotropes, as exemplified by the separation of the azeotropic mixture of acetone, benzene, and chloroform into its constituent pure components, thereby reducing the amount of azeotropic waste normally associated with separation of azeotropic mixtures.

ACKNOWLEDGMENTS

This work was supported by the U.S. Department of Energy under grant DE-FG02-94ER14447 and the MIT Undergraduate Research Opportunities Program.

APPENDIX A: MODELING AND DESIGN EQUATIONS FOR THE MVC

Although the bulk of the analysis conducted in this chapter is based on the assumption of limiting conditions (LC) (i.e., of infinite reflux/reboil ratios, infinite number of trays), it should be recognized that LC are not really achievable in a practical column. The LC analysis really only serves to delimit and characterize the separations and performance levels feasible in the MVC. Hence, presented in this appendix are the modeling and design equations for an MVC, with which it is possible to predict the operating characteristics of the MVC under non-LC conditions.

The concept of the middle vessel still pot composition moving in the composition simplex away from the net product $\boldsymbol{x}^P(\xi)$ and within the vector cone delimited by the distillate product $\boldsymbol{x}^D(\xi)$, bottoms product $\boldsymbol{x}^B(\xi)$, and the middle vessel composition $\boldsymbol{x}^M(\xi)$, as described by equations (11) and (12), are both still valid; however, the values of $\boldsymbol{x}^D(\xi)$, $\boldsymbol{x}^B(\xi)$, and, consequently, the value of $\boldsymbol{x}^P(\xi)$ must now be determined via a tray-by-tray calculation within the MVC. Finally, if we account for a constant holdup in each of the trays in the rectifying and stripping sections, equations (11) and (12) are modified to obtain

$$\frac{d\boldsymbol{x}^c(\xi)}{d\xi} = \boldsymbol{x}^c(\xi) - \boldsymbol{x}^p(\xi), \tag{A1}$$

$$\frac{d\boldsymbol{x}^c(\xi)}{d\xi} = \lambda[\boldsymbol{x}^c(\xi) - \boldsymbol{x}^D(\xi)] + (1-\lambda)[\boldsymbol{x}^c(\xi) - \boldsymbol{x}^B(\xi)], \tag{A2}$$

where $\boldsymbol{x}^C(\xi)$ denotes the composition of the overall holdup in the column (inclusive of holdup in the middle vessel, and each of the trays in both the rectifying and stripping sections).

From the overall mass balance in the middle vessel, we thus obtain

$$\frac{dM}{dt} = L_d + V_b - V_d - L_b, \tag{A3}$$

and the overall component mass balance yields

$$\frac{dM\boldsymbol{x}^M}{dt} = L_d\boldsymbol{x}^1 + V_b\boldsymbol{y}^{NB} - V_d\boldsymbol{y}^M - L_b\boldsymbol{x}^M. \tag{A4}$$

Mass balances written for each of the trays in the rectifying section (envelope C in Fig. 9.1) yield

$$H_{nd}\frac{d\boldsymbol{x}^{nd}}{dt} = L_d\boldsymbol{x}^{nd+1} + V_d\boldsymbol{y}^{nd-1} - V_d\boldsymbol{y}^{nd} - L_d\boldsymbol{x}^{nd} \qquad \forall 1 \leq nd \leq ND \tag{A5}$$

where H_{nd} is the holdup in the nd^{th} tray (assumed constant), $\boldsymbol{y}^0 = \boldsymbol{y}^M$ is the middle vessel vapor composition, and with the total condenser assumption, $\boldsymbol{x}^{ND+1} = \boldsymbol{y}^{ND} = \boldsymbol{y}^D$. Mass balance around the condenser (envelope F) yields

$$V_d = L_d + D. \tag{A6}$$

A similar set of equations can also be obtained for the stripping section of the MVC (envelopes E, G),

$$H_{nb}\frac{d\boldsymbol{x}^{nb}}{dt} = L_b\boldsymbol{x}^{nb+1} + V_b\boldsymbol{y}^{nb-1} - V_b\boldsymbol{y}^{nb} - L_b\boldsymbol{x}^{nb} \qquad \forall 1 \leq nb \leq NB, \tag{A7}$$

$$L_b = V_b + B, \tag{A8}$$

where H_{nb} is the holdup in the nb^{th} tray, $\boldsymbol{x}^{NB+1} = \boldsymbol{x}^M$ is the middle vessel liquid composition, and with the total reboiler assumption, $\boldsymbol{y}^0 = \boldsymbol{x}^1 = \boldsymbol{x}^B$. Finally, the vapor-liquid equilibrium (VLE) relationship on each tray and in the middle vessel is given by

$$\boldsymbol{y}^i = \boldsymbol{y}^i[\boldsymbol{x}^i, T(\boldsymbol{x}^i, P), P] \qquad \forall\, i \in \{\{nd\}, \{nb\}, M\}, \tag{A9}$$

where $\boldsymbol{y}^{\mathrm{I}}$ is modeled using the NRTL local-composition activity coefficient model. Solving for the set of equations (A3) through (A9), it is then possible to obtain the value of $\boldsymbol{x}^B(\xi)$ and $\boldsymbol{x}^D(\xi)$ for a given point in ξ. Keeping in mind that

$$C\boldsymbol{x}^c = M\boldsymbol{x}^M + \sum_{nd}^{ND} H_{nd}\boldsymbol{x}^{nd} + \sum_{nb}^{NB} H_{nb}\boldsymbol{x}^{nb}, \tag{A10}$$

it is then possible to solve equations (A1) through (A9) to obtain $\boldsymbol{x}^M(\xi)$, the motion of the middle vessel composition with ξ, as a function of the defined operating parameters. In general, we are able to define operating parameters such that the reflux ratio, reboil ratio, and λ can be specified intrinsically in the VVF-MVC. Correspondingly, there is one degree of freedom less in the CVF-MVC. A degree of freedom analysis is provided by Cheong and Barton (1999b).

Finally, in the presence of finite reflux ratios and reboil ratios defined accordingly as $R_d = L_d/D$ and $R_b = V_b/B$, and with tray holdup times in the rectifying (h_{nd}) and stripping (h_{nb}) sections of the MVC defined, respectively, as $h_{nd} = H_{nd}/D$ and $h_{nb} = H_{nb}/B$, equations (A5) to (A8) can be rearranged as

$$h_{nd}\frac{d\boldsymbol{x}^{nd}}{dt} = R_d\boldsymbol{x}^{nd+1} + (R_d + 1)\boldsymbol{y}^{nd-1} - R_d\boldsymbol{x}^{nd} - (R_d + 1)\boldsymbol{y}^{nd} \qquad \forall 1 \leq nd \leq ND, \tag{A11}$$

$$h_{nb}\frac{d\boldsymbol{x}^{nb}}{dt} = (R_b + 1)\boldsymbol{x}^{nb+1} + R_b\boldsymbol{y}^{nb-1} - (R_b + 1)\boldsymbol{x}^{nb} - R_b\boldsymbol{y}^{nb} \qquad \forall 1 \leq nb \leq NB. \tag{A12}$$

APPENDIX B: SIMPLE DISTILLATION RESIDUE CURVES AND RESIDUE CURVE MAPS

The simple distillation residue curve is defined as the locus of the liquid composition remaining from a simple distillation process. Thus, for a given point on the residue curve, its tangent is given by the tie-line that connects that point (i.e., liquid residue composition) to the corresponding vapor composition that is in equilibrium with the liquid. These corresponding vapor compositions can also be plotted on the composition simplex to form the distillate curve. The collection of residue curves in the composition simplex thus forms the residue curve map for that given set of components.

The simple distillation process can be expressed mathematically as

$$\frac{d(H\boldsymbol{x})}{dt} = -Vy(\boldsymbol{x}), \tag{B1}$$

or with an appropriate substitution of a warped time variable τ,

$$d\tau = \frac{H}{V}dt, \tag{B2}$$

we obtain the more familiar equation for the residue curves as a function of the warped time τ,

$$\frac{d\boldsymbol{x}}{d\tau} = \boldsymbol{x} - \boldsymbol{y}(\boldsymbol{x}), \tag{B3}$$

with which Doherty and Perkins (1978) interpreted the residue curve maps as a phase portrait diagram for the system of differential equations defining the simple distillation process and studied the topological nature of residue curves extensively. Van Dongen and Doherty (1985a, 1985b) then went on to show that as the R_d or $R_b \to \infty$ in a column, the liquid composition profile in a column approaches the shape of a simple distillation residue curve. They also state that at R_d or $R_b = \infty$, the liquid composition profile is exactly the same as a simple-distillation residue curve. Cheong and Barton (1999a) clarified that the shape of the liquid composition profile in a distillation tray column approaches the shape of the simple distillation residue curve, but they will never be mathematically equivalent.

Van Dongen and Doherty (1985a) also demonstrated the equivalence of the residue curves to the column composition profile at infinite reflux/reboil ratios by making some first-order approximations resulting in a set of equations equivalent to the residue curve equation for the rectifying section of the column,

$$\frac{d\boldsymbol{x}'}{dh'} = \boldsymbol{x}' - y(\boldsymbol{x}'), \tag{B4}$$

where h' is the height along the rectifying section of the column measured from the middle feed tray. Equivalently, for the stripping section of the column,

$$\frac{d\boldsymbol{x}}{dh} = \boldsymbol{x} - \boldsymbol{y}(\boldsymbol{x}), \tag{B5}$$

where h is the height up the stripping section of the distillation column measured from the bottom of the column. Comparing equations (B4) and (B5) with (B3), it can be observed that the distillation column operating at infinite reflux is a space-analog of the time-distributed residue curve. Whereas movement along the simple distillation residue curve [phase portrait diagram of equation (B3)] represents a progress or reversal in warped time τ, movement along the phase portrait diagrams of equations (B4) and (B5) is a movement up or down the distillation column at infinite reflux. The symmetry of equations (B3) to (B5) thus implies that the phase portrait diagrams will be similar. Hence, the simple distillation residue curve map can be used equivalently to predict the composition profile of a column under infinite reflux. This result is used in our prediction of the products derived from an MVC operating at limiting conditions (i.e., infinite reflux).

Note that the above prediction is based on a first-order approximation of the column composition profile. Hence, the actual column composition profile will never be equivalent to the residue curve, although it may approach the residue curve asymptotically at limiting conditions; however, it remains true that the limits of the distillation line sequence at limiting conditions (i.e., composition of the top and bottom trays of the column) unambiguously approach the fixed points (i.e., alpha and omega limit sets, respectively) of the residue curve (Cheong and Barton, 1999a).

REFERENCES

Ahmad, B. S., and Barton, P. I. (1996). Homogeneous multicomponent azeotropic batch distillation. *AIChE J.* 42:3419.

Barolo, M., Guarise, G. B., Ribon, N., Rienzi, S. A., and Trotta, A. (1996a). Some issues in the design and operation of a batch distillation column with a middle vessel. *Comput. Chem. Eng.* 20(S):S37.

Barolo, M., Guarise, G. B., Rienzi, S. A., Trotta, A., and Macchietto, S. (1996b). Running batch distillation in a column with a middle vessel. *Ind. Eng. Chem. Res.* 35(12):4612.

Barolo, M., and Botteon, F. (1997). Simple method of obtaining pure products by batch distillation. *AIChE J.* 43(10):2601.

Barolo, M., Guarise, G. B., Rienzi, S. A., and Trotta, A. (1998). Understanding the dynamics of a batch distillation column with a middle vessel. *Comput. Chem. Eng.* 22(S):S37.

Bernot, C., Doherty, M. F., and Malone, M. F. (1990). Feasibility and separation sequences in multicomponent batch distillation. *Chem. Eng. Sci.* 45(5):1207.

Cheong, W., and Barton, P. I. (1999a). Azeotropic distillation in a middle vessel batch column. I: Model formulation and linear separation boundaries. *Ind. Eng. Chem. Res.* 38(4):1504.

Cheong, W., and Barton, P. I. (1999b). Azeotropic distillation in a middle vessel batch column. II: Nonlinear separation boundaries. *Ind. Eng. Chem. Res.* 38(4):1531.

Cheong, W., and Barton, P. I. (1999c). Azeotropic distillation in a middle vessel batch column. III: Model validation. *Ind. Eng. Chem. Res.* 38(4):1549.

Davidyan, A. G., Kiva, V. N., Meski, G. A., and Morari, M. (1994). Batch distillation in a column with a middle vessel. *Chem. Eng. Sci.* 49(18):3033.

Doherty, M. F., Perkins, J. D. (1978). On the dynamics of distillation processes—I: The simple distillation of multi-component non-reacting, Homogeneous liquid mixtures. *Chem. Eng. Sci.* 33:281.

Hasebe, S., Kurooka, T., Aziz, B. B. A., Hashimoto, I., and Watanabe, T. (1996). Simultaneous separation of light and heavy impurities by a complex batch distillation column. *J. Chem. Eng. Japan* 29:1000.

Hasebe, S., Noda, M., and Hashimoto, I. (1997). Optimal operation policy for multi-effect batch distillation system. *Comput. Chem. Eng.* 21(S):S1221.

Lelkes, Z., Lang, P., Benadda B., and Moszkowicz, P. (1998). Feasibility of extractive distillation in a batch rectifier. *AIChE J.* 44(4):810.

Meski, G. A., and Morari, M. (1995). Design and operation of a batch distillation column with a middle vessel column. *Comput. Chem. Eng.* 19(S):S597.

Reinders, W., and De Minjer, C. H. (1940). Vapour-liquid equilibrium in ternary systems. *Recl. Trans. Chim.* 59:207.

Robinson, C. S., and Gilliland, E. R. (1950). *Elements of Fractional Distillation*, ed. 4, p. 388. McGraw-Hill Book Co., New York.

Safrit, B. T., Westerberg, A. W., Diwekar, U., and Wahnschafft, O. M. (1995). Extending continuous conventional and extractive feasibility insights to batch distillation. *Ind. Eng. Chem. Res.* 34(10):3257.

Safrit, B. T., and Westerberg, A. W. (1997a). Improved operational policies for batch extractive distillation columns. *Ind. Eng. Chem. Res.* 36(2):436.

Safrit, B. T., and Westerberg, A. W. (1997b). Algorithm for generating the distillation regions for azeotropic multicomponent mixtures. *Ind. Eng. Chem. Res.* 36(5):1827.

Safrit, B. T., and Westerberg, A. W. (1997c). Synthesis of azeotropic batch distillation separation systems. *Ind. Eng. Chem. Res.* 36(5):1841.

Skogestad, S., Wittgens, B., Litto, R., and Sørensen, E. (1997). Multivessel batch distillation. *AIChE J.* 43(4):971.

Van Dongen, D. B., and Doherty, M. F. (1985a). Design and synthesis of homogeneous azeotropic distillations. 1: Problem formulation for a single column. *Ind. Eng. Chem. Fundam.* 24:454.

Van Dongen, D. B., and Doherty, M. F. (1985b). On the dynamics of distillation process—VI: Batch distillation. *Chem. Eng. Sci.* 40(11):2087.

Wahnschafft, O. M., Koehler, J. W., Blass, E., and Westerberg, A. W. (1992). The product composition regions of single-feed azeotropic distillation columns. *Ind. Eng. Chem. Res.* 31(10):2345.

Wahnschafft, O. M., Le Rudulier, J. P., and Westerberg, A. W. (1993). A Problem decomposition approach for the synthesis of complex separation process with recycles. *Ind. Eng. Chem. Res.* 32(6):1121.

Wigdado, S., and Seider, W. D. (1996). Azeotropic distillation. *AIChE J.* 42(1):96.

Wittgens, B., Litto, R. Sørensen, E., and Skogestad, S. (1996). Total reflux operation of multivessel batch distillation. *Comput. Chem. Eng.* 20(S):S1041.

CHAPTER 10

Using Heuristics to Systematically Identify Process Flow Diagram Alternatives in the Context of Pollution Prevention

David W. Pennington

ORISE Research Fellow, Systems Analysis Branch, Sustainable Technology Division, National Risk Management Research Laboratory, U.S. Environmental Protection Agency, Cincinnati, OH 45268

Abstract

The identification and comparison of design modifications are not typically practiced consistently or routinely, particularly in the context of pollution prevention. A structured heuristic (rule of thumb) approach is demonstrated in this chapter that can help designers systematically identify pollution prevention opportunities. The heuristics were derived from basic chemical engineering principles and encoded into a prototype expert system, P2TCP. The reaction and separation systems of a process are first analyzed independently and then potential interactions are considered. Prior specification of a process flow diagram is not required, facilitating the identification of alternatives during conceptual and retrofit design. The key concepts of the approach are illustrated using a hypothetical example based on the production of allyl chloride.

INTRODUCTION

The synthesis or analysis of a process flow diagram is a highly skilled and creative task, typically involving expert knowledge to identify options and mathematical analysis techniques for subsequent evaluation. Once the design problem has been specified, the next step is the generation of alternatives. There are many different methods for generating chemical process

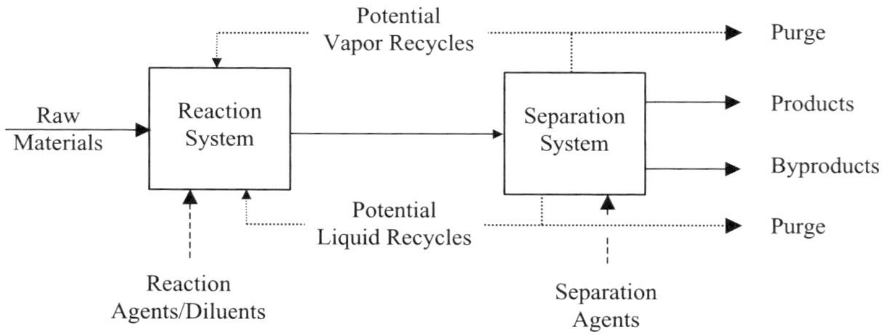

Figure 10.1: Potential streams reaction and separation systems.

alternatives (Cano-Ruiz and McRae, 1998). As time constraints exist, tools that help generate and evaluate such alternatives are extremely beneficial. In this chapter we focus on the generation of process design alternatives for continuous chemical processes (Fig. 10.1) using a structured approach based on heuristics derived from basic chemical engineering principles.

As the task of synthesis or analysis is knowledge intensive (heavily dependent on experience and qualitative information that is often uncertain, imprecise, and incomplete), this makes the use of knowledge-based expert systems that rely heavily on heuristics attractive aids for designers. A heuristic is a rule of thumb, simplification, or educated guess that reduces or limits the search for solutions in domains that are difficult and poorly understood. Unlike algorithms, heuristics do not guarantee optimal (or even feasible) solutions and are commonly used with no theoretical guarantee.

A chemical process design can be divided into a number of generic sections to provide a structured basis for the identification of pollution prevention opportunities. For example, detailed design theory can help to determine pollution prevention opportunities at different levels of the "onion model" in Figure 10.2 (Smith and Petela, 1991a, 1991b, 1992). Nevertheless, the use of this theory and available simulation tools can involve a significant number of skilled man-hours, and there is no guarantee that a systematic approach will be adopted within each ring of the onion or that interactions between the rings will be properly addressed in a structured manner. Using software tools can help reduce these limitations.

Douglas (1985, 1988, 1992) proposed the hierarchical procedure (Table 10.1) and associated synthesis heuristics that can help identify pollution prevention opportunities in a structured manner. In addition to its use in conceptual design, several retrofit studies have been performed using this approach (Mizsey and Fonyo, 1990; Fonyo et al., 1994; Rossiter et al., 1993).

Using the hierarchical procedure and the economic-oriented heuristics presented by Douglas (1985, 1988, 1992) a base case Process Flow Diagram (PFD) can be rapidly synthesized. The minimum design knowledge required is embodied in these heuristics, and hence the base case is not intended

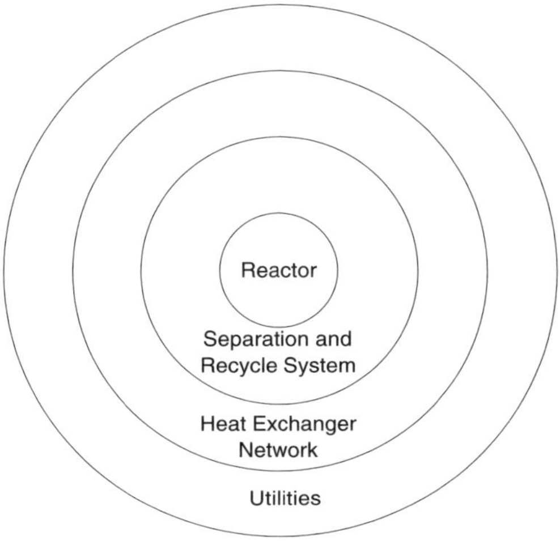

Figure 10.2: The onion model (Smith and Petela, 1991a).

to be a final or detailed design. Importantly, a number of decisions will be made in the hierarchy that can be reconsidered to identify alternative process options. The scope of these alternatives relates to the depth of design knowledge considered and the degree to which iterations of the hierarchical procedure are performed.

Limitations of the heuristics presented by Douglas (1985) may include their economic orientation (although maximizing economic performance and minimizing pollution are often covariant) and the minimalist level of design knowledge represented. As the recommendations can be fairly

TABLE 10.1
Hierarchical design decision levels (Douglas, 1988)

Hierarchy level
1. Input information
2. Input-output structure of the flow sheet
3. Recycle structure of the flow sheet
4. General structure: phase splits (a) Vapor recovery system (b) Liquid recovery system (c) Solid recovery system
5. Heat exchanger network

general, the use of these heuristics and the hierarchical procedure to identify P2 opportunities can be considered similar to using guideword or brainstorming techniques, albeit more structured. A more in-depth analysis might be achieved by using detailed design procedures in conjunction with such a structured hierarchy, analogous to a combination of the work of Smith and Petela (1991a, 1991b, 1992) and Douglas (1985, 1988, 1992). Advantages over the use of "simplistic" design heuristics can include (Pennington, 1997)

- Representation of reaction kinetics and thermodynamics to help identify effects of changes in operating conditions, heat mode, and reactor type
- Consideration of other separation unit operations in addition to phase splitting and distillation
- Identification of trade-offs between energy consumption and waste generation

In the hierarchical procedure, process systems or the rings of the onion are sequentially considered. For example, the reaction system is commonly specified before the consideration of separation techniques (Mizsey and Fonyo, 1990; Smith and Linnhoff, 1988). Process conditions are often frozen to optimize selectivity and conversion in the reaction system to minimize process development time (Ingleby et al., 1986). As the systems interact, this traditional iterative approach may not result in the development of an optimum solution in terms of overall process waste generation and energy consumption. Therefore it may not be appropriate to select among alternatives until all layers of the onion have first been considered (Mizsey and Fonyo, 1990; Smith and Linnhoff, 1988; Rossiter and Ingleby, 1986).

Process flow diagram alternatives can first be identified independently for the reaction and separation systems using basic chemical engineering principles (Pennington, 1999; Pennington and Yue, 2000). Interactions between the reaction and separation systems can then be evaluated to identify a shorter list of potential process flow diagram options. This adaptation of the hierarchical procedure and the use of detailed design theory to help determine pollution prevention opportunities are demonstrated in this chapter. The approach is illustrated here for a hypothetical example based on the production of allyl chloride and was represented in a computer-based prototype expert system, P2TCP (Pennington, 1997).

More detailed design theory (e.g., the consideration of mixture properties to refine the identification of potential separation options) and the use of mathematics-oriented tools (e.g., design simulation, heat integration, mathematical programming approaches to design synthesis) can help to further screen preliminary flow diagram options initially identified using heuristics. Similarly, taking into account different chemical reaction paths, alternative separation and reaction agents, the waste treatment infrastructure, life-cycle analysis, metrics to environmentally compare alternatives (the subject of chapter 3) and fugitive emissions can all augment the process of identifying pollution prevention opportunities (Cano-Ruiz and McRae, 1998). The consideration and illustration of all of these valuable issues are obviously beyond the scope of this chapter.

REACTION SYSTEM ANALYSIS

BACKGROUND

The theory of reactor design is well established (e.g., Dudukovic, 1995; Levenspiel, 1972; Folger, 1992; Hopper et al., 1994; Smith, 1995) and can be used to help identify pollution prevention (P2) opportunities (e.g., Smith and Petela, 1991b; Douglas, 1988; Smith, 1995; I. Chem. E., 1995). P2 may be achieved by using a different reaction path (not considered here) or by changes to the system design for a given reaction path. Nevertheless, despite the availability of information, performing a systematic analysis of a reaction system to identify potential modifications and conflicts in terms of each operating variable often remains limited by the resources required (time, money, in-house knowledge). Opportunities may therefore be overlooked.

Using software tools can reduce the resources required to design reaction systems. As with most synthesis tools, operating conditions are determined for a predefined goal, usually maximum selectivity or conversion. Hence, the identified design and operating conditions may represent the optimum for the reaction system, but this may not correspond with the optimum in terms of overall process waste generation and energy consumption. To identify the overall process optimum, interactions with the separation system usually need to be considered. To facilitate consideration of these interactions, tools are required to help determine the effects of changes within reaction systems on the exit stream and energy requirements in a broad, qualitative manner.

Changes in operating variables and reactor type can result in significant reductions in waste generation and energy consumption (Smith, 1995; I. Chem. E., 1995). Selectivity and the extent of conversion of raw materials are typically influenced by concentration, temperature, and pressure. This requires an ability to describe reaction schemes in terms of parallel and series interactions (e.g., Levenspiel, 1972; Smith, 1995). Other specifications include reaction rate orders, activation energies, heats of reaction, and specific heat capacities. In heterogeneous systems, rates can be difficult to analyze; however, if the reaction step is rate limiting, then effects of concentration and temperature will be associated with the reaction only. Reaction is often the rate-limiting step, particularly in low-temperature systems (Smith, 1995).

A number of reaction systems require the use of a catalyst. Changes in catalyst, and hence relative reaction rates, can be an effective method to reduce waste generation and energy consumption; however the selection of a catalyst is typically empirical and case specific (Smith, 1995), and hence not considered in this chapter.

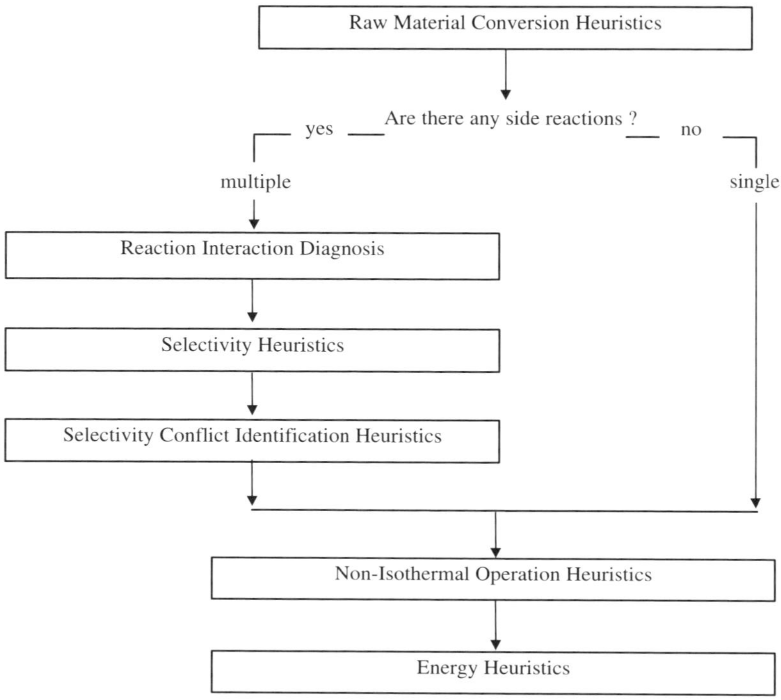

Figure 10.3: Reaction system analysis procedure (Pennington, 1997, 1999).

APPROACH

A structured approach was devised (Fig. 10.3) based on qualitative design theory, represented in the form of heuristics, to help identify the effects of changes within reaction systems on the exit stream and energy requirements (Pennington, 1997, 1999). This structured knowledge base is applicable for homogeneous reaction systems, and the results facilitate the consideration of interactions with the separation system to help determine an overall optimum process design. An initial core of knowledge is represented that may have some limitations in terms of general applicability and scope. For example, decreasing the pressure in a gas-phase reaction may not always result in a reduction in unrecovered raw materials. Research is therefore required to extend the approach to nonhomogeneous systems and to help identify opportunities for novel techniques like reactive distillation.

CASE STUDY

The identification of process design alternatives using heuristics is illustrated using a hypothetical example based on the production of allyl chloride

(Pennington, 1997). Allyl chloride (AC) is a bulk chemical produced globally by the vapor phase chlorination of propylene (Kirk-Othmer Encyclopaedia of Chemical Technology, 1991):

$$\underset{\text{propylene}}{CH_2{=}CH{-}CH_3} + Cl_2 \rightarrow \underset{\text{allyl chloride}}{CH_2{=}CH{-}CH_2Cl} + HCl.$$

Three key byproducts are generated. Hydrochloric acid (HCl) is a byproduct generated in the main reaction, 1,2-dichloropropane (1,2-DCP) in a parallel side reaction and 1,3-dichloropropene (1,3-DCP) in a combined side reaction:

$$\underset{\text{propylene}}{CH_2{=}CH{-}CH_3} + Cl_2 \rightarrow \underset{\text{1,2-dichloropropane}}{CH_2Cl{-}CHCl{-}CH_3},$$

$$\underset{\text{allyle chloride}}{CH_2{=}CH{-}CH_2Cl} + Cl_2 \rightarrow \underset{\text{1,3-dichloropropene}}{CHCl{=}CH{-}CH_2Cl} + HCl.$$

All of the reactions in the allyl chloride process are irreversible, exothermic, and first order in terms of each reactant.

The reaction system of the allyl chloride process was analyzed using a structured set of heuristics (Pennington, 1997, 1999) and modeled for confirmation using the simulation package ASPEN PLUS (Aspen Tech., Inc. USA). The heuristics and the model results helped identify the effects of changes on selectivity, raw material conversion, and energy generation. A summary is presented in Tables 10.2 and 10.3. Parameters used in the full analysis are presented in Tables 10.4 and 10.5. Rate constants are expressed in the Arrhenius form. Specific heat capacity data were predicted using the ASPEN PLUS process simulation package and are based on the geometric average between 298 K and 823 K, the range of potential operating conditions.

TABLE 10.2
Comparison of simulation model and P2TCP predictions for affects on the isothermal reaction system energy (Pennington, 1997)

Increased variables	*Inlet heater duty*		*Direct heat exchange*		*Heat in outlet stream*	
	Model	*P2TCP**	*Model*	*P2TCP**	*Model*	*P2TCP**
Inlet temperature	↑	—	↑	—		
Conversion			↑	**3.72a, 3.73a**	↑	**3.75a**
Selectivity			↓	**3.79b**	↑	**3.711c**
Reactor temperature					↑	**3.713a**

*Heuristic number (see Table 10.3).

TABLE 10.3
Pertinent prototype energy requirement heuristics for reaction system (Pennington, 1997, 1999)

No.	*If*	*Then*
3.72a	Exothermic main reaction and isothermal	Increase in conversion increases reactor energy output
3.73a	Exothermic side reaction and isothermal	Decrease in conversion decreases reactor energy output
3.75a	$(Cp'' - Cp') > 0$	Increase in conversion increases exit stream heat content
3.79	Isothermal, exothermic main reaction and exothermic side reaction	(a) If $\Delta H_{r,\text{main}} < \Delta H_{r,\text{side}}$, then increased selectivity results in increased generation or decreased utilization
		(b) If $\Delta H_{r,\text{main}} > \Delta H_{r,\text{side}}$, then increased selectivity results in increased generation or decreased utilization
3.711a	Main $(Cp'' - Cp') > 0$ and side < 0	Increased selectivity results in an increase in exit stream heat content
3.711b	Main $(Cp'' - Cp') < 0$ and side > 0	Increased selectivity results in a decrease in exit stream heat content
3.711c	Not 3.711a or 3.711b	Consider conflict of effects related to heat capacity on exit stream heat content
3.713a	$(Cp'' - Cp') > 0$ and change reactor temperature	Increase in temperature will increase the exit stream heat content relative to reactor heat duty

Cp - Combined stoichiometric heat capacity relative to the stoichiometric coefficient of a reactant; Cp″ - products; Cp′ - reactants.

TABLE 10.4
Allyl chloride process reaction parameters (Sawyer and Martell, 1992)

Equation no.	*Frequency factor (A) (kmol/s-m^3-MPa^2)*	*Activation energy (E) (kJ/kmol)*	*Heat of reaction (ΔH_{ri}) (kJ/kmol @ 298K)*
6.19	89 324	63 270	−111 653
6.20	5.07	15 957	−184 228
6.21	1.2E+08	24 773	−213 537

TABLE 10.5
Heat capacity constants

Chemical	*Heat capacity (kJ/mol-K)*
Propylene	98 093
Chlorine	35 795
Allyl chloride	125 250
1,2-DCP	158 200
1,3-DCP	137 900
HCl	44 664

The effects identified for isothermal systems using the two tools correspond. Changes in inlet temperature were not considered for isothermal systems, as there is no overall change in energy. It was assumed in the simulation study, and verified using the heuristics, that the effects of changes to the reaction system on the heat transferred in isothermal systems are analogous to changes in the exit stream heat content for nonisothermal systems. Unlike isothermal systems, an increase in the inlet temperature can increase the exit stream heat content. It should be noted that the exit stream heat content can vary for both types of system due to changes in heat capacity associated with composition. Associated effects are analogous for isothermal and nonisothermal operation.

For nonisothermal operation, heat transfer from the reactor was assumed to be constant. Only the heat content of the exit stream was considered to be variable. If the reactor duty is not constant, then a combination of the changes identified by the heuristics for both isothermal and nonisothermal operation would be observed in simulations.

SEPARATION SYSTEM ANALYSIS

BACKGROUND

A number of approaches are presented in the literature that can be used to help identify alternatives for the separation system of a process in the context of pollution prevention (e.g., Jaksland, 1996; Jaksland et al., 1995; Barnicki and Fair, 1990, 1992; Barnicki, 1991; Douglas, 1995; Huang and Fan, 1990). Considerations include (the last four are considered in this chapter)

- Optimization of existing separation unit operations (e.g., Smith and Petela, 1992; Douglas, 1988; Smith, 1995; I. Chem. E., 1995)
- Use of different separation techniques for a given split
- Use of additional separation

- Modification of the composition and concentrations in the stream, leaving the reaction system to enable the use of different separation techniques
- Optimization to account for other interactions between the reaction and separation systems, including changes associated with recycled components

The identification of potential separation techniques and sequences is based heavily on design experience (Pennington, 1997; Pennington and Yue, 2000; Smith, 1995; Jaksland, 1996; Jaksland et al., 1995; Barnicki and Fair, 1990, 1992; Barnicki, 1991; Douglas, 1995; El-Halwagi and Petrides, 1994; McCabe et al., 1993; Rousseau, 1987; King, 1980). To completely separate a mixture with 6 components with a choice of 10 kinds of unit operation for each binary split, there are 4,200,000 potential separation combinations (Barnicki and Fair, 1990). Not all of these combinations will be feasible, and it is therefore desirable to select a subset using preliminary screening heuristics.

The separation of a mixture is typically achieved by exploiting differences in physical or chemical properties. Hence, potential techniques for a given split are often identified on the basis of these differences (Jaksland, 1996; Barnicki, 1991). This knowledge is often an oversimplification with limited validation but is commonly used for preliminary screening to help identify potentially feasible alternatives.

Methodologies for sequencing splits based on distillation and given input stream composition data are well established. A sequence may be determined by sequentially selecting splits using heuristics until all required separations have been considered or by using available quantitative approaches in which all alternatives are taken into account and ranked. A quantitative methodology for the determination of distillation sequences based on minimizing total vapor load demonstrated more consistent results than the step-wise (hierarchical), single-pass approaches based on heuristics in an economic context (Porter and Momoh, 1991). Established heuristic and mathematical approaches may not, however, reflect the optimum in terms of minimum waste generation and energy consumption. Similar limitations are associated with heuristic-based methodologies for the synthesis of sequences in which nondistillation separation techniques are also taken into account (Jaksland, 1996; Barnicki, 1991).

APPROACH

A prototype screening level methodology was developed to help designers determine potential options for the separation system without having to predefine the reaction system (Fig. 10.4) (Pennington, 1997; Pennington and Yue, 2000). After comparison with the reaction system insights and the consideration of potential interactions, the number of alternatives can be further reduced and then simulated.

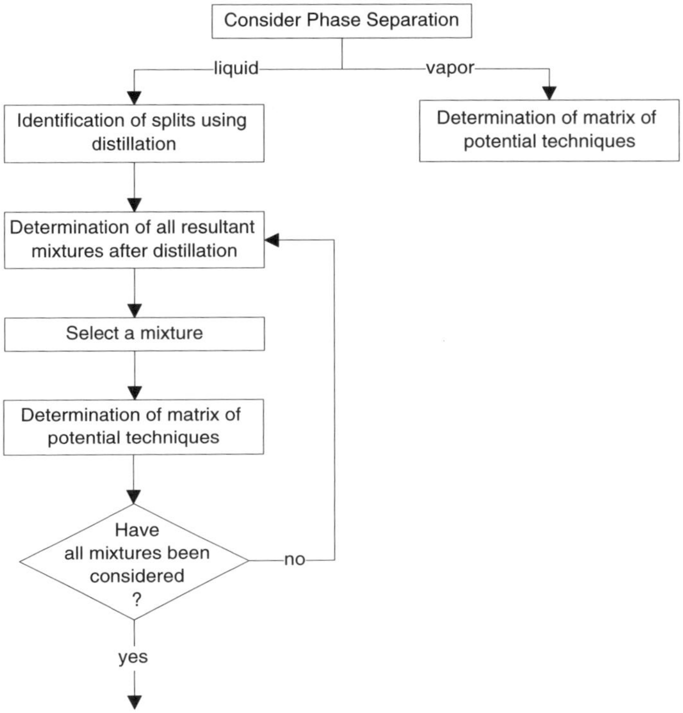

Figure 10.4: Overview of proposed methodology for the identification of separation technique options (Pennington and Yue, 1999).

Unlike typical approaches, the separation alternatives are not constrained to a case-specific mixture definition:

- All components that may be in the stream leaving the reaction system of a process are considered (e.g., the use of a reactant in excess to ensure the complete conversion of another is not preassumed)
- Case-specific mass fractions in the stream are not assumed

All potential splits that could be required are addressed. The splits required to separate a mixture are dependent on the predefined destination of each component (product, coproduct, waste, recycle). This information helps reduce the number of alternatives. Required separations that can be performed initially using preferred techniques are determined to further reduce the number of alternatives requiring consideration.

Whenever practical, the stream leaving a reaction system is typically split into two phases (Douglas, 1988; Barnicki and Fair, 1990; Barnicki, 1991).

TABLE 10.6
Component destination after phase splitting based on boiling point (Pennington and Yue, 1999)

Component destination	*Normal boiling point range (°C)*
Vapor stream	<-20
Vapor-liquid distributed	$-20 <$ Boiling Point < 50
Liquid stream	>50

Phase splitting the initial mixture is therefore considered a preferred technique, although this default assumption requires further evaluation in the context of pollution prevention. Separation alternatives are then identified independently for the resultant liquid and vapor mixtures.

Based partially on Barnicki and Fair (1990) Barnicki (1991), normal boiling point differences are used to determine the suitability of phase splitting and resultant mixtures after phase splitting (Table 10.6) (Pennington and Yue, 2000). Using a vacuum or operating at higher pressures may increase relative volatility differences but is not addressed, as phase splitting may no longer be preferential. As an initial default, distributed components are included in the liquid phase mixture only and a sharp split is assumed. If a sharp split is not practical or if it is more desirable to include a component in the vapor stream based on destination, the distributed components can be considered in the vapor mixture. If phase separation of a mixture results in a liquid stream with a low flow rate relative to that of the vapor (<5% mole), then this phase splitting will not typically be favored (Barnicki and Fair, 1990; Barnicki, 1991). The mixture is treated as a vapor only under these circumstances. Similarly, if a phase split would result in a trace vapor stream, then the mixture is considered as a liquid only. As no concentrations are assumed, designers are required to be aware of these situations.

No preferential techniques exist for the separation of vapor mixtures (Barnicki and Fair, 1990, 1992). For the separation of liquid mixtures, simple distillation is typically favored (Douglas, 1988; Barnicki and Fair, 1990; Barnicki, 1991). From a pollution prevention perspective, no mass separation agent is required in simple distillation, and heat streams can usually be integrated. Liquid splits that can be performed using simple distillation initially and the resultant mixtures are therefore identified as default preferential techniques. As the optimal sequence is a function of mixture composition, concentrations, and heat integration, the selection of a sequence is not performed at this screening stage.

Distillation is considered the preferred technique for a split if separation can be achieved with a low reflux and a low number of stages (Smith, 1995). These conditions may be met if (Barnicki and Fair, 1990; Barnicki, 1991)

1. The relative volatility (α) of two adjacent boiling components is greater than 1.5

2. The split does not involve a dilute separation (i.e., it does not have a distillate or bottoms product less than approximately 15% mole of the feed).

Distillation may still be considered feasible if α is greater than 1.1; however, other techniques may be competitive (Douglas, 1988; Smith, 1995). Relative volatility is calculated using Raoult's Law and the Antoine equation (Pennington, 1997; Pennington and Yue, 2000). This represents separation of a binary mixture into pure components at atmospheric pressure.

After identifying initial splits that can be performed using preferred techniques, a matrix of alternatives is determined for each remaining liquid mixture, requiring that separation and potential sequences are identified. A similar procedure is adopted for separations required in the vapor stream. Potentially feasible techniques are presented in a matrix for each component pair in each mixture. Only pairs that have different destinations (individual products, waste, recycle) are considered. Jaksland (1996), Jaksland et al. (1995), Smith (1995), Barnicki and Fair (1990), and Barnicki (1991) have presented heuristics appropriate for this identification task (Pennington, 1997, 1999). These screening heuristics do not require the consideration of mixture composition or specific mass separating agents; however, as the heuristics are typically based on pure component or structural differences, not all separation techniques are addressed and their validation remains limited. A summary of the heuristics adopted for the purposes of illustration in this chapter is presented in Tables 10.7 and 10.8.

CASE STUDY

A typical design for the separation system of the allyl chloride process is presented in Figure 10.5.

A thermal split is used to separate the stream leaving the reaction system. The allyl chloride (AC) product is subsequently separated from the byproducts (1,2-DCP and 1,3-DCP) using simple distillation. An excess of propylene is used to ensure the complete conversion of chlorine in the reaction system. Hence, the vapor stream after the phase split contains predominantly propylene and the coproduct, HCl gas. The HCl gas is absorbed from the vapor stream using water to generate a dilute acid coproduct. Remaining traces of HCl are removed using a sodium hydroxide solution, which results in the generation of an aqueous waste. Before recycling the propylene, it is essential to remove all traces of water. Traces of water, introduced by the separation techniques used, are removed using compression and adsorption.

An analysis was performed using the proposed procedure and heuristics to help determine alternative options for the separation system and to consider potential interactions with the reaction system (Pennington, 1997; Pennington and Yue, 2000). All components that could be in the exit stream

TABLE 10.7
Heuristics for the identification of potential techniques for component pairs in liquid mixtures (Pennington, 1999)

Property	*Identification criteria*	*Potential techniques*	*Special consideration notes*
Relative volatility*	>1.1	Distillation	Bulk separations
Freezing point (K)	Difference >20	Crystallization	If freezing point is less than 288 K, then refrigeration may be required.
Kinetic diameter or van der Waals volume (Å)†	Ratio >1.1	Liquid-liquid extraction	Additional separating agents and regeneration required. Microporous membranes and adsorption are limited for bulk separations.
		Stripping	
		Microporous membranes	
		Adsorption	
Dipole moment (C/m)	Ratio >1.1	Liquid-liquid extraction	Additional separating agents and regeneration required. Adsorption limited for bulk separations.
	Difference‡ >0.1	Adsorption	
		Stripping	

*Relative volatility is determined at each pure component boiling point and the maximum used.
†Kinetic diameter is a combined measure of cross-sectional area and shape characteristics of a molecule (Barnicki and Fair, 1990). Data may be limited, in which case van der Waals volume may be used.
‡If dipole moment of a component is zero, then the criteria are based on the property difference.

TABLE 10.8
Heuristics for the identification of potential techniques for component pairs in vapor mixtures (Pennington, 1999)

Property	*Identification criteria*	*Potential techniques*	*Special consideration notes*
Functional group	Difference*	Chemical absorption	May give a sharp split. Additional separating agents and regeneration are required.
Relative volatility†	>2	Cryogenic, high pressure, and vacuum distillation	May give a sharp split. Used for bulk separations.
Kinetic diameter or van der Waals volume (Å)‡	Ratio >1.1	Molecular sieve adsorption	May give a sharp split. Dilute separation. Additional separating agents and regeneration required.
Relative volatility†	>7	Condensation	May not provide sharp split.
Boiling point (K)	Difference >40		
Chemical family	Organics from an inert gas (air, helium, etc.)	Equilibrium-limited adsorption	Used for separation of dilute components, except pressure swing adsorption. May give sharp splits. Bed regeneration required.
Chemical families	Gases or compounds which reflect current industrial practice	Equilibrium-limited adsorption	As above.

*Different functional groups that may result in chemical complexing with a bond energy between 10 and 60 kJ/mol. A list was presented by (Barnicki and Fair, 1992).
†Relative volatility is determined at each pure component boiling point and the maximum used.
‡Kinetic diameter is a combined measure of cross-sectional area and shape characteristics of a molecule (Barnicki and Fair, 1992). Data may be limited, in which case van der Waals volume may be used.

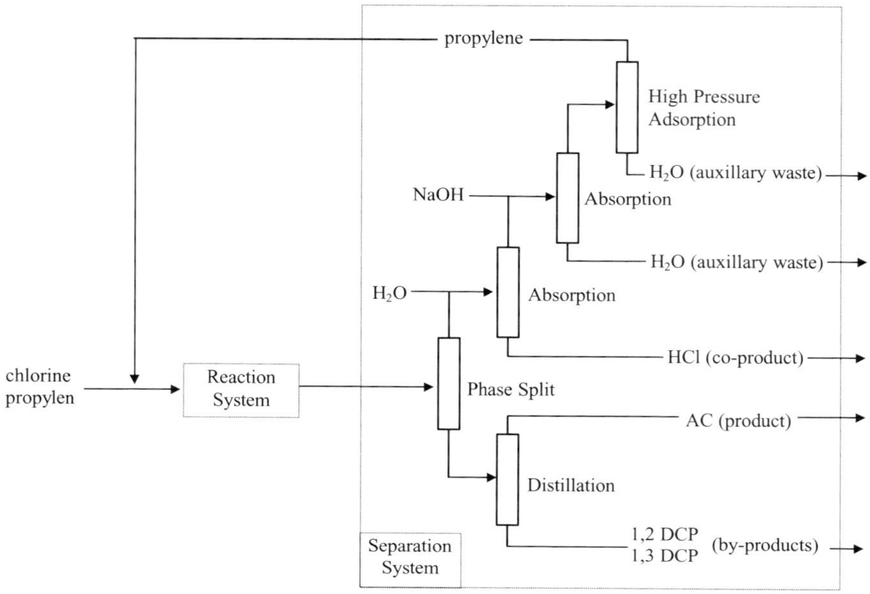

Figure 10.5: Allyl chloride process flowsheet.

of the reaction system were considered and no concentrations were assumed. The physical property and chemical structure data used to determine the feasibility of thermal separation processes are presented in Table 10.9. The Antoine coefficients in Table 10.9 and the physical property data in Table 10.10 helped identify potential separation alternatives after phase splitting.

Based on the heuristics, the stream leaving the reaction system of the allyl chloride process can be separated into two phases, as in the existing process. This is expected to be a bulk separation for all reasonable process options. Therefore, the use of a phase split for the initial mixture was confirmed. Allyl chloride, the product, was determined to be a distributed component. As an initial assumption that is readdressed later, allyl chloride was first considered to separate into the liquid stream.

It is necessary to separate AC in the liquid mixture from the byproducts (1,2-DCP and 1,3-DCP). It was identified using the heuristics that this separation may be readily achieved in a single step using distillation, reflecting typical industrial practice; however if a high selectivity can be achieved in the reaction system, then this bulk separation technique may not be practical. For this situation, it was determined that all other types of separation techniques considered (crystallization, liquid-liquid extraction, stripping, microporous membranes, adsorption, and liquid-liquid

TABLE 10.9
Physical property data to help determine the feasibility of thermal separation techniques

Component	Boiling point* (K)	Antoine coefficients† (temperature – K, pressure – Pa)		
		A	*B*	*C*
AC	318	21.235	2838.133	-27.888
1,2-DCP	369	21.55	3386.804	-32.081
1,3-DCP	366	21.429	3437.146	-29.725
HCl	188	21.831	1882.878	-5.879
Propylene	225	21.196	2037.812	-15.634
Chlorine	239	21.127	2084.631	-22.165

*AIChE (A. I. Chem. E., 1995).
†Regression based on data from ASPEN PLUS for 180–400 K.

extraction) are potential alternatives that could be used to separate off trace contaminants.

Separation of the raw materials (propylene and chlorine) from the co-product HCl in the vapor stream may be required, depending on the reaction system specifications. Again all five categories of techniques considered using the heuristics (chemical absorption, cryogenic, high pressure and vacuum distillation, molecular sieve adsorption, equilibrium-limited adsorption, and condensation) for both possible splits were identified as possibilities.

TABLE 10.10
Physical property data for the identification of liquid and vapor separation techniques

Component	Freezing point* (K)	van der Waals volume* (m^3/kmol)	Dipole moment* ($x10^{-30}$ C/m)	Acid/base absorption group†	Chemically adsorbed component†
AC	138	0.042	6.5	—	—
1,2-DCP	173	0.055	4.87	—	—
1,3-DCP	191	0.051	5.8	—	—
HCl	159	0.016	—	yes	yes
Propylene	88	0.034	—	—	yes
Chlorine	172	0.024	—	—	yes

*AIChE (A. I. Chem. E., 1995).
†Chemicals identified by Barnicki and Fair (1992) and Barnicki (1991) that may be adsorbed or absorbed.

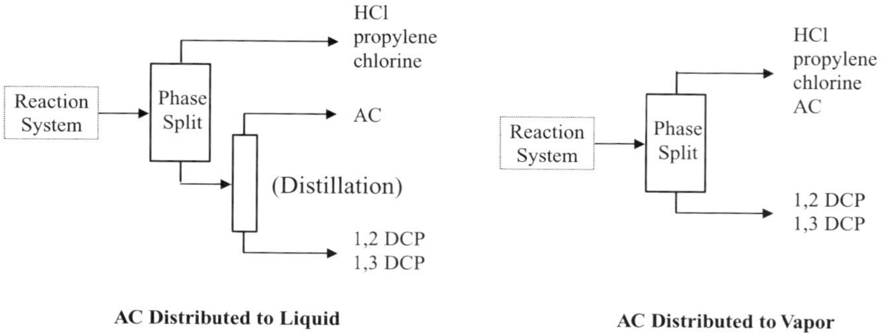

Figure 10.6: Potential alternatives for separation after phase splitting.

Allyl chloride was considered earlier by default to be only in the liquid stream after phase splitting. If this default decision is rejected, then the resultant vapor stream would consist of allyl chloride, chlorine, propylene, and HCl. If this alternative phase split is sharp, then no further separation of the liquid stream would be required. Splits between three further pairs of components in the vapor stream would need to be considered. Again all categories of techniques taken into account using the heuristics were determined for these splits to be potential options.

Based on the heuristics and a list of components ranked by boiling point data, it was demonstrated that allyl chloride may be readily separated from the vapor stream in a single step using distillation. This will always be a bulk separation, as HCl and allyl chloride are both main products. It was therefore assumed that other techniques or sequences would not be competitive for the full range of reasonable process operating conditions. Hence, two likely alternatives that correspond to the different allyl chloride destination options after phase splitting are identified (Fig. 10.6). Separation alternatives for the remaining tertiary vapor mixture after distillation are identical irrespective of the allyl chloride destination.

An excess of propylene is typically used to ensure complete conversion of chlorine (considered in the next section). After phase splitting, the hypothetical separation of propylene and the coproduct, HCl, would result in the generation of an aqueous waste (as in Fig. 10.5). Absorption, used to remove HCl, was among the options identified by the heuristics; however, water, the extraneous absorbent introduced, was not considered in the analysis presented. Hence, the important additional step of adsorption that is used to ensure the removal of this water was not identified.

Using an alternative for the typical unit operations in Figure 10.5 may reduce energy consumption and waste generation. Distillation and condensation, which are bulk separation techniques, were identified as potential options. At lower flow rates of propylene, equilibrium adsorption could be more promising. This technique has been demonstrated in industry for the similar separation of benzene from HCl; however, commercial reactors are adiabatic and propylene is used as a diluent in the reaction system. Therefore a lower flow rate may not be feasible unless an alternate diluent is identified (see subsequent discussion).

INTERACTIONS BETWEEN THE REACTION AND SEPARATION SYSTEMS

APPROACH

The reaction and separation systems of a process can be analyzed independently, as demonstrated. Interactions between the two systems can then be taken into account to identify a shorter list of potential flow diagram options (Pennington, 1997, 1999). The interactions considered using the heuristics include

- Effects on the separation system of the composition and concentrations in the stream leaving the reaction system.
- Effects of using recycles, changes in their flow rates, and possible contaminants on the reaction and separation systems.

CASE STUDY

The effects of interactions between the separation and reaction systems of the allyl chloride process were considered using heuristics in terms of

- The use of raw materials in excess
- The decision to recycle
- The extent of conversion
- Selectivity
- Separation efficiency

The decision to recycle a component, the extent of conversion of each raw material, and selectivity all affect flow rates through the separation system and which techniques are then appropriate. Associated changes may reduce waste generation and energy consumption, as illustrated for the recycle of propylene in the previous section.

Commercial reactors usually are adiabatic. An excess of propylene is typically used as a diluent because it has a higher heat capacity than chlorine, it ensures complete conversion of chlorine, which is a high concern from a safety perspective, and it reduces the byproduct 1,3-DCP; however, the vapor separation system used to recover propylene from HCl has a high energy consumption, results in the generation of aqueous waste, and introduces a potential safety hazard (water) into the process (Fig. 10.5). As suggested earlier, an alternate diluent may facilitate the use of a lower flow rate of propylene and a different separation system. As one of the alternatives, it was determined that HCl contamination of the propylene recycle would act as an inert; hence a diluent in the reaction system (Pennington, 1997). The partial recycle of HCl with propylene may therefore help reduce the waste generation and energy consumption of the process design considered by

- Reducing the propylene flow rate through the recycle loop
- Enabling the use of alternative separation techniques with less stringent separation efficiencies

SUMMARY

The prototype of a structured heuristic-based approach is presented in this chapter that can help designers of continuous chemical processes systematically identify process flow diagram alternatives to inherently reduce waste generation and energy consumption. Advantages of using such an approach include

- The analysis performed is not resource intensive and does not require the prior specification or simulation of a process flow diagram. It can therefore help identify pollution prevention opportunities during conceptual as well as retrofit design.
- The heuristics can be derived systematically using fundamental chemical engineering principles. This approach can therefore provide a greater depth of analysis and more structure than keyword approaches or other similar brainstorming techniques.
- Each system (reaction and separation) of a process can be analyzed independently, not sequentially, and then potential interactions are considered to help ensure that all options are identified within the search space and scope of the heuristics used.

As with any tools, the results of expert systems should not be accepted blindly, and there is room for considerable further development. As well as efforts to validate and extend the current core of knowledge used, focus on the development of heuristics to identify opportunities for the use of advanced technologies (e.g., combined reaction-separation techniques) is considered a high priority.

ACKNOWLEDGMENTS

The author gratefully acknowledges the advice from Dr. J. F. Porter, Dr. G. McKay, and Prof. P.L. Yue of the Chemical Engineering Department, Hong Kong University of Science and Technology.

REFERENCES

A. I. Chem., E. (1995). Physical and thermodynamic properties of pure chemicals: Data compilation. *Design Institute for Physical Property Data*. Taylor and Francis, New York.

Barnicki, S. D. (1991). Separation System Synthesis: A Knowledge-Based Approach. Ph.D. thesis, The University of Texas at Austin.

Barnicki, S. D., and Fair, J. R. (1990). Separation system synthesis: A knowledge-based approach: 1. Liquid mixture separations. *Ind. Eng. Chem. Res.* 29:421–432.

Barnicki, S. D., and Fair, J. R. (1992). Separation system synthesis: A knowledge-based approach: 2. Gas/vapor mixtures. *Ind. Eng. Chem. Res.* 31:1679–1693.

Cano-Ruiz, H., and McRae, G. J. (1998). Environmentally conscious chemical process design. *Annu. Rev. Energy Environ.* 23:499–536.

Douglas, J. M. (1985). A hierarchical decision procedure for process synthesis. *AIChE J.* 31:353–362.

Douglas, J. M. (1988). *Conceptual Design of Chemical Processes*. McGraw-Hill Publishing Company, Singapore.

Douglas, J. M. (1992). Process synthesis for waste minimization. *Ind. Eng. Chem. Res.* 31:238–243.

Douglas, J. M. (1995). Synthesis of separation system flow sheets. *AIChE J.* 41(12):2522–2536.

Dudukovic, M. P. (1995). Chemical reaction engineering: The environment and pollution prevention. *Hem. Ind.* 49:451–460.

El-Halwagi, M. M., and Petrides, D. P. (1994). Pollution prevention via process and product modifications. *AIChE Symposium Series*, Vol. 90, AIChE, New York.

Folger, H. S. (1992). *Elements of Chemical Reactor Engineering*, ed. 2. Prentice-Hall, Englewood Cliffs, NJ.

Fonyo, Z., Kurum, S., and Rippin, D. W. T. (1994). Process development for waste minimization: The retrofitting problem. *Comp. Chem. Eng.* 18:S591–S595.

Hopper, J. R., Yaws, C. L., Vichailak, M., and Ho, T. C. (1994). Pollution prevention by process modification: Reactions and separations. *Waste Management* 14(3–4):187–202,

Huang, Y. L., and Fan, L. T. (1990). An adaptive heuristic-based system for synthesis of complex separation sequences. *Artificial Intelligence in Process Engineering*, pp. 311–361. Academic Press, New York.

I. Chem., E. (1995). Environmental Training for the Process Industries. Safety Health and Environment Department. Institution of Chemical Engineers, UK.

Ingleby, S., Rossiter, A. P., and Douglas, J. M. (1986). Economic evaluation of integrated reactor systems. *Chem. Eng. Res. Des.* 64:241–248.

Jaksland, C. A., Gani, R., and Lien, K. M. (1995). Separation process design and synthesis based on thermodynamic insights. *Chem. Eng. Sci.*, 50:511–530.

Jaksland, C. (1996). Separation Process Design and Synthesis Based on Thermodynamic Insights. Ph.D. thesis, Dept. of Chemical Engineering. Technical University of Denmark.

King, C. J. (1980). *Separation Processes*. McGraw-Hill, New York.

Kirk-Othmer Encyclopaedia of Chemical Technology (1991). John Wiley & Sons, New York.

Levenspiel, O. (1972). *Chemical Reaction Engineering*, ed. 2. Wiley, New York.

McCabe, W. L., Smith, J. C., and Harriott, P. (1993). *Unit Operations of Chemical Engineering*. McGraw-Hill, New York.

Mizsey, P., and Fonyo, Z. (1990). Toward a more realistic overall process synthesis: The combined approach. *Comp. Chem. Eng.* 14(11):1213–1236.

Pennington, D. W. (1997). A pollution prevention tool for continuous chemical processes. Ph.D. thesis, UMI Dissertation Abstracts, Hong Kong University of Science & Technology.

Pennington, D. W. (1999). Chemical process design using heuristics in the context of pollution prevention. *J. Clean Products Processes* 1(1):170–179.

Pennington, D. W., and Yue, P. L. (2000). A screening-level prototype for the synthesis or analysis of separation systems to support identification of inherently cleaner chemical processes. *J. Clean Products Processes* 2(2):82–91.

Porter, K. E., and Momoh, S. O. (1991). Finding the optimum sequence of distillation columns: An equation to replace the "rules of thumb" (heuristics). *Chem. Eng. J.* 46:97–108.

Rossiter, A. P., and Ingleby, S. (1986). Integrated reactor systems: Design for variable operating conditions. *Chem. Eng. Res. Des.* 64:365–371.

Rossiter, A. P., Spriggs, H. D., and Klee, P. (1993). Apply process integration to waste minimization. *Chem. Eng. Prog.* 89:30–36.

Rousseau, R. W. (1987). *Handbook of Separation Process Technology*. Wiley, New York.

Sawyer, D. T., and Martell, A. E. (1992). *Industrial Environmental Chemistry*. Plenum Press. New York.

Smith, R., and Linnhoff, B. (1988). The design of separators in the context of overall processes. *Chem. Eng. Res. Des.* 66:195–228.

Smith, R., and Petela, E. (1991a). Waste minimisation in the process industries. Part 1: The problem. *Chem. Eng.* 31:24–25.

Smith, R., and Petela, E. (1991b). Waste minimisation in the process industries. Part 2: Reactors. *Chem. Eng.* 12:17–23.

Smith, R., and Petela, E. (1992). Waste minimisation in the process industries. Part 3: Separation and recycle systems. *Chem. Eng.* 13:24–28.

Smith, R. (1995). *Chemical Process Design*. McGraw-Hill, New York.

CHAPTER 11

Shortcut Approaches to the Recovery and Allocation of Multicomponent Volatile Organic Compounds

Gautham Parthasarathy and Mahmoud M. El-Halwagi

Department of Chemical Engineering, Auburn University, Auburn, AL 36849

Abstract

Volatile organic compounds (VOCs) are widely used in chemical process industries. In many cases, their recovery and recycle can lead to economic and environmental benefits. Whereas previous works have focused on recovering VOCs from gases, little attention has been paid to integrating recovery with potential reuse of the recovered VOCs. In many cases, the condensation of VOC-laden gases to meet air-pollution regulations should be accompanied by the design of an optimal allocation scheme that makes optimum usage of the recovered VOCs. This chapter presents shortcut approaches to the simultaneous design of condensation and allocation systems for VOCs. First, we review the literature in the area of designing VOC condensation networks. Next, we pose the problem of recovery and allocation along with its challenges. Then, we present solution strategies for special cases covering a wide range of scenarios for industrial VOC recovery and reuse. A case study is used to demonstrate the potential benefits of the approach.

INTRODUCTION

VOCs are widely used in chemical process industries as solvents, material utilities, surface-coating and cleaning agents, chemical reagents, and intermediate chemicals. There are various definitions of VOCs. Most definitions are based on relating their vapor pressure to temperature. For instance, a VOC may be defined as any hydrocarbon whose vapor pressure at 100°C is greater than that of water (101 kPa or 760 mm Hg) (Heinsohn and

Kabel, 1999). Alternatively, a VOC may be defined as any hydrocarbon whose vapor pressure at 25°C is greater than 0.014 kPa (11 mm Hg). Sometimes, certain compounds with low reactivity, such as methane, are excluded from these definitions, leading to the term nonmethane hydrocarbon, which is used synonymously with VOC. Most VOCs are classified as hazardous air pollutants (HAPs) and are subject to stringent clean-air act requirements, primarily because of their contributions to ozone depletion. In addition to their environmental impact, VOCs are typically valuable chemicals. Therefore, there is a strong need to develop a systematic framework to synthesize optimal VOC recovery and allocation systems that meet all environmental specifications while reconciling the economic factors of recovery and reuse.

VOC recovery techniques are broadly classified into those based on separations using mass separating agents (e.g., absorption, adsorption), membrane separations (e.g., pervaporation), or heat-induced separations (e.g., condensation). In this chapter we focus on the use of condensation using refrigerants to recover the multiple VOCs that are in the gaseous phase. Other VOC recovery techniques have been described (Ruddy and Carroll, 1993; Moretti and Mukhopadhyay, 1993; Pollack et al., 1996; Dunn and El-Halwagi, 1996; El-Halwagi, 1997). In the following section we review the design techniques for VOC condensation networks. Later we address the more general problem of simultaneous condensation and reuse of VOCs.

DESIGN OF VOC CONDENSATION NETWORKS

An important category in VOC-recovery systems deals with the separation of VOCs via condensation in Heat-Induced Separation Networks (HISENs) in which condensation is caused by the application of cooling (Dunn and El-Halwagi, 1994a, 1994b; El-Halwagi et al., 1995). A more general system is the Energy-Induced Separation Network (EISEN) that employs pressurization/depressurization in addition to cooling to induce separation (Dunn et al., 1995). There are systematic techniques for synthesizing and optimizing the design of these condensation networks (Dunn and El-Halwagi, 1994a, 1994b; El-Halwagi et al., 1995; Richburg and El-Halwagi, 1995).

First, let us state the design task for synthesizing a VOC condensation network (Dunn and El-Halwagi, 1994a, 1994b). Later we will generalize the problem statement to include the reuse of the condensed VOCs. Given are N_G VOC-laden gaseous streams, containing N_c VOCs, with known flow rates, G_i, where i is an index for the gaseous streams that range between 1 and N_G. For each stream i, the supply and target compositions of each component k (where k is an index for the VOCs and ranges between 1 and N_C) are given as $y^s_{i,k}$ and $y^t_{i,k}$, respectively. The target composition of each stream represents an upper bound on the actual outlet composition and is based on environmental, safety, or technical constraints. Hence, the outlet composition of the kth component in the ith stream is unknown and must satisfy the following constraint:

$$y^{out}_{i,k} \le y^t_{i,k}, \qquad \text{where } i = 1, 2, \ldots, N_G \quad \text{and} \quad k = 1, 2, \ldots, N_C. \quad (1)$$

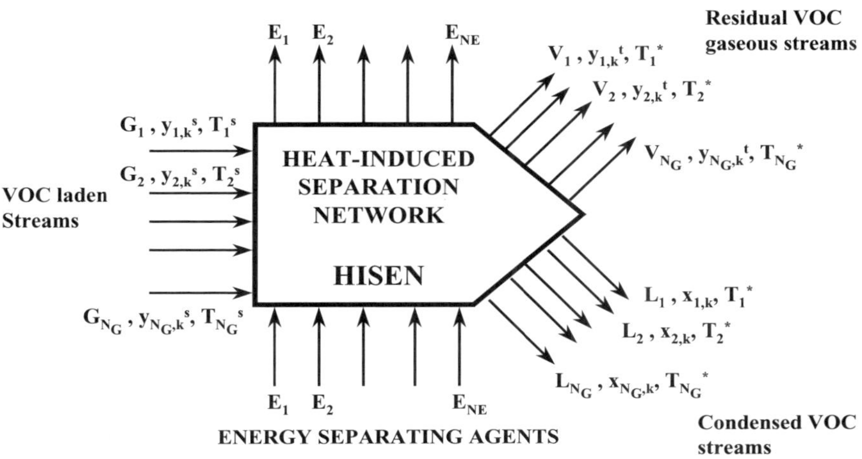

Figure 11.1: Problem of designing an HISEN for VOC condensation (El-Halwagi et al., 1995).

Available for service are several coolants/refrigerants that can be used as energy-separating agents (ESAs) for condensation. Also given are the supply temperature, T_i^S, and pressure, P_i^S, for each gaseous stream. The objective is to develop a network for condensation of the VOCs such that the gaseous streams satisfy the predefined constraints on them. The problem is represented in Figure 11.1.

Figure 11.2 shows a typical configuration of a VOC recovery system. These configurations involve up to three heat exchangers for dehumidification, heat integration, and external cooling, respectively. The initial step is to cool the stream to a temperature slightly above the freezing point of water (i.e., ~273 K). This results in dehumidification of the gas, which prevents detrimental icing effects in subsequent stages. Next, the stream is cooled to T^* to recover the VOCs. The temperature T^* is an optimization variable and determines the extent of VOC condensation. To utilize the cooling capacity of the gaseous stream at T^*, it is recycled back to the system for heat integration. The remaining cooling duty is accomplished by an external coolant/refrigerant.

The design of an HISEN can be carried out using a graphical pinch-based representation of the problem. The essence of this approach is to first transform the VOC-recovery task into a problem of cooling the gas to a temperature T^*. This temperature is determined by using vapor-liquid equilibrium data. Once the separation problem is converted into the corresponding heat transfer problem, a pinch diagram is constructed to optimize VOC recovery (Fig. 11.3). On this enthalpy-temperature diagram, the VOC-laden stream is represented as the hot composite stream, whereas the cooled stream is represented as the cold composite stream. Using this diagram, one can determine the minimum operating cost of the system and then trade it off with fixed cost. This procedure leads to the complete design of the system.

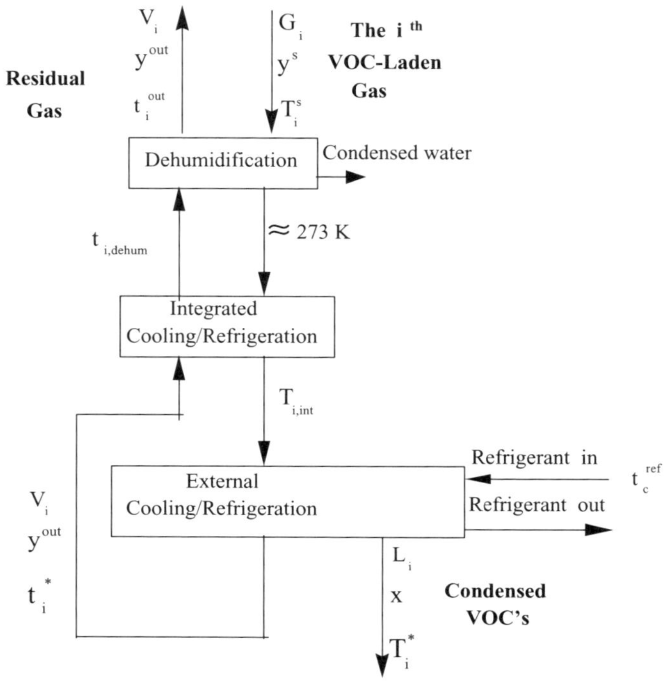

Figure 11.2: A typical structure of a VOC condensation system (Dunn and El-Halwagi, 1994a; Richburg and El-Halwagi, 1995).

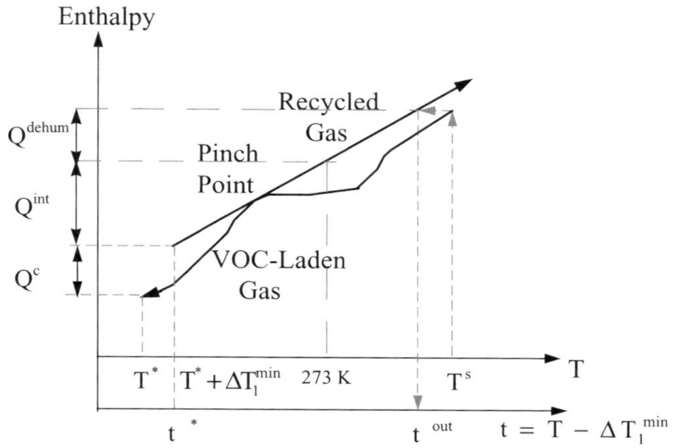

Figure 11.3: Minimizing condensation cost via pinch analysis (Richburg and El-Halwagi, 1995).

This graphical technique can be generalized by formulating the design task as a mathematical optimization program that can be solved *globally* (Dunn and El-Halwagi, 1994a, 1994b; El-Halwagi et al., 1995). Furthermore, the effect of pressure can be included to design an energy-induced separation network or EISEN that employs pressurization, depressurization, and cooling to effect condensation (Dunn et al., 1995).

ALLOCATION OF RECOVERED VOCs

As mentioned earlier, condensation of the VOCs must be integrated with their reuse in the process. Therefore, separation has to be integrated with decisions on how much VOC to recover and how the recovered species will be used. Toward this goal, we need to invoke the more general concept of *mass integration* that is a holistic approach dealing with the optimum generation, separation, and routing of streams and species. It provides a systematic framework for simultaneously considering the recovery and the allocation of species and streams. For recent reviews on mass integration, the reader is referred to El-Halwagi (1997) and El-Halwagi and Spriggs (1998). Mass integration is based on fundamental principles of chemical engineering combined with systems analysis using graphical and optimization-based tools. The first step in conducting mass integration is the development of a global mass allocation representation of the whole process from a species viewpoint (e.g., El-Halwagi and Spriggs, 1998; El-Halwagi, 1997; El-Halwagi et al., 1996; Garrison et al., 1995). For each targeted species (e.g., each VOC), there are sources (streams that carry the species) and process sinks (units that can accept the species). Properties of sources (e.g., flow rate, composition, pressure, temperature, etc.) can be modified by adding new units that intercept the streams prior to being fed to the process sinks and condition their properties to the desired values. This is performed in a waste interception network (WIN). Examples of this include the VOC condensation network. It is worth noting that a process is characterized not only by mass aspects, but also by energy aspects. Dunn and Srinivas (1997) introduced the problem of synthesizing Heat-Induced Waste Minimization Networks (HIWAMIN) in which mass and energy integration are simultaneously addressed. In addition to the interception and allocation of VOCs, the energy-separating agents are considered for the dual usage as condensation media or as heating/cooling utilities. Therefore, competing mass and energy objectives can be systematically reconciled.

We are now in a position to address the problem of simultaneously condensing and allocating the recovered VOCs.

PROBLEM STATEMENT

The problem is represented in Figure 11.4. In addition to the problem of designing a VOC condensation network (i.e., an HISEN) described previously,

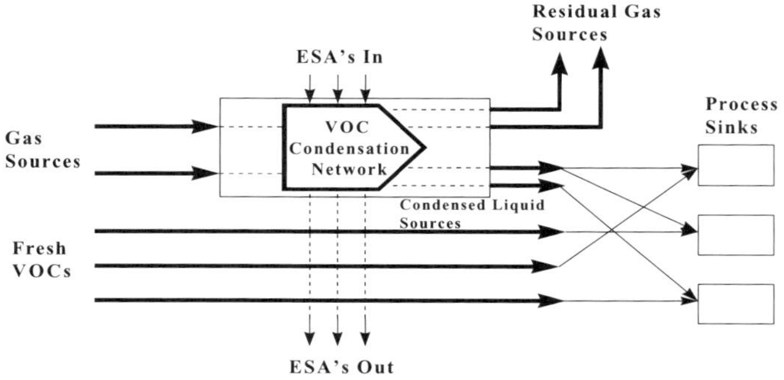

Figure 11.4: Problem statement.

the condensed VOCs are to be considered for recycle/reuse within the process. Several process units may be considered as sinks to which condensed VOCs can be recycled. Let the total number of sinks selected for the problem statement be given by N_{units}. For these sinks to use the recycled VOCs, the following flow rate and composition constraints must be met:

$$L_j^{Sink,lower} \leq L_j^{Sink} \leq L_j^{Sink,upper}, \qquad j = 1, 2, \ldots, N_{\text{units}}, \tag{2}$$

where $L_j^{Sink,lower}$ and $L_j^{Sink,upper}$ are given values of lower and upper bounds on admissible flow rate to unit j and L_j^{Sink} is the total flow rate of VOC fed into unit j, and

$$z_{j,k}^{lower} \leq z_{j,k}^{in} \leq z_{j,k}^{upper}, \qquad j = 1, 2, \ldots, N_{\text{units}} \quad \text{and} \quad k = 1, 2, \ldots, N_c, \tag{3}$$

where $z_{j,k}^{lower}$and $z_{j,k}^{upper}$ are given values of lower and upper bounds on admissible composition of kth VOC to unit j and $z_{j,k}^{in}$ is the composition of kth VOC fed into unit j.

To supplement the use of recovered VOC, fresh VOCs can also be purchased and mixed with recovered VOCs to meet the various sink requirements. The objective is to develop optimal strategies for recovery, allocation, and reuse of the VOCs to satisfy the requirements of all process units (sinks) and meet all prespecified environmental and process constraints at a minimum cost.

To address this problem, the designer should

- Identify the extent of separation for each VOC-laden gaseous stream.
- Choose the optimal refrigerant.
- Design an optimal condensation system to effect the determined extent of separation.
- Determine how much of the recovered VOCs can be recycled/reused and where their destination should be.

- Determine what optimum quantities of fresh VOCs should be purchased and how they should be mixed and allocated.

In addition to the aforementioned design tasks explicitly stated in the previous section, there are also some subtle difficulties. Although the target compositions of the various VOC-laden gaseous sources are given, it is possible to go beyond these target compositions and condense more of the sources so as to recover and recycle liquid VOCs. The decision on how far to condense each gaseous source depends on both the condensation and mixing costs. Hence, the target compositions are not solely dictated by environmental regulations. Instead, if there is an economic merit, additional separation may be undertaken. Another difficulty is that there exists a multicomponent nonideal equilibrium among the various VOCs present in each source. This nonlinear interaction calls for the simultaneous consideration of all species in the design procedure. The composition and flow rate of each condensed source depend on the initial composition, flow rate, temperature, and pressure of the respective gaseous source. The heat duties involved in the condensation depend on supply temperatures of the various gaseous sources. A trade-off is required between the costs involved in condensation and the cost of fresh VOC sources so as to come up with the final optimal network at a minimum cost. The problem can be solved rigorously via involved mathematical optimization (Parathasarathy and El-Halwagi, 2000). Nonetheless, the objective of this chapter is to devise shortcut approaches to solving the problem by invoking physical insights.

SOLUTION STRATEGY

We propose a five-step solution strategy to solve this challenging problem:

I. Identify bounds on the condensation temperature for each stream.

II. Decompose the recovery and the allocation problems by selecting an iterative condensation temperature for each stream.

III. Solve the decomposed VOC recovery problem.

IV. Solve the decomposed VOC allocation problem.

V. Evaluate total annualized cost of recovery and allocation and check stopping criterion. If minimum total annualized cost of recovery and allocation is attained, then stop. Otherwise, return to step II.

The following description expounds on the details of each step.

STEP I: BOUNDING OF SEARCH SPACE FOR CONDENSATION TEMPERATURES

The objective of this step is to identify bounds on the condensation temperature for each stream so as to reduce the computational effort. Therefore, the step entails the identification of an upper bound and a lower bound on the legitimate condensation temperature for each gaseous source i (referred to as T_i^{lower} and T_i^{upper}, respectively). The lower bound can be determined by examining several temperatures that limit further cooling of the stream. Examples include the lowest temperature of the coolants/refrigerants ($T^{ref,\ \mathrm{min}}$), the highest temperature at which a VOC freezes ($T^{freezing,\ \mathrm{max}}$), and the highest temperature at which detrimental byproducts (e.g., hydrates) start to form ($T_i^{byproducts,\ \mathrm{max}}$). The highest of all of these temperatures will prevent the feasibility of further cooling and will establish a lower bound on all viable condensation temperatures; i.e.,

$$T_i^{lower} = \max\left(T^{ref,\ \mathrm{min}},\ T^{freezing,\ \mathrm{max}},\ T_i^{byproducts,\ \mathrm{max}}\right), \qquad i = 1, 2, \ldots, N_G. \quad (4)$$

On the other hand, the determination of the upper bound on condensation temperature is more involved. This bound should be based on the least extent of separation needed to achieve the target compositions of the gaseous sources. Therefore, we need to understand the vapor-liquid equilibrium of the system. Hence, we first generate data on the nonideal multicomponent equilibrium that exist among the VOCs. This can be achieved through simulation, literature, or experimentation. The data generated should include gas phase compositions, liquid phase compositions, and condensed liquid flow rates at different condensing temperatures. For each gaseous source being condensed, there exist unique paths in the gas and liquid phases, along which the compositions and corresponding flow rates change as we lower temperature. This can be plotted as "condensation routes" for both the liquid and gaseous phases. For instance, for two VOCs and a carrier gas, the condensation route can be plotted on a triangular composition diagram as shown in Figure 11.5.

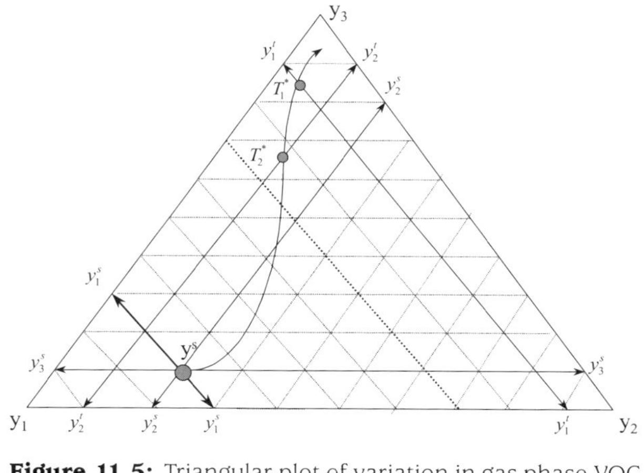

Figure 11.5: Triangular plot of variation in gas phase VOC composition with temperature (the index i has been dropped for simplicity).

In Figure 11.5, the vertices corresponding to y_1 and y_2 represent pure VOCs 1 and 2, respectively. Let y_3 correspond to pure carrier gas. The curve represents the gas-phase condensation route for the VOC-laden source and is pointed in the direction of decreasing temperature and toward the carrier gas vertex. This curve represents variation of gas phase compositions of the two VOCs from supply to outlet compositions. Corresponding to $y^t_{i,1}$ and $y^t_{i,2}$, the temperatures to which the gaseous source must be condensed are given by T_1^* and T_2^*. To satisfy constraint (1) for both components, we choose the lower temperature (corresponding to the limiting VOC), which is taken as T_i^{upper}. The same rationale for selecting the upper bound temperature applies to any number of multiple components.

STEP II: ITERATIVE PROBLEM DECOMPOSITION

In this step we decompose the recovery and the allocation problems by selecting an iterative condensation temperature for each stream. One way of discretization is to select a representative number of temperatures within the range bounded by T_i^{lower} and T_i^{upper}. Once an iterative temperature is selected, the following equations can be employed to determine flow rate and composition information for the gas and the condensate.

Material Balances

Overall Material Balances:

$$G_i = V_i + L_i, \qquad i = 1, 2, \ldots, N_G, \tag{5}$$

where V_i is the flow rate of the gaseous stream leaving the system and L_i is the flow rate of the condensed VOC stream.

Component Material Balances:

$$G_i y^s_{i,k} = V_i y^{out}_{i,k} + L_i x_{i,k}, \qquad i = 1, 2, \ldots, N_G \quad \text{and} \quad k = 1, 2, \ldots, N_C. \tag{6}$$

Vapor-Liquid Equilibrium

$$\Theta_k P_i y^{out}_{i,k} = \gamma_k P^{sat}_k (T_i^*) x_{i,k}, \qquad i = 1, 2, \ldots, N_G \quad \text{and} \quad k = 1, 2, \ldots, N_C, \tag{7}$$

where Θ_k and γ_k are vapor fugacity coefficient and liquid activity coefficient of the kth VOC, respectively. For a given system with known pressure, the previous equation can be solved to provide a relationship, Ω_i, between vapor composition and temperature; i.e.,

$$y^{out}_{i,k} = \Omega_{i,k}(T_i^*), \qquad i = 1, 2, \ldots, N_G \quad \text{and} \quad k = 1, 2, \ldots, N_C. \tag{8}$$

Because no condensation occurs prior to reaching the dew point,

$$\begin{aligned} &y_{i,k} = y^s_{i,k} \quad \text{for} \quad T_i^* > T_i^{dew}, \\ &y_{i,k} = \Omega_{i,k}(T_i^*) \quad \text{for} \quad T_i^* \leq T_i^{dew}, \quad i = 1, 2, \ldots, N_G \quad \text{and} \quad k = 1, 2, \ldots, N_C. \end{aligned} \tag{9}$$

By solving these equations, we obtain the flow rate and composition data for the outlet gas and the condensate.

STEP III: SOLUTION OF THE DECOMPOSED CONDENSATION PROBLEM

Once an iterative condensation temperature is selected, we can solve the decomposed VOC recovery problem using the techniques described in the foregoing section "Design of VOC Condensation Networks." The result of this step is the sizing, configuration, and cost estimation of the optimum condensation network.

STEP IV: SOLUTION OF THE DECOMPOSED ALLOCATION PROBLEM

From step II, the liquid phase flows and compositions corresponding to the iterative condensation temperature have been identified. The condensed sources, along with any external purchased VOC-laden sources (either pure VOC or a mixture of VOCs), can be optimally allocated to satisfy the requirements of any sinks that may be present. This optimal allocation can be achieved by solving a linear mixing and allocation problem with the objective of minimizing the cost of external fresh purchased VOC. In this problem, the optimal strategies for mixing and routing condensed sources from the allocation network, as well as fresh VOC sources to the various sinks, must be considered. The need of the sinks, coupled with the cost of the sources, will determine the extent of mixing. The objective function is given as

$$\min = \sum_{i=N_G+1}^{N_G+N_{fresh}} c_i L_i. \tag{10}$$

The following constraints describe the mixing and allocation problem:

Source Splitting Material Balance:

$$L_i = \sum_{j=1}^{N_S} l_{i,j}, \qquad i = 1, 2, \ldots, N_G + N_{fresh}. \tag{11}$$

Sink Mixing Material Balance:

$$L_j^{Sink} = \sum_{i=1}^{N_G+N_{fresh}} l_{i,j}, \qquad j = 1, 2, \ldots, N_S. \tag{12}$$

Sink Inlet Component Material Balance:

$$L_j^{Sink} \cdot z_{j,k}^{in} = \sum_{i=1}^{N_G+N_{fresh}} l_{i,j} \cdot x_{i,k}, \qquad j = 1, 2, \ldots, N_S,\ k = 1, 2, \ldots, N_C. \tag{13}$$

The allocation problem is represented schematically in Figure 11.6.

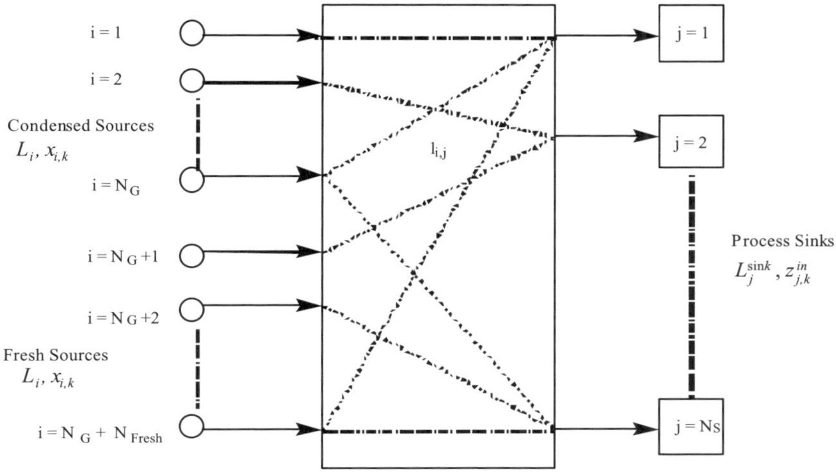

Figure 11.6: Optimal mixing and allocation network.

The above mixing and allocation problem can also be solved using graphical tools, such as the triangular source-sink mapping diagram for visualization insights of the system (Parthasarathy and El-Halwagi, 1997).

STEP V: TOTAL COST EVALUATION

In this step we add up the annualized costs of the condensation system and the fresh sources to yield the total annualized cost of the system at the iterative temperature selected in step II. The procedure is repeated until all iterative temperatures are employed and the least-cost system is selected as the optimal solution.

SPECIAL CASES

The largest computational effort in the aforementioned solution strategy is the iteration over the condensation temperatures. We now describe two special cases that alleviate the need for iterations:

1. High recovery of all VOCs.
2. No economic incentive to recover VOCs beyond their target compositions.

SPECIAL CASE I: HIGH RECOVERY

This case deals with high recovery of the VOCs from the gaseous streams. In this case, almost all of the VOCs in the entering gaseous streams become the condensate. Hence, equations (5) and (6) can be simplified into

$$G_i \approx L_i, \qquad i = 1, 2, \ldots, N_G, \tag{14}$$

$$y^s_{i,k} \approx x_{i,k}, \qquad i = 1, 2, \ldots, N_G \quad \text{and} \quad k = 1, 2, \ldots, N_C, \tag{15}$$

with both G_i and $y^s_{i,k}$ calculated on a carrier-free basis. Consequently, the design problems for the condensation network and the allocation are disengaged, and there is no need for iterations over the condensation temperatures. Therefore, the previously described solution strategy is significantly simplified.

Alternatively, the problem can also be solved graphically. To illustrate, consider a gaseous source with three VOCs: A, B, and C. Based on supply temperature, pressure, and composition, as well as thermodynamic data, the condensation route can be generated as discussed earlier. Let us consider the target compositions for three species: y^t_A, y^t_B, and y^t_C, as shown in Figure 11.7. To determine the optimum temperature of the gas leaving the condensation network, we compare the target compositions on the condensation route (the direction of the arrow is that of decreasing temperature). Therefore, the whole stream has to be cooled until the most stringent target composition (the farthest on the condensation route) is reached. On Figure 11.7 this corresponds to y^t_B. The point of intersection of y^t_B and the condensation route corresponds to the optimum outlet gas temperature, T^*. We cannot stop condensation prior to this temperature, otherwise the constraint on the outlet composition for VOC B will not be met. Also, there is no need to cool any further, as this is a high VOC recovery case, for which additional condensation offers no economic incentive. Therefore, T^* is the optimal outlet gas temperature, and the gas compositions at this point are the optimal outlet gas compositions, y^{out}_A, $y^{out}_B = y^t_B$, and y^{out}_C, and can be determined

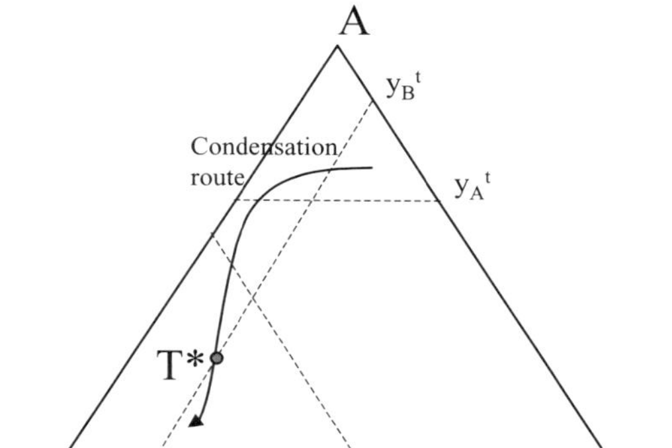

Figure 11.7: Identification of optimal outlet gas temperature and compositions (carrier-free basis).

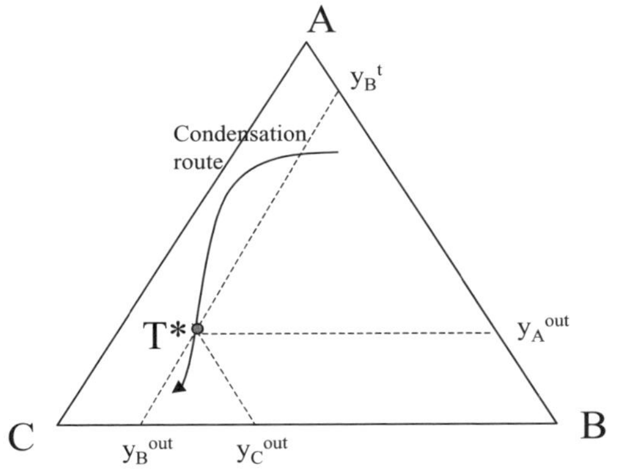

Figure 11.8: Optimal outlet gas composition for the three VOCs (carrier-free basis).

from the ternary diagram (Fig. 11.8). With this information, the optimal HISEN can be synthesized using the graphical pinch diagram (Richburg and El-Halwagi, 1995) as discussed earlier.

SPECIAL CASE II: NO ECONOMIC INCENTIVE FOR EXCEEDING TARGETED RECOVERY

In this special case there is no economic incentive to recover VOCs beyond their gas-phase target compositions. Examples of when this case is encountered include

a. Very dilute VOC-laden sources.

b. Value of fresh liquid VOCs is very low compared to the cost of condensation.

In such cases, the optimum condensation temperature becomes equal to T_i^{upper} determined in step I of the solution strategy, and the need for iteration is eliminated.

CASE STUDY

We consider a magnetic tape manufacturing process shown in Figure 11.9 (Dunn and El-Halwagi et al., 1994a, 1994b). Magnetic tape is manufactured via a coating process where a solvent mixture containing magnetic pigments, binders, dispersants, and lubricants is applied to a polyester base film. The solvent being used here is a mixed solvent consisting of three VOCs, namely, cyclohexanone (CH), methyl isobutyl ketone (MIBK), and methyl ethyl ketone (MEK). There is one sink where the solvent is added to the

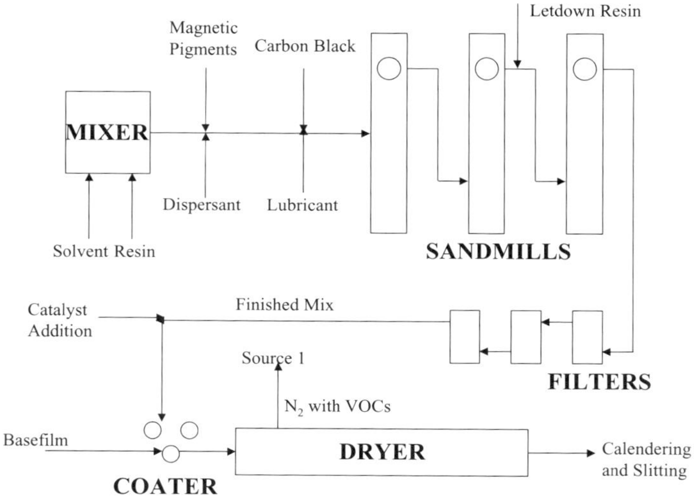

Figure 11.9: Magnetic tape manufacturing process.

process, namely the mixer. The basefilm containing VOCs is now dried in a multiple zone oven. The VOC-laden gaseous emission streams exit the coating line dryers. The data about the sources, as well as the requirements of the sink, are given in Appendix 1. It is desired to reduce the load of the three VOCs in the gaseous source by a certain fraction before discharging them. This fraction is expressed in the form of desired final target compositions for the VOCs in the gaseous source given in Appendix 1.

There are VOC losses and purges in the process from which the VOCs cannot be recovered via condensation. It is desired to integrate this process and come up with an optimal condensation and allocation that can satisfy the requirements of the sink as well as predefined process constraints on the gaseous source. There is one external refrigerant, namely, liquid nitrogen, whose cost and operating data are listed in Appendix 1.

APPLICATION OF SPECIAL CASE 1 TO CASE STUDY

This case deals with high recovery of the VOCs from the gaseous streams. In this case, it is required to achieve at least 99% removal of all VOCs from gas phase. Data on the entering hot VOC-laden gaseous source are given by

Total flow rate = 3000 kg/hr.

Temperature = 50°C.

$y^{s}_{CH} = 0.02.$

$y^s_{MEK} = 0.03.$

$y^s_{MIBK} = 0.005.$

$y^s_{N_2} = 0.945.$

First, we develop the multicomponent, nonideal vapor liquid equilibrium that governs the condensation of the available hot VOC-laden source from the dryer. For the source, the dew point is calculated as 28.2°C. Here we use the Wilson equation for calculating the liquid phase activity coefficients. The vapor phase is assumed to be ideal, which is a reasonable assumption at low pressures. Subsequently, the value of T^* to condense to attain 99% recovery can be determined from the triangular diagram as shown in Figure 11.10, which is an enlargement of the area near the MEK vertex.

Reaching the target compositions of cyclohexanone, MEK and MIBK require a T^* of −23°C, −65°C, and −39°C. Hence, T^* is selected as −65°C with MEK being the limiting component. In Figure 11.10, the values of y^{mod}_{MEK} correspond to the carrier-free gas phase composition of MEK. As the temperature is lowered progressively, the condensation route tends toward the pure MEK vertex, and the values taken by y^{mod}_{MEK} tend toward one. The condensed

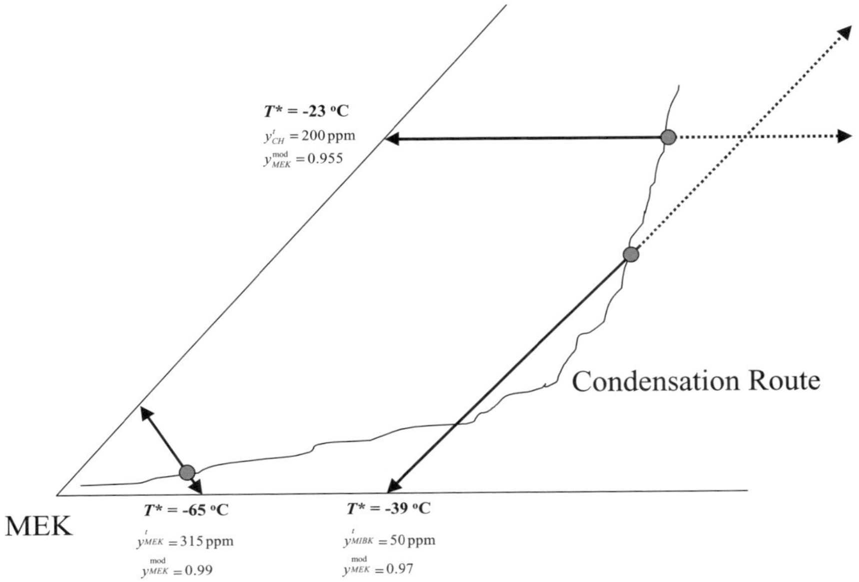

Figure 11.10: Enlargement of area near the MEK vertex of triangular diagram.

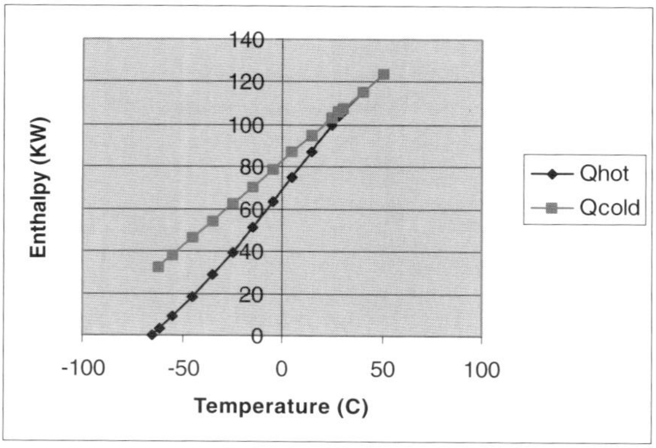

Figure 11.11: HISEN pinch diagram for T^* of $-65°C$.

VOC source has a flow rate of 165 kg/hr and composition

$$x_{CH} = 0.365,$$

$$x_{MEK} = 0.545,$$

$$x_{MIBK} = 0.090.$$

Next, we optimize the condensation network using the HISEN pinch analysis as shown in Figure 11.11.

The HISEN pinch diagram (Fig. 11.11) can be plotted for the source at T^* of $-65°C$ from which we can obtain the various heat duties Q^c, Q^{int}, and Q^{dehum}. Knowing the various heat duties, we can size and cost the different exchangers and determine the optimal HISEN condensation network. The total annualized cost of the network is \$225,170. The optimal condensation network for the source is given in Figure 11.12.

Once condensation is completed, we have to meet the objective of minimizing the cost of external purchased VOC as described by equations (10) to (13). By solving this optimization program, we determine the optimum cost of external purchased VOC to be \$103,950/year. Therefore, the total cost of the above system is calculated as \$329,120/year.

APPLICATION OF SPECIAL CASE 2

In this case, let us assume that there is no economic incentive to recover the VOCs beyond their gas-phase target compositions. Data on supply and target compositions for all VOCs in the entering VOC-laden gaseous source are given as follows:

Total flow rate = 3000 kg/hr.

Temperature = 50°C.

$y^s_{CH} = 0.02.$

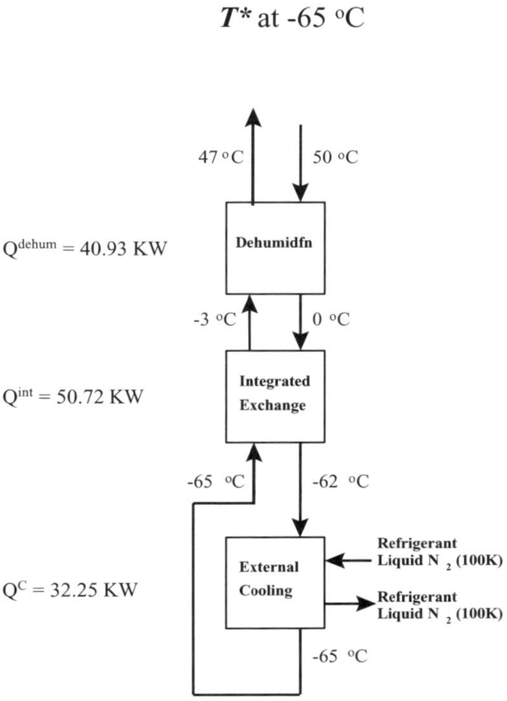

Figure 11.12: Optimal condensation network for special case 1.

$y^s_{MEK} = 0.03.$

$y^s_{MIBK} = 0.005.$

$y^s_{N_2} = 0.945.$

The target compositions for the three VOCs are given by

$y^t_{CH} = 0.001.$

$y^t_{MEK} = 0.026.$

$y^t_{MIBK} = 0.004.$

As mentioned earlier, in this special case the optimum condensation temperature corresponds to the minimum extent of separation (upper bound on condensation temperatures) that satisfies all target compositions. This can be obtained as shown in Figure 11.13.

As can be seen from Figure 11.13, the optimal condensation temperature is $-5°C$ and CH is the limiting component. In Figure 11.13, the values of y^{mod}_{MEK}

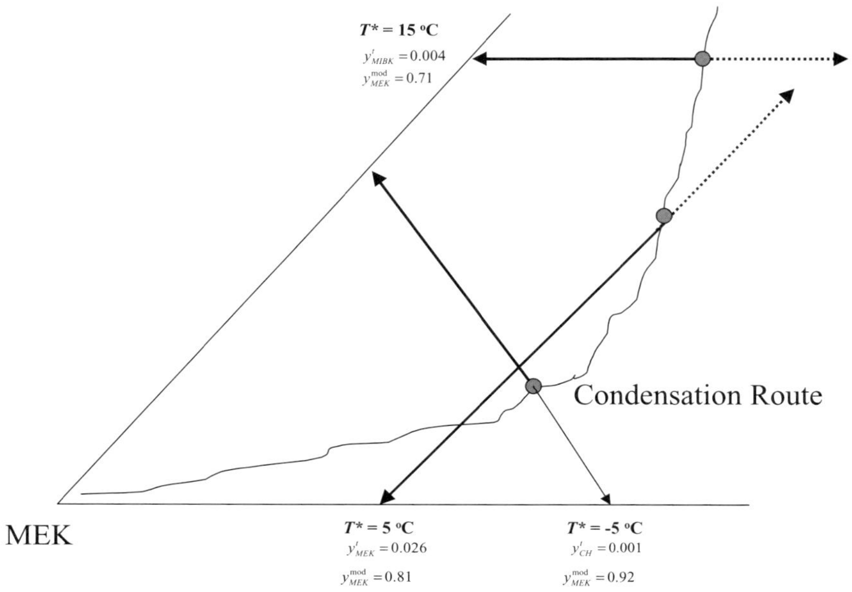

Figure 11.13: Enlargement of area near the MEK vertex of triangular diagram for selection of optimum condensation temperature.

correspond to the carrier-free gas phase composition of MEK. The resultant condensed source has a flow rate of 93.83 kg/hr and composition of

$$x_{CH} = 0.599,$$

$$x_{MEK} = 0.283,$$

$$x_{MIBK} = 0.118.$$

Next, we synthesize the optimal HISEN condensation network. The total annualized cost of the network is $114,960. The optimal condensation network for the source is given in Figure 11.14.

Once the condensation system is designed, we solve the optimal allocation program (equations 10–13) to get a minimum cost of external purchased VOC as $153,355/year, and the total cost of the above system is calculated as $268,315/year.

CONCLUSIONS

This chapter has a systematic solution procedure toward solving the problem of multicomponent nonideal VOC recovery and reuse. It links the recovery of VOCs with their optimal allocation to meet the requirements of different sinks in the process. The approach is within the framework of mass

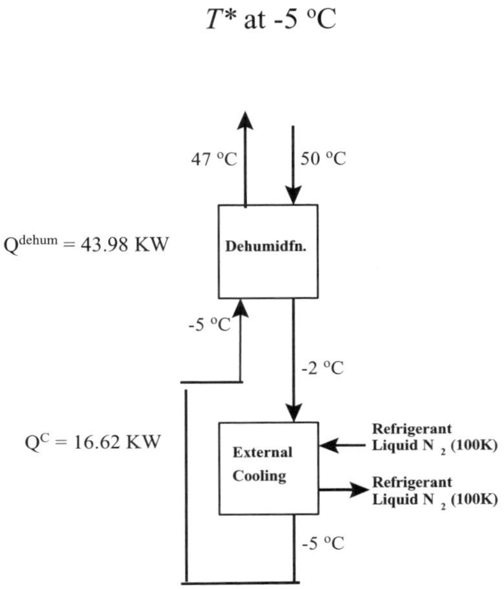

Figure 11.14: Optimal condensation network for special case 2.

integration, which is a holistic approach to the systematic and optimal global allocation of different species. The solution is derived from the recognition of a key variable, the condensation temperature of each stream that allows decomposition of the original problem into two subproblems, each of which can be solved independently. Two special cases are defined, and solution approach for each of these is described. These special cases cover a range of industrial VOC recovery and reuse scenarios for which no iterations in the design procedure are needed. A case study dealing with a magnetic tape manufacturing process is considered. The analysis is found to bring about cost savings besides meeting all required constraints and besides integrating the current process optimally.

NOMENCLATURE

A_i	area of heat exchanger calculated, m^2
c_i	cost of ith fresh VOC source, \$/kg
C_p	specific heat for gaseous sources, kJ/kg · K
FCOST	fixed cost, \$
G_i	total flow rate (or carrier-free flow rate) of ith gaseous source, kg/s
L_i	total flow rate of ith condensed or purchased liquid VOC, kg/s

$L_j^{\sin k}$	total required inlet flow rate for j^{th} sink, kg/s
$l_{i,j}$	individual flow rate from i^{th} liquid source (condensed or purchased) to j^{th} sink, kg/s
N_C	total number of targeted VOCs
N_E	total number of candidate coolants/refrigerants
N_{fresh}	total number of purchased fresh liquid VOCs
N_G	total number of hot gaseous sources
N_S	total number of process sinks
P_i^s	supply pressure of the ith hot gaseous source, atm
P_k^{sot}	saturation pressure of kth VOC
Q^c	external cooling utility requirements for the gaseous source, kW
REF_COST	cost of refrigerant, \$/kJ removed
T_i	temperature of ith hot gaseous source, K
t_i	temperature of ith cold stream, K
T^{ref}	cooling temperature of refrigerant, K
T_i^*	condensing temperature of the ith gaseous source, K
$T_k^{freezing}$	freezing temperature of kth VOC, K
T^{minsep}	temperature corresponding to minimum separation of all VOCs, K
TAC	total annualized cost, \$/year
U	overall heat transfer coefficient for the gas source, kW/m^2 K
V_i	total flow rate (or carrier-free flow rate) of ith residual gas source following condensation, kg/s
$x_{i,k}$	overall (or carrier-free) composition of kth VOC in i^{th} condensed or purchased liquid VOC source
$y_{i,k}^{out}$	overall (or carrier-free) outlet composition of kth VOC in i^{th} gaseous source
$y_{i,k}^s$	overall (or carrier-free) supply composition of kth VOC in i^{th} gaseous source
$y_{i,k}^t$	overall (or carrier-free) target composition of kth VOC in i^{th} gaseous source
$z_{j,k}^{in}$	required overall (or carrier-free) inlet composition of kth VOC in j^{th} sink

GREEK

γ_k	liquid phase activity coefficient of kth VOC
$\Delta T_{\min}$	minimum allowable temperature driving force between hot and cold streams, K
$\Delta T_{\log mean}$	log mean temperature driving force, K
Ω	VLE function relating vapor phase VOC composition to temperature
θ	number of operational seconds per year
Θ_k	vapor fugacity coefficient of kth VOC
τ	years of depreciation, years

SUBSCRIPTS

i	index for hot gaseous sources and for fresh purchased liquid VOC sources
j	index for process sinks
k	index for targeted VOC
min	refers to minimum

SUPERSCRIPTS

c	cooling
dehum	dehumidification
dew	dew point
freezing	freezing point
in	inlet
int	integrated
lower	lower bound
min	minimum
min sep	minimum separation
out	outlet
ref	refrigerant
s	supply composition
sat	refers to saturation
Sink	refers to sink
t	target composition
upper	upper bound
*	at final condensation conditions

APPENDIX 1: BACKGROUND EQUATIONS AND DATA

UNIT SIZING, FIXED, AND OPERATING COST EQUATIONS

We consider up to three heat exchangers for each gaseous source, one each for dehumidification, heat integration, and external cooling/refrigeration. The various heat duties for these can be obtained from the pinch diagram (see Fig. 11.2). For instance, for the ith VOC-laden gaseous source,

For dehumidification:

$$Q_i^{dehum} = U_i^{dehum} \cdot A_i^{dehum} \cdot \Delta T_{\log mean,i}^{dehum},$$

where

$$\Delta T_{\log mean,i}^{dehum} = \frac{\left(T_i^s - t_i^{out}\right) - \left(273 - t_i^{dehum}\right)}{\log\left(\dfrac{T_i^s - t_i^{out}}{273 - t_i^{dehum}}\right)}.$$

Once the area of the heat exchanger is evaluated, its fixed cost, $FCOST_i^{dehum}$, can be evaluated:

$$FCOST_i^{dehum} = FCOST_i^{dehum}\left(A_i^{dehum}\right).$$

For the heat-integration exchanger,

$$Q_i^{\text{int}} = U_i^{\text{int}} \cdot A_i^{\text{int}} \cdot \Delta T_{\log mean,i}^{\text{int}},$$

where

$$\Delta T_{\log mean,i}^{\text{int}} = \frac{\left(273 - t_i^{dehum}\right) - \left(T_i^{\text{int}} - t_i^*\right)}{\log\left(\dfrac{273 - t_i^{dehum}}{T_i^{\text{int}} - t_i^*}\right)},$$

the fixed cost is given by

$$FCOST_i^{\text{int}} = FCOST_i^{\text{int}}\left(A_i^{\text{int}}\right).$$

For the external cooling/refrigeration section, total heat duty is given by

$$Q_i^c = U_i^c \cdot A_i^c \cdot \Delta T_{\log mean,i}^c,$$

where

$$\Delta T_{\log mean,i}^c = \frac{\left(T_i^{\text{int}} - T^{ref}\right) - \left(t_i^* - T^{ref}\right)}{\log\left(\dfrac{T_i^{\text{int}} - T^{ref}}{t_i^* - T^{ref}}\right)},$$

and the fixed cost is evaluated through

$$FCOST_i^c = FCOST_i^c\left(A_i^c\right).$$

The total fixed cost for the heat exchangers cooling/condensing the ith gaseous source is given by

$$FCOST_i = FCOST_i^{dehum} + FCOST_i^{\text{int}} + FCOST_i^c.$$

Operating cost of condensation is primarily attributed to external cooling/refrigeration

$$O \cdot C_i = Q_i^c \cdot REF\_COST \cdot \theta_i.$$

The total annualized cost can now be calculated for the ith gaseous source as

$$TAC_i = \frac{F \cdot C_i}{\tau_i} + O \cdot C_i.$$

BASIS OF COST CALCULATIONS

The following equations were used to calculate the total annualized costs for the condensation network:

Fixed Cost of Heat Exchanger = 3000* (area in $m^2)^{0.6}$ \$ (this is for any exchanger, namely, for dehumidification, heat integration, or external cooling)

$U^{dehum} = 0.1$ KW/m$^2\cdot$ K.

$U^{int} = 0.1$ KW/m$^2\cdot$ K.

$U^{c} = 0.2$ KW/m$^2\cdot$ K.

Depreciation over 3 years.

Operating Cost for External Cooling using N_2 Refrigeration Systems = \$180/MMJ.

Removed, Operating Temperature = 100 K.

FRESH VOC COSTS

Cost of fresh CH = \$0.05/kg.

Cost of fresh MEK = \$0.08/kg.

Cost of fresh MEK = \$0.10/kg.

REFERENCES

Dunn, R. F., and El-Halwagi, M. M. (1994a). Selection of optimal VOC-condensation systems. *Waste Mgt.* 14:103–113.

Dunn, R. F., and El-Halwagi, M. M. (1994b). Optimal design of multi-component VOC-condensation systems. *J. Haz. Mtls.* 38:187–206.

Dunn, R. F., Zhu, M., Srinivas, B. K., and El-Halwagi, M. M. (1995). Optimal design of energy induced separation networks for VOC recovery. *AIChE Symp. Ser.* 90:74–85.

Dunn, R. F., and El-Halwagi, M. M. (1996). *Design of Cost-Effective VOC Recovery Systems*. TVA Pub., Muscle Shoals, AL.

Dunn, R. F., and Srinivas, B. K. (1997). Synthesis of heat-induced waste minimization networks (HIWAMINs). *Adv. Environ. Res.* 1:275–301.

El-Halwagi, M. M. (1997). *Pollution Prevention through Process Integration: Systematic Design Tools*. Academic Press, San Diego.

El-Halwagi, M. M., Srinivas, B. K., and Dunn, R. F. (1995). Synthesis of heat-induced separation networks. *Chem. Eng. Sci.* 50:81–97.

El-Halwagi, M. M., Hamad, A. A., and Garrison, G. W. (1996). Synthesis of waste interception and allocation networks. *AIChE J.* 42:3087–3101.

El-Halwagi, M. M., and Spriggs, H. D. (1998). Solve design puzzles with mass integration. *Chem. Eng. Prog.* 94:25–44.

Garrison, G. W., Hamad, A. A., and El-Halwagi, M. M. (1995). Synthesis of waste interception networks. AIChE Annual Meeting, Miami.

Heinsohn, R. J., and Kabel, R. L. (1999). *Sources and Control of Air Pollution*. Prentice-Hall, Englewood Cliffs, NJ.

Moretti, E. C., Mukhopadhyay, N. (1993). VOC control: Current practices and future trends. *Chem. Eng. Prog.* 89:20–26.

Parthasarathy, G., and El-Halwagi, M. M. (2000). Optimum mass integration strategies for condensation and allocation of multicomponent VOCs. *Chem. Eng. Sci.* 55:881–895.

Parthasarathy, G., and El-Halwagi, M. M. (1997). Mass integration for multicomponent non-ideal systems. AIChE Annual Meeting, Los Angeles.

Pollack, N. R., Tramposch, W. G., and Sengupta, U. (1996). The VOC abatement market: An overview of regulatory, technology, and market trends. AIChE Annual Meeting, Chicago.

Richburg, A., and El-Halwagi, M. M. (1995). A graphical approach to the optimal design of heat-induced separation networks for VOC recovery. *AIChE Symp. Ser.* 91:256–259.

Ruddy, E. N., and Carroll, L. A. (1993). Select the best VOC control strategy. *Chem. Eng. Prog.* 89:28–35.

PART FIVE

MACROSCOPIC DESIGN AND ENVIRONMENTAL IMPACT ASSESSMENT

CHAPTER 12

Designing Processes for Environmental Problems

Yan Fu and Urmila Diwekar

Environmental Institute, Carnegie Mellon University, Pittsburgh, PA 15213

Douglas Young and Heriberto Cabezas

National Risk Management Research Laboratory, U.S. Environmental Protection Agency, Cincinnati, OH 45268

Abstract

Designing for the environment requires consideration of environmental impacts. The Generalized Waste Reduction (WAR) algorithm is a methodology that allows the user to evaluate the potential environmental impact of the design of a chemical process. In this methodology, chemicals are assigned values that describe their potential to have an adverse affect on the environment relative to other chemicals. These relative values cover eight different potential environmental impact categories; however, environmental impacts must also be weighted and balanced against other concerns, such as cost and long-term sustainability. These multiple, often conflicting, goals pose a challenging and complex optimization problem, requiring multiobjective optimization. This chapter addresses the problem of quantifying and analyzing the various objectives involved in process design for the environment. A detailed case study will be presented that will illustrate the concepts leading to environmentally friendly and cost-effective designs.

1 INTRODUCTION

Currently, environmentally friendly or green processes are designed based on new concepts in process engineering, such as process integration. These concepts embody a number of closely related methodologies for designing new processes and retrofitting existing ones by taking into consideration the performance of the entire process as a whole (Rossiter, 1994). As part of this process integration, pollution prevention has been incorporated into the initial stages of process design and development. This has been shown to improve both the technical and the economic performances of chemical processes.

Process design involves a number of steps: chemical synthesis (the chemical pathway from reactants to products), process synthesis, which involves decisions about the specific unit operations in the process and decisions about the process connectivity, and process simulation. In recent years, researchers have realized the importance of including the chemical synthesis step into the overall process design. Unlike the traditional process design, where engineers are simply looking for low-cost options, the new concepts in process design encourage including environmental considerations into the design steps.

In general, process simulators provide local scale models to predict the mass and energy flows around a chemical process. Cost models combined with these performance models can define the profitability of the process. The goals in terms of profitability are relatively easy to define, and researchers in academia and industries have used simulators and modeling tools to achieve economic optimization; however, incorporating environmental considerations into process design optimization problems is a formidable task. For instance, consider a base process design that emits 1000 kg/hr of pollutants whereas an alternate process design emits 2000 kg/hr of pollutants. Depending on the types of pollutants emitted by each process, it is quite conceivable that the alternate design could be the environmentally friendlier one. That is, the environmental impact of emitting 2000 kg/hr of certain pollutants could be less than emitting 1000 kg/hr of different pollutants, which would make the alternate design preferable from an environmental viewpoint.

To address the need of incorporating environmental impact considerations into process design, Cabezas et al. (1997) developed a potential environmental impact (PEI) balance as a generalization of the WAste Reduction (WAR) algorithm first introduced by Hilaly and Sikdar (1994). They had introduced the concept of a pollution balance that was the precursor to the PEI balance. The PEI balance quantifies the environmental impact of the pollutants in a process and ultimately serves as the basis for quantitative indicators that represent the environmental friendliness or unfriendliness of a given process. Cano-Ruiz and McRae (1998) provide a comprehensive review of techniques used to incorporate environmental considerations into process design.

To incorporate environmental impacts into an optimization problem, they must be weighted and balanced against other concerns, such as cost and long-term sustainability. These multiple, often conflicting, goals pose a challenging and complex optimization problem, requiring multiobjective optimization. This chapter addresses the problem of quantifying and analyzing the various objectives involved in process design for the environment. Toward this goal, a novel multiobjective optimization framework is presented here. This framework is based on a new and efficient algorithm for multiobjective optimization and is implemented in the ASPEN simulator. The multiobjective optimization framework will provide a set of non-dominant designs where a further improvement for one objective will be at the expense of another. This approach will find a set of potentially optimal designs where trade-offs can be explicitly identified, unlike cost-benefit analysis, which converts all objectives into one objective: currency. This proposed approach is particularly valuable in a multiobjective situation where there are a large number of desirable and important objectives that are not easily translated into dollars (e.g., human health and life, environmental impacts, and plant flexibility).

The objective of this chapter is to use optimization techniques to design chemical manufacturing processes that have a maximum economic benefit and a minimum environmental impact. The use of the PEI balance and environmental indexes from the WAR algorithm will also be demonstrated. A brief synopsis of the Generalized WAR algorithm and an outline of our methodology for estimating PEI will be presented. The methodology for multiobjective optimization using the indexes from the WAR algorithm and engineering economics is discussed. The methodology is illustrated with a case study of the production of benzene that was simulated with the ASPEN simulator. (Use of the ASPEN simulator does not imply approval or endorsement by the U.S. Environmental Protection Agency.) Finally, a discussion of the results and some conclusions are presented.

2 POTENTIAL ENVIRONMENTAL IMPACT THEORY

The PEI of a given quantity of material is defined as the expected effect that this material would have on the environment if it were emitted into the environment. This implies that a release of material has not occurred, but if it did, it would have a quantifiable effect which is termed the *potential environmental impact*. This quantity is probabilistic and will in reality deviate from the observed values for particular cases. It should be noted that PEI is a conceptual quantity that cannot be directly measured; however, it can be estimated from measurable quantities (Cabezas et al., 1997; Cabezas et al., 1999; Young and Cabezas, 1999), such as stream flow rates, stream compositions, and environmental impact parameters, such as toxicity.

To apply these concepts to process design, Cabezas et al. (1999) proposed that an environmental impact balance be incorporated into the traditional process design methodology to account for environmental effects. The PEI balance is described by the expression

$$\frac{\partial I^{(cp)}}{\partial t} = \dot{I}_{in}^{(cp)} - \dot{I}_{out}^{(cp)} + \dot{I}_{gen}^{(cp)}, \qquad (1)$$

where $I^{(cp)}$ is the amount of PEI inside the chemical process, $\dot{I}_{in}^{(cp)}$ and $\dot{I}_{out}^{(cp)}$ are the input and output rates of potential environmental impact to the chemical process, and $\dot{I}_{gen}^{(cp)}$ is the rate of generation of PEI inside the process.

$\dot{I}^{(cp)}_{gen}$ represents the creation or consumption of PEI by chemical reactions inside the chemical process.

For steady state processes, equation (1) reduces to

$$0 = \dot{I}^{(cp)}_{in} - \dot{I}^{(cp)}_{out} + \dot{I}^{(cp)}_{gen}. \tag{2}$$

To make explicit use of equation (2) in chemical process design, Mallick et al. (1996) and Cabezas et al. (1997) developed a first-order approximation to estimate the various terms in this expression from measurable quantities. The expressions are

$$\dot{I}^{(cp)}_{in} = \sum_i^{EnvCat} \alpha_i \dot{I}^{(in)}_i = \sum_i^{EnvCat} \alpha_i \sum_j^{Streams} \dot{M}^{(in)}_j \sum_k^{Comps} x_{kj} \psi^s_{ki} + \cdots, \tag{3}$$

$$\dot{I}^{(cp)}_{out} = \sum_i^{EnvCat} \alpha_i \dot{I}^{(out)}_i = \sum_i^{EnvCat} \alpha_i \sum_j^{Streams} \dot{M}^{(out)}_j \sum_k^{Comps} x_{kj} \psi^s_{ki} + \cdots, \tag{4}$$

where the sum over i is taken over all environmental impact categories (e.g., human toxicity, global warming, etc.), α_i is the relative weighting of impact category i, $\dot{I}^{(l)}_i$ is the PEI input ($l = in$) or output ($l = out$) associated with impact category i, the sum over j is taken over all streams into or out of the process, $\dot{M}^{(l)}_j$ is the mass flow rate of stream j into or out of the process, the sum over k is taken over chemical components, x_{kj} is the mass fraction of chemical k in stream j, and ψ^s_{ki} is the specific PEI of chemical k associated with environmental impact category i.

There are two classes of PEI indexes used to evaluate the environmental friendliness of process designs: output indexes and generation indexes. The two most important output indexes ($\dot{I}^{(cp)}_{out}$ and $\hat{I}^{(cp)}_{out}$) describe the rate of PEI output and the output PEI per mass of product. The rate of PEI output $\dot{I}^{(cp)}_{out}$ is simply given by equation (4). The PEI output per mass of product is given by

$$\hat{I}^{(cp)}_{out} = \frac{\dot{I}^{(cp)}_{out}}{\sum_p \dot{P}_p}, \tag{5}$$

where $\dot{P}_p$ is the mass flow rate of the product stream p. The output index described by equation (5) is normalized to the production rate of the manufacturing process. There are two similarly defined PEI generation indexes ($\dot{I}^{(cp)}_{gen}$ and $\hat{I}^{(cp)}_{gen}$) given by the following expressions:

$$\dot{I}^{(cp)}_{gen} = \dot{I}^{(cp)}_{out} - \dot{I}^{(cp)}_{in}, \tag{6}$$

$$\hat{I}^{(cp)}_{gen} = \frac{\dot{I}^{(cp)}_{gen}}{\sum_p \dot{P}_p}, \tag{7}$$

where $\dot{I}^{(cp)}_{gen}$ is the rate of PEI generation inside the process and $\hat{I}^{(cp)}_{gen}$ is the PEI generated per mass of product. In the WAR algorithm, these four indexes are used to evaluate chemical manufacturing processes. The PEI indexes can be calculated from process flow information using the mass flow of the input and output streams and the compositions of those streams (from equations (3) and (4). This process flow data can be obtained through the use of a process simulator or even actual plant data. The intended use of this methodology is to compare the PEI indexes of alternative process designs. This will provide an indication as to which design is the most environmentally friendly. The lower the value of the indexes, the more environmentally friendly the process design.

The PEI indexes can be evaluated in terms of the total amount of PEI in the output streams, which would include the PEI of the product and nonproduct streams, or they can be evaluated in terms of the PEI present in the nonproduct streams. The latter evaluation provides more pertinent information. The nonproduct evaluation considers only components in those streams that are considered to be waste streams. This technique does not penalize a process design for producing a chemical that has a relatively high PEI or a process design that produces more of the product than alternative process designs. This technique only assesses the PEI of the waste (nonproduct) streams. Thus, alternative process designs are evaluated and compared on the PEI of their respective waste streams. The nonproduct evaluation will be employed in this case study; it can be denoted by a superscript (*NP*) in place of (*cp*) in equations (1) to (7).

3 CHEMICAL ENVIRONMENTAL IMPACTS

The specific PEI of a chemical *k* associated with environmental impact category i, ψ^s_{ki}, is estimated from measures of chemical environmental impact such as LD_{50}, LC_{50}, etc. These measures are used to establish environmental impact values or *Scores* for each chemical. The *Scores* for each environmental impact category, *i*, are normalized by the *Average Score* for each environmental impact category according to the expression

$$\psi^s_{ki} = \frac{(Score)_{ki}}{\langle (Score)_{ki} \rangle_i}, \tag{8}$$

where $(Score)_{ki}$ is a measure of chemical environmental impact, such as $1/LD_{50}$, $1/LC_{50}$, etc., and $\langle (Score)_{ki} \rangle_i$ is the arithmetic average of all the nonzero *Scores* in impact category *i*. The normalization of equation (8) ensures that all of the ψ^s_{ki} are numerically of the same order of magnitude so that they can be added across different impact categories.

The WAR algorithm uses the following techniques for estimating the PEI of a chemical. These estimations are used to describe relative potency of the chemicals in eight PEI categories. The impact categories used within the WAR algorithm are not inclusive of all possible impact categories, but do represent those categories that are considered to represent the most significant environmental concerns to the chemical manufacturing industry. Impact categories that have not been incorporated into the WAR algorithm include land use, resource depletion, noise, odor, etc.

TABLE 12.1
The potential environmental impact categories used within the WAR algorithm

Local toxicological		*Global*	*Regional*
Human	*Ecological*	*atmospheric*	*atmospheric*
Human toxicity potential by ingestion	Aquatic toxicity potential	Global warming potential	Acidification potential
Human toxicity potential by inhalation and dermal exposure	Terrestrial toxicity potential	Ozone depletion potential	Photochemical oxidation potential

The environmental impacts measures used are based on the work of Young (in preparation) and Young and Cabezas (1999). The *Score* measures of environmental impacts used in equation (8) fall into four general categories: local human toxicity, local ecological toxicity, regional atmospheric, and global atmospheric (Table 12.1).

There are four local toxicological categories: human toxicity potential by ingestion (HTPI); human toxicity potential by exposure, dermal and inhalation (HTPE); aquatic toxicity potential (ATP); terrestrial toxicity potential (TTP). There are two global atmospheric categories: global warming potential (GWP) and ozone depletion potential (ODP). There are two regional atmospheric categories: acidification or acid rain potential (AP) and photochemical oxidation or smog formation potential (PCOP).

The *scores* for human toxicity potential by ingestion (HTPI) and terrestrial toxicity potential (TTP) for a chemical i were determined by

$$(Score_i)_{HTPI} = (Score_i)_{TTP} = \frac{1}{(LD_{50})_i}, \tag{9}$$

where LD_{50} is the oral dose of chemical i that produced a 50 % death rate in test rats (National Institute for Occupational Safety and Health, 1999; Lewis, 1996; National Library of Medicine, 1999). The scores for aquatic toxicity potential (ATP) for a chemical i were determined by

$$(Score_i)_{ATP} = \frac{1}{(LC_{50})_i}, \tag{10}$$

where LC_{50} is the concentration that produced a 50 % death rate in the test species, *Pimephales promelas* (fathead minnows) (U.S. Environmental Protection Agency, 1998; U.S. Environmental Protection Agency, 1999; National Library of Medicine, 1999; Prager, 1995; Verschueren, 1996). The scores for human toxicity by exposure (HTPE) were obtained from time-weighted averages of threshold limit values (National Institute for Occupational Safety and Health, 1998; American Conference of Governmental Industrial Hygienists, 1994; U.S. Department of Labor, 1998).

The scores for the two regional atmospheric and the two global atmospheric categories were obtained from the work of (Heijungs et al., 1992) and directly inserted into equation (8). The ARP was obtained by dividing the rate of release in the atmosphere H^+ of promoted by chemical i to the same rate promoted by SO_2. The PCOP was obtained by dividing reaction rate of a unit mass of chemical i with hydroxyl radical ($OH\bullet$) by the same rate for ethylene. The GWP was obtained by dividing the absorption of infrared radiation by a unit mass of a chemical i by the absorption by a unit mass of CO_2 over a span of 100 years. The ODP was obtained by dividing the rate at which a unit mass of chemical i reacts with ozone to form molecular oxygen to the same rate for CFC-11 (trichlorofluoromethane).

4 MULTIOBJECTIVE OPTIMIZATION

The WAR algorithm defined earlier provides a first approximation toward defining environmental objectives; however, the environmental impact indexes and economic objectives are so different in terms of evaluations, quantification, and magnitude that the choice of relative weights for environmental and economic impacts depends upon the decision-makers' perspectives. Thus, it is necessary to provide decision-makers with the complete economic-environmental surface so that they would know the full range of alternatives and understand the trade-offs among the objectives implied by each alternative before making their selection.

As is well known, mathematics cannot isolate a unique optimum when there are multiple competing objectives. Mathematics can, at most, aid designers to eliminate design alternatives dominated by others, leaving a number of alternatives in what is called the Pareto set (Hwang et al., 1980). For each of these designs it is impossible to improve one objective without sacrificing the value of another relative to some other design alternative in the set. From among the dominating solutions, it is then a value judgment by the customer to select which design is the most appropriate. At issue is an effective means of finding the members of the Pareto set for a design problem, especially when there are more than two or three objectives; the analysis per design requires significant computations to complete, and there are an almost uncountable number of design alternatives.

The Generalized WAR algorithm expresses environmental objectives in terms of potential impact indexes combined together using weighing factors, α_i. This formulation used in the WAR algorithm can be easily expressed in terms of the weighting method for multiobjective optimization problem, where different weights are assigned to obtain the Pareto surface. The generalized formulation is shown below.

The weighting method based on the WAR algorithm

$$Min.\dot{I}_{out}^{(NP)} = \sum_{i=1}^{EnvCat} \alpha_i \dot{I}_i^{(NP)} \tag{11}$$

mass and energy balance constraints bounds with decision variables.

The basic strategy of the weighting method is to transform the multiobjective problem into a series of single-objective problems by giving weighting factors to each objective. The Pareto set can be derived by solving the large number of optimization problems created by modifying the weighting factors of each of the objectives; however, the major disadvantage of using the weighting method is its inefficiency that arises because of the linear combination of objectives; it is also difficult to control the region of the nondominated surface on which the decision-maker is heavily favored.

The constraint method is another technique for generating an approximation of the Pareto set. The basic strategy is also to transform the multiobjective problem into a series of single objective problems. A purely algorithmic solution approach (Cohon, 1978) is to select one of the objectives to maximize (e.g., profit) while each of the other objectives (e.g., environmental impacts) is turned into an inequality constraint with a parametric right-hand side ($\varepsilon_1, \ldots, \varepsilon_p$). It is important to note that each optimal design in the Pareto set that is derived from a combination of $\alpha_i\,(i = 1, \ldots, EnvCat)$ by the weighting method can also be generated from a corresponding combination of $\varepsilon_i\,(i = 1, \ldots, EnvCat)$ using the constraint method. One method is a mapping of the other. For example, the upper bound of ε_i used in the constraint method corresponds to $\alpha_i = 0$ in the weighting method, and the lower bound of ε_i used in the constraint method corresponds to $\alpha_i = +\infty$ in the weighting method. In other words, the upper and lower bounds of ε_i cover the entire range between $0 - +\infty$ for α_i. The constraint method offers the advantages of better control over exploration of the nondominated set and of locating points anywhere along the nondominated surface. Note profit has been included in the formulation shown below.

The constraint method based on profit

$$Max.\ Profit \tag{12}$$

subject to

$$\dot{I}_i^{(NP)} \leq \varepsilon_i, \qquad i = 1, \ldots, EnvCat$$

mass and energy balance constraints bounds with decision variables.

The Pareto set for various environmental impacts and economic objectives involves solutions of a large number of optimization problems. For example, if there are 6 objectives and 5 of them are evaluated over 10 levels, we would have to solve 100,000 optimization problems. To circumvent this problem, a new and efficient multiobjective optimization algorithm has been developed (Fu and Diwekar, in preparation). Figure 12.1 shows the generalized multiobjective framework for finding maximum profit and minimum environmental impacts designs; however, for the sake of brevity, only the representative results in terms of the bounds of the different objectives will be presented. This is equivalent to approximating the Pareto set by obtaining optimal designs with two values of $\varepsilon_i\,(i = 1, \ldots, EnvCat$ and $i = Profit)$ for each objective function ($\dot{I}_i, i = 1, \ldots, EnvCat$ and $Profit$). For the new algorithm, this represents the first step toward obtaining the Pareto set.

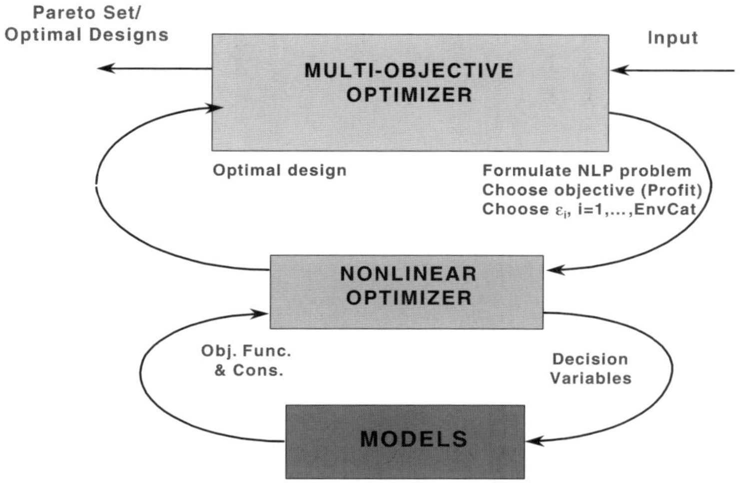

Figure 12.1: An efficient multiobjective optimization framework.

The following case study illustrates the advantages of the multiobjective optimization framework to obtain profitable and environmentally friendly designs. The results are presented in terms of lower and upper bounds of the different objectives as a representative of the complete Pareto set.

5 CASE STUDY: BENZENE PRODUCTION

The hydrodealkylation (HDA) of toluene to produce benzene, often classified as a benchmark example in chemical process synthesis, is used in this study. The flowsheet is described below.

5.1 FLOWSHEET

The HDA process has been extensively studied by Douglas (1988) using a hierarchical design/synthesis approach. The problem presented and solved here is based on the flowsheet structure analyzed by Diwekar et al. (1992), which involved the selection of the flowsheet configuration and some of the operating conditions that maximize profit.

The primary reaction of the HDA process is

$$C_6H_5CH_3 + H_2 \longrightarrow C_6H_6 + CH_4. \tag{13}$$

In addition to this desired reaction, an undesired reaction,

$$2C_6H_6 \longleftrightarrow C_6H_5 - C_6H_5 + H_2, \tag{14}$$

also occurs. The primary reaction is exothermic. These transformations are carried out in an adiabatic reactor. The furnace shown in Figure 12.2 is used

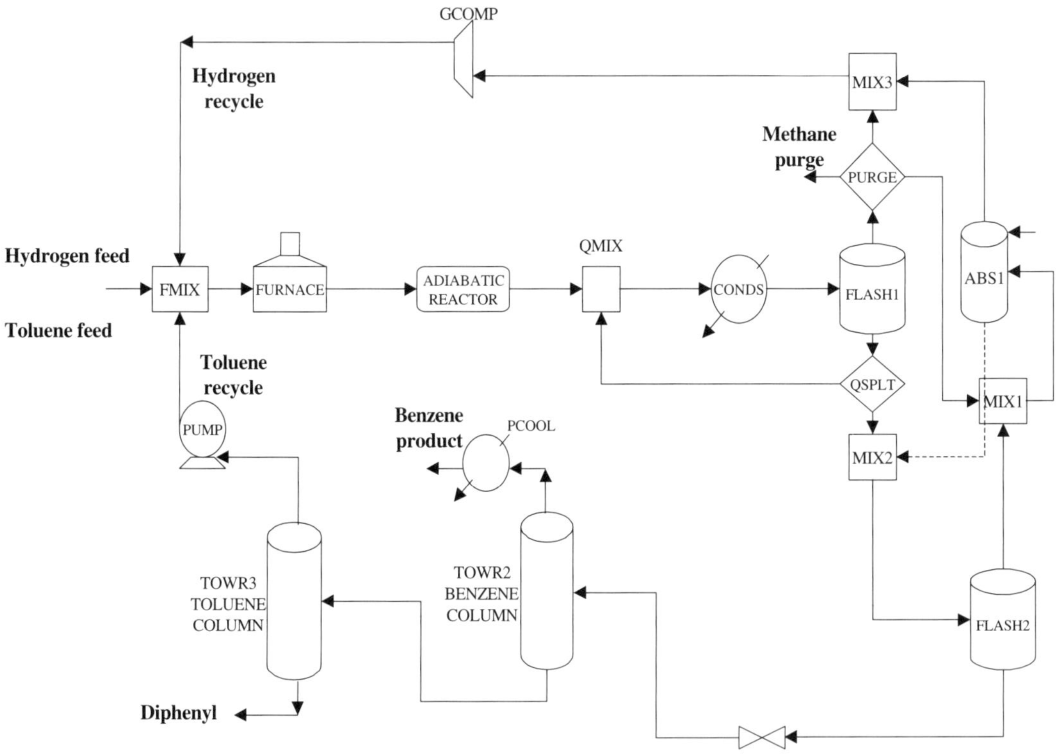

Figure 12.2: The process flowsheet for the production of benzene through the hydrodealkylation (HDA) of toluene.

to preheat the feed. These homogeneous gas-phase reactions occur in the range of 894 K and 974 K. A molar ratio of at least 5:1 hydrogen to aromatics is maintained to prevent coking. The reactor effluents must be quenched to 894 K to prevent coking in the heat exchanger following the reactor.

The HDA flowsheet is shown in Figure 12.2. In this process, benzene is formed by the reaction of toluene with hydrogen, equation (13). The hydrogen feed stream has a purity of 95% (the rest is methane) and is mixed with a fresh inlet stream of toluene, a recycled toluene stream, and a recycled hydrogen stream. The feed mixture is heated in a furnace before being fed to an adiabatic reactor. The reactor effluent contains unreacted hydrogen and toluene, benzene (the desired product), diphenyl (also known as biphenyl), and methane. It is quenched and subsequently cooled in a flash separator to condense the aromatics from the noncondensable hydrogen and methane. The vapor stream from the flash unit contains hydrogen that is recycled. The liquid stream contains traces of hydrogen and methane that are separated from the aromatics in a secondary flash unit. The liquid stream from the secondary flash unit consisting of benzene, diphenyl, and toluene is separated in two distillation columns. The first column separates the product, benzene, from diphenyl and toluene, whereas the second separates the diphenyl from toluene, which is recycled back into the reactor.

The important parameters, molar flow rate of the hydrogen feed, molar flow rate of the toluene feed, furnace temperature, and conversion of the adiabatic reactor, have been chosen as the decision variables for this multi-objective analysis. The bounds and initial values on these decision variables are shown in Table 12.2. The different objectives include maximizing the annualized profit and minimizing the different environmental impacts of

TABLE 12.2
Environmental impact indexes for the components in the HDA process

	HTPI	*HTPE*	*ATP*	*TTP*	*GWP*	*ODP*	*PCOP*	*AP*
Hydrogen	0	0	0	0	0	0	0	0
Methane	0	0	0.057	0	0.0035	0	0.014	0
Benzene (product)	0	0	0	0	0	0	0	0
Toluene	0.078	2.2E-06	0.065	0.078	0	0	1.2	0
Diphenyl (pollutant)	0.12	0.0016	0.88	0.12	0	0	0	0
Diphenyl (byproduct)	0	0	0	0	0	0	0	0

AP — acid rain potential;
ATP — aquatic toxicity potential;
GWP — global warming potential;
HTPE — human toxicity potential by dermal/inhalation exposure;
HTPI — human toxicity potential by ingestion;
ODP — ozone depletion potential;
PCOP — photochemical oxidation potential;
TTP — terrestrial toxicity potential.

the output as calculated by the WAR algorithm subjected to the following constraints:

- The purity of the benzene product is at least 95 %
- The benzene production rate must be maintained at 120 kmol/hr
- The adiabatic reactor must have a volume less than 500 m^3
- The hydrogen feed must have a purity of 95 %

This process is modeled by using the ASPEN simulator. The highly nonlinear nature of the flowsheet calls for a nonlinear multiobjective optimization algorithm to find the optimal designs.

5.2 ENVIRONMENTAL OBJECTIVES

The PEI indexes for each of the chemicals presented in the HDA process are listed in Table 12.2. The ODP indexes and ARP indexes for all components are zero; hence there is no need to include them as separate objectives. Furthermore, the indexes for all components of human toxicity potential by ingestion (HTPI) and terrestrial toxicity potential (TTP) are equivalent, and the optimal solutions for minimizing or maximizing them are the same; hence, they are listed together. This study includes six objectives total: HTPI or TTP, HTPE, ATP, GWP, PCOP, and an economic objective defined in the next subsection.

5.3 ECONOMIC OBJECTIVE

The economic objective is represented in terms of the annualized profit. The cost model (Diwekar et al., 1992) is represented by linear fixed-charge costs. The data for the model are provided in Table 12.3. The annualized profit is given by

$$Profit(\$10^3/year) = \sum_{i=1}^{7} C_i^* F_i - \sum_{i=1}^{3} C_{ui}^* U_i - \sum_{i=1}^{5} (C_{eqfi} + C_{eqvi}^* V_{eqi}). \quad (15)$$

The notation is provided in Table 12.3. In this illustrative example, the cost for waste treatment has been neglected but would be included in a detailed economic optimization problem.

5.4 THE MULTIOBJECTIVE OPTIMIZATION FRAMEWORK

The multiobjective optimization problem is represented in terms of the constraint method shown in equation (12). Figure 12.3 shows the schematic diagram of the multiobjective optimization framework for the HDA process

TABLE 12.3
Cost data for HDA process

Streams		*Cost/price (\$/kmol)*	*Symbols*	*Symbols for molar flow rate of streams ($\times 10^3$ kmol/year)*
Benzene product	$\geq$ 95% benzene	19.90	C_1	F_1
Diphenyl (byproduct)		11.84	C_2	F_2
Methane purge	(heating value)	3.37	C_3	F_3
Hydrogen purge	(heating value)	1.08	C_4	F_4
Hydrogen feed	95% hydrogen, 5% methane	2.50	$-C_5$	F_5
Toluene feed	100% toluene	14.00	$-C_6$	F_6
Diphenyl (pollutant)		0.00	$-C_7$	F_7

Utilities	*Costs*	*Symbols*	*Symbols for amount of utilities ($\times 10^{-3}$ kJ/year)*
Heating (stream)	$\$8.0 \times 10^6/kJ$	C_{u1}	U_1
Cooling (water)	$\$0.7 \times 10^6/kJ$	C_{u2}	U_2
Fuel	$\$4.0 \times 10^6/kJ$	C_{u3}	U_3

Investments	*Fixed-charge costs (\$$10^3$/year)*	*Symbols*	*Linear coefficient (\$$10^3$/year)*	*Symbols*
Compressor	7.155	C_{eqf1}	$0.815 \times$ bhp (kW)	$C^{ast}_{eqv1} V_{eq1}$
Benzene column	16.30	C_{eqf2}	$1.55 \times$ no. of trays	$C^{ast}_{eqv1} V_{eq2}$
Toluene column	3.90	C_{eqf3}	$1.12 \times$ no. of trays	$C^{*}_{eqv1} V_{eq3}$
Furnace	34.5	C_{eqf4}	$1172 \times$ heat duty (10^6 kJ/year)	$C^{*}_{eqv1} V_{eq4}$
Adiabatic reactor	74.30	C_{eqf5}	$1.257 \times$ reactor volume (m_3)	$C^{*}_{eqv1} V_{eq5}$

implemented in the public version of the ASPEN™ simulator. The multiobjective optimization algorithm formulates the nonlinear programming (NLP) problems for obtaining each point in the Pareto set using the new algorithm (Fu and Diwekar, in preparation). The new multiobjective optimization algorithm minimizes the number of NLPs problems to be solved, along with optimizing the various objectives. The ASPEN™ input file of the HDA process defines the detailed information for the process, including stream and block connectivities, calculation sequences, loop convergence parameters, etc. The multiobjective optimization algorithm and the NLP optimizer are implemented in the ASPEN™ simulator as unit operation blocks. The decision variables are transferred from the NLP optimizer to the simulator

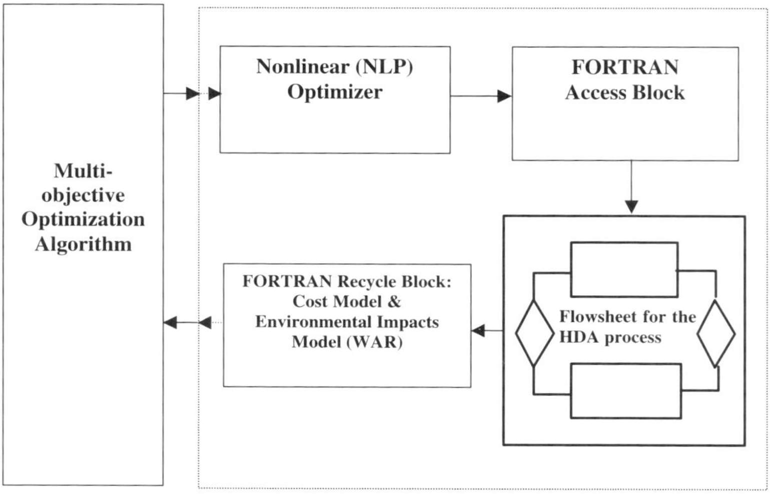

Figure 12.3: A schematic diagram of the multiobjective optimization framework for the HDA process using the public version of ASPEN.

model using the FORTRAN access block. Once ASPEN™ finishes evaluating the design according to certain decision variables, the FORTRAN recycle block calculates the cost and environmental impacts of this design. The NLP optimizer determines new decision variables, and the inner NLP optimization cycle repeats until the optimal solution is found. The outer multiobjective optimization algorithm then defines a new single-objective problem to be solved by the inner cycle. The iterations continue until a reasonable approximation to the Pareto set is obtained (in minimum cycles of NLPs).

In the current study we are only presenting 10 optimal designs, representing the first approximation towards the Pareto set.

6 RESULTS AND DISCUSSIONS

In the HDA process, benzene is the desirable product obtained by the HDA of toluene. The raw materials are toluene and hydrogen. Diphenyl, which is also formed during this process, can either be sold as a byproduct or treated as a pollutant. The following section presents two cases: one in which diphenyl is treated as a pollutant and one where diphenyl is treated as a byproduct.

6.1 DIPHENYL AS A POLLUTANT

In this case, diphenyl is considered to be a pollutant, which is a distinction given by the WAR algorithm to all nonproduct compounds. Diphenyl is

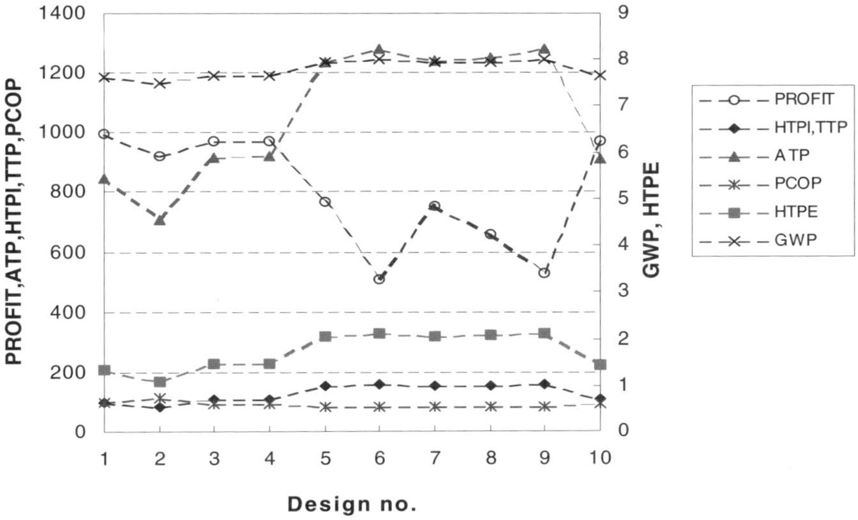

Figure 12.4: Objectives of the 10 different designs for the HDA process with diphenyl as a pollutant.

one of the major contributors to the PEIs as shown in Table 12.2. Because diphenyl is a pollutant, it will not be considered to contribute to the annualized profit, which can be seen from Table 12.3, where its price is quoted as $0/kg-mol.

Table 12.4 shows 10 different design cases that were obtained by minimizing and maximizing each of the six objectives (profit and five different environmental impacts); duplicate designs were removed. Table 12.4 shows the values of the four decision variables, the six objectives, and the output flow rates of the five components for each of ten different designs.

Figure 12.4 shows different environmental impacts and the profit of these 10 designs with diphenyl as a pollutant. Figure 12.4 shows that the trend of environmental impacts ATP, HTPI, TTP, GWP, and HTPE all follow a similar pattern. Diphenyl is the biggest contributor to the environmental impacts ATP, HTPI, TTP, and HTPE. The diphenyl effluent stream is significant enough to dominate those impact categories. Although, the dominant contributor to the GWP is methane and not diphenyl, from Table 12.4 it can be seen that the output flow rate of methane follows a similar trend to the output flow rate of diphenyl; hence, the GWP also displays the same trend as the ATP, HTPI, TTP, and HTPE. The indexes for PCOP for the 10 optimum designs seem to be insensitive to the different designs. The dominant contributor to the PCOP, on the other hand, is toluene. Recall, toluene is a raw material for the HDA process, and therefore profit is lower for designs that require a high toluene flow.

TABLE 12.4
Results of 10 different designs for the case of treating diphenyl as a pollutant

	Design no.									
	1	*2*	*3*	*4*	*5*	*6*	*7*	*8*	*9*	*10*
Decision variables										
Hydrogen feed (kmol/hr)	248.71	255	246.82	246.77	245	255	245.55	248.73	255	246.52
Toluene feed (kmol/hr)	127.87	125.63	128.99	129.02	133.81	134.49	133.81	133.72	134.51	128.73
Conversion of adiabatic reactor	0.62395	0.5783	0.642	0.6425	0.7	0.7	0.7	0.7	0.7	0.64054
Furnace temperature (F)	1095	1100	1100	1100	1100	1105	1105	1100	1100	1100
Objectives										
Profit (10^3/year)	996.7	922.99	970.09	969.76	767.55	508.508	752.389	658.131	530.152	973.224
HTPI,TTP	101.752	83.597	110.85	111.12	152.33	158.065	152.663	154.439	158.074	109.886
HTPE	1.33723	1.076	1.4665	1.4704	2.0492	2.12734	2.05375	2.07801	2.12746	1.45302
ATP	848.34	706.44	918.7	920.81	1236.3	1279.72	1238.8	1252.03	1279.8	911.193
GWP	7.60787	7.4967	7.666	7.6678	7.9351	8.01215	7.93713	7.9461	8.01319	7.65057
PCOP	100.727	112.29	96.681	96.571	85.648	86.8944	85.6951	85.9503	86.904	96.8563
Total output flow rate for each component (kg/hr)*										
Hydrogen	228.309	242.87	223.45	223.32	214.77	233.043	215.859	222.279	233.004	223.298
Methane	2178.66	2146.8	2195.3	2195.8	2272.4	2294.43	2272.95	2275.51	2294.73	2190.88
Toluene	59.9778	70.372	56.274	56.172	45.779	46.5821	45.8125	46.0012	46.5867	56.4803
Diphenyl	814.806	655.58	893.59	895.96	1248.7	1296.3	1251.46	1266.24	1296.38	885.373
Benzene	9093.56	9071.1	9103.9	9104.2	9129	9132.8	9125.81	9103.4	9134.34	9091.7

*Total output flow rate for each component is the sum of mass flow rates from the three output streams, including the methane purge stream, benzene product stream, and the diphenyl stream.

Identifying a trend helps to combine similar trend objectives as one objective, thereby reducing the problem size considerably. Hence, we can combine the PEIs ATP, HTPI, TTP, GWP, and HTPE into one objective. The PCOP and the profit remain as separate objectives. Using trends in this manner reduces this six-objective optimization problem to a three-objective optimization problem. The complete Pareto set of this reduced optimization problem will not be presented here.

Figure 12.4 also depicts the trade-offs between the different objectives. This helps identify the best designs (i.e., those designs that are profitable as well as environmentally friendly). From Figure 12.4 it can be seen that design Nos. 1 and 2 are the best designs. The profit of design No. 1 is the highest among all 10 designs, and its environmental impacts (except PCOP) are lower than design Nos. 3 to 10. Design No. 2 has the best environmental performance (except PCOP), and its profit is only 7.4% less than design No. 1. Design Nos. 6 and 9 are the worst designs because they have the lowest profit and the highest environmental impacts.

6.2 DIPHENYL AS A BYPRODUCT

In this case study, diphenyl is assumed to be a byproduct that can be sold for $11.84/kg-mole (shown in Table 12.3). According to the WAR algorithm, the environmental impact values for product streams are set to zero; thus, in this case the environmental impact values for the diphenyl stream are set to zero (shown in Table 12.2).

Table 12.5 presents the results of 10 different designs for the HDA process considering diphenyl as a byproduct. These results were obtained by minimizing and maximizing each of the six objective functions using the nonlinear optimizer and removing the duplicate designs.

Similar to Figure 12.4, Figure 12.5 shows the trends of different objectives and the trade-offs between them for the case where diphenyl is treated as a byproduct. The trends for the PEIs are different than the previous study. For example, the ATP and GWP display similar trends (see Fig. 12.5). In this study, methane dominates those impact categories. The PCOP, HTPI, TTP, and HTPE impact categories behave similarly, as toluene is the major contributor to all of these categories. Therefore, in this case the ATP and GWP impact categories can be combined into one objective; the PCOP, HTPI, TTP, and HTPE impact categories combined into a second objective; and the profit defines the third objective. Note that the two systems that have been described in this chapter reduced to three objectives; however, these reduced objectives were defined differently.

Multiobjective framework helps to discard bad designs (e.g., design No. 8) that have lower profits and higher environmental impacts. Among the better designs are the ones that have (1) higher profit, lower environmental impacts, (2) lower environmental impacts at the expense of small profit loss, (3) higher profit at the expense of a very small amount of higher environmental impacts, or (4) lower environmental impacts for some categories

TABLE 12.5
Results of different designs for the case of treating diphenyl as a byproduct

	Design no.									
	1	*2*	*3*	*4*	*5*	*6*	*7*	*8*	*9*	*10*
Decision variables										
Hydrogen feed (kmol/hr)	248.71	248.21	246.76	246.73	246.77	245	255	255	248.52	246.52
Toluene feed (kmol/hr)	127.87	133.83	129.06	128.64	129.02	133.81	134.49	125.63	133.84	128.73
Conversion of adiabatic reactor	0.62395	0.7	0.643	0.6387	0.6425	0.7	0.7	0.57828	0.7	0.64054
Furnace temperature (F)	1095	1100	1100	1100	1100	1100	1105	1100	1100	1100
Objectives										
Profit (10^3/year)	1059.3	776.7	1038	1043	1039	863.55	608.17	974.14	768.93	1041.3
HTPI,TTP	4.6871	3.594	4.382	4.443	4.39	3.5775	3.6403	5.5004	3.5958	4.4138
HTPE	0.0001	1E-04	1E-04	1E-04	1E-04	0.0001	0.0001	0.0002	0.0001	0.0001
ATP	128.05	132.7	128.8	128.5	128.8	132.48	133.79	126.91	132.76	128.52
GWP	7.6079	7.95	7.67	7.646	7.668	7.9351	8.0121	7.497	7.9516	7.6506
PCOP	100.73	85.96	96.47	97.27	96.57	85.648	86.894	112.31	85.987	96.856
Total output flow rate for each component (kg/hr)*										
Hydrogen	228.309	221.03	223.23	223.79	223.32	214.766	233.043	242.853	221.604	223.298
Methane	2178.66	2276.6	2196.4	2189.5	2195.8	2272.37	2294.43	2146.9	2277.08	2190.88
Toluene	59.9778	45.993	56.075	56.852	56.172	45.779	46.5821	70.3859	46.0136	56.4803
Diphenyl	814.806	1264	898.41	877.29	895.96	1248.68	1296.3	655.39	1265.43	885.373
Benzene	9093.56	9114.8	9104.8	9092.6	9104.2	9129	9132.8	9071.42	9113.85	9091.7

*Total output flow rate for each component is the sum of mass flow rates from the three output streams including the methane purge stream, benzene product stream, and the diphenyl stream.

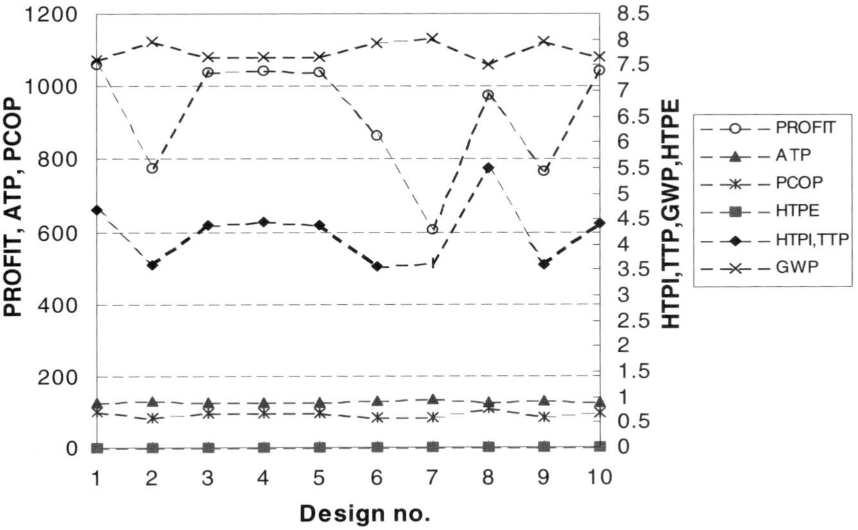

Figure 12.5: Objectives of the 10 different designs for the HDA process with diphenyl as a byproduct.

at the expense of a small amount of higher environmental impacts for other categories. It is then a value judgment of the decision-makers as to which design among the better designs is the most appropriate one.

As with the previous system, the trend for profit follows the trends for some environmental impacts: HTPI, TTP, PCOP, and HTPE; however, the desirable and undesirable designs are different from the earlier case. In this system the undesirable designs can clearly be eliminated; however, the best designs are difficult to identify. For example, design No. 8 can be eliminated as compared to design No. 1 because No. 8 has higher environmental impacts and lower profit.

6.3 COMPARISON OF TWO CASE STUDIES

The results listed in Tables 12.4 and 12.5 can be compared to identify similar designs. These similar designs are tabulated in Table 12.6. For those designs that are found in both cases, three trends were observed: (1) As expected, the profit is higher with diphenyl as a byproduct than that with diphenyl as a pollutant. (2) When diphenyl is treated as a pollutant, the environmental impact categories HTPI, TTP, HTPE, and ATP are higher. This is due to the fact that diphenyl is the dominant component for these categories. (3) The environmental impacts GWP and PCOP are the same for both cases. Because the environmental impact indexes of the GWP and PCOP for diphenyl are zero, diphenyl does not contribute to either the GWP or the PCOP.

TABLE 12.6
Same designs found in both case studies

Design No.	*Diphenyl as a pollutant*	*Diphenyl as a byproduct*
1	1	1
2	4	5
3	5	6
4	6	7
5	2	8
6	10	10

These two studies illustrate the use of multiobjective optimization framework for obtaining desirable designs and eliminating undesirable designs. The profit does not always conflict with the environmental objectives. In fact, designs can be obtained that are relatively benign and cost effective.

Although the current paper presents only 10 designs for each case study, the complete Pareto set can provide more insight into the trends and trade-offs, and as a result can provide better designs; however, to get the more accurate Pareto surface, a large number of optimization problems need to be solved within the lower and upper bound of each objective, shown in Table 12.7, which will be the focus of the future work. Note that the average CPU time to obtain a single optimal solution is of the order of 1 minute on an open VMS DEC 400 workstation. In the future, the complete Pareto set will be obtained by sampling uniformly over objective space within the lower and upper bounds of different environmental impacts (shown in Table 12.7) so that it can ensure that representative designs for all regions of the Pareto set are being considered.

TABLE 12.7
Bounds for each objective of two case studies

	Diphenyl as a pollutant		*Diphenyl as a byproduct*	
Objectives	*Lower bound* ε_{low}	*Upper bound* ε_{up}	*Lower bound* ε_{low}	*Upper bound* ε_{up}
Profit ($\$10^3$/year)	508.508	996.7	608.17	1059.3
HTPI,TTP	83.597	158.074	3.5775	5.5004
HTPE	1.076	2.12746	0.0001	0.0002
ATP	706.44	1279.8	128.05	133.79
GWP	7.4967	8.01319	7.6079	8.0121
PCOP	85.648	112.29	85.648	112.31

7 CONCLUSIONS

This chapter presented a methodology to quantify and integrate environmental and economic objectives in all stages of process design. The WAR algorithm provided a mechanism to quantify the PEI of the process designs. The problem of designing for the environment is posed as a multiobjective optimization problem involving several conflicting and synergistic objectives. An optimization framework has been presented that allows for the efficient determination of the Pareto surface for a multiobjective problem. The Pareto surface provides a means to assess the trade-offs among the different objective functions. A case study illustrated the methodology to augment traditional chemical process simulators, such as ASPEN, with the capacity to design for environmentally friendly and cost-effective processes. The chapter provided a first step toward the sustainable design in terms defining and quantifying the environmental and economic impacts and integrating it into the design process; however, to address the issues of inaccuracies or lack of data in quantifying the environmental indexes, the problem of uncertainties must be dealt with. In the future, the framework will be extended to include the uncertainty impacts.

8 ACKNOWLEDGMENTS

The authors wish to thank the National Risk Management Research Laboratory within the United States Environmental Protection Agency for moral and financial support of this work, in particular, E. Timothy Oppelt, Subhas Sikdar, and Teresa Harten.

9 SYMBOLS

$I^{(cp)}$	the amount of PEI in a chemical process (units of PEI)
$\dot{I}_i^{(in)}$	the rate of PEI of category i entering a chemical process (units of PEI/time)
$\dot{I}_i^{(NP)}$	the rate of PEI of category i leaving a chemical process including only nonproduct streams (units of PEI/time)
$\dot{I}_{gen}^{(cp)}$	the rate of generation of PEI within a chemical process (units of PEI/time)
$\hat{I}_{gen}^{(cp)}$	the PEI generated within a chemical process per mass of product streams leaving the process (PEI/mass of products)
$\dot{I}_{in}^{(cp)}$	the rate of PEI entering a chemical process (units of PEI/time)
$\dot{I}_{out}^{(cp)}$	the rate of PEI leaving a chemical process (units of PEI/time)
$\hat{I}_{out}^{(cp)}$	the PEI leaving a chemical process per mass of product streams leaving the process (PEI/mass of products)
$\dot{M}_j^{(in)}$	the mass flow rate of stream j into a process (units of mass/time)

$\dot{M}_j^{(out)}$ the mass flow rate of stream j out of a process (units of mass/time)
$\dot{P}_P$ the mass flow rate of product stream p (units of mass/time)
$(Score)_{ki}$ a characteristic quantity of chemical k used to determine a PEI value of that chemical for impact category i (units vary for each category)
$\langle(Score)_{ki}\rangle_i$ the average value of all k chemicals, with nonzero values, in category i (units vary for each category)
x_{kj} the mass fraction of chemical k in stream j (units of mass of chemical k/mass of stream j)

GREEK SYMBOLS

α_i the weighting factor for impact category i (dimensionless)
ψ_{ki}^s the specific PEI of chemical k for impact category i (units of PEI/mass of chemical k)

REFERENCES

American Congress of Governmental Industrial Hygienists (1994). *Threshold Limit Values and Biological Exposure Indices for 1994-1995*. Cincinnati, OH.

Cabezas, H., et al. (1997). Pollution prevention with chemical process simulators: The generalized waste reduction (WAR) algorithm. *Comp. Chem. Eng.* 21:S305–S310.

Cabezas, H., et al. (1999). Pollution prevention with chemical process simulators: The generalized waste (WAR) algorithm—full version. *Comp. Chem. Eng.* 23:625.

Cano-Ruiz, J. A., and McRae, G. J. (1998). Environmentally conscious chemical process design. *Annu. Rev. Energy Environ.* 23:499–536.

Cohon, J. (1978). *Multiobjective Programming and Planning*. Academic Press, San Diego, CA.

Diwekar, U. M., et al. (1992). An MINLP process synthesizer for a sequential modular simulator. *Ind. Eng. Chem. Res.* 31:313–322.

Douglas, J. M. (1988). *Conceptual Design of Chemical Processes*. McGraw-Hill, New York.

Fu, Y., and Diwekar, U. M. (in preparation). A New and Efficient Multi-objective Optimization Algorithm Based on the Hammersley Sequential Sampling Technique. To be submitted to Operation Research.

Heijungs, R., et al. (1992). *Environmental Life Cycle Assessment of Products—Guide*. Centre of Environmental Science, Leiden, Netherlands.

Hilaly, A. K., and Sikdar, S. K. (1994). Pollution balance: A new method for minimizing waste production in manufacturing processes. *J. Air Waste Manage. Assoc.* 44:1303–1308.

Hwang, C. L., et al. (1980). Mathematical programming with multiple objectives: A tutorial. *Comp. Oper. Res.* 7:5–31.

Lewis, R. J. (1996). *Sax's Dangerous Properties of Industrial Materials*, ed. 9. Van Nostrand Reinhold, New York.

Mallick, S. K., et al. (1996). A pollution reduction methodology for chemical process simulators. *Ind. Eng. Chem. Res.* 35:4128.

National Institute for Occupational Safety and Health (1998). *NIOSH Pocket Guide to Chemical Hazards*. Washington, DC.

National Institute for Occupational Safety and Health (1999). *Registry of Toxic Effects of Chemical Substances (RTECS)*. Washington, DC.

National Library of Medicine (1999). Hazardous Substances Data Bank (HSDB), Bethesda, MD.

Prager, J. C. (1995). *Environmental Contaminant Reference Databook*. Van Nostrand Reinhold, New York.

Rossiter, A. P. (1994). Process Integration and Pollution Prevention. *Pollution Prevention via Process and Product Modifications*, Vol. 303(90) (M. M. El-Halwagi and D. Petrides), pp. 12–22. AIChE, New York.

U.S. Department of Labor Occupational Safety and Health Administration (1998). *29 Code of Federal Regulations 1910: Occupational Safety and Health Standards*. Washington, DC.

U.S. Environmental Protection Agency (1998). Aquatic Toxicity Information Retrieval (AQUIRE) Database. Duluth, MN.

U.S. Environmental Protection Agency (1999). ECOSAR 0.99e. Washington, DC.

Verschueren, K. (1996). *Handbook of Environmental Data on Organic Chemicals*, ed. 3. Van Nostrand Reinhold, New York.

Young, D. M. (in preparation) Estimating Rat Oral LD_{50} and Fathead Minnow LC_{50} by Molecular Methods. To be submitted to Environmental Toxicology.

Young, D. M., and Cabezas, H. (1999). Designing sustainable processes with simulation: The waste reduction (WAR) algorithm. *Comp. Chem. Eng.* 23:1477–1491.

CHAPTER 13

Environmental Impact and Environmental Risk Minimization via Process Design and Optimization

C. G. Vassiliadis, S. K. Stefanis, and E. N. Pistikopoulos

Centre for Process Systems Engineering,
Department of Chemical Engineering,
Imperial College, London, U.K.

Abstract

The need to include environmental impact criteria in the design phase of process manufacturing systems has been widely accepted by industry, governmental, and legislation bodies and the scientific community. This has led to major international initiatives to establish formal environmental impact assessment approaches, most of which, however, are rather qualitative in the form of general guidelines. In this work we present an optimization-based environmental risk minimization strategy that rigorously incorporates the various components of the environmental risk (i.e., the mechanism of occurrence of unexpected events), usually related to equipment failures, the severity of their consequences, the detailed process model, and process specifications/constraints, as well as explicit reliability and maintenance characteristics. A system effectiveness vector is defined, which is characterized by both the environmental criteria and the profitability of the system, the optimization of which identifies the optimal process design, process operation, and preventive maintenance strategies leading to environmental risk minimization. The approach is illustrated with a process example problem.

Keywords: environmental risk, system effectiveness, process design

1 INTRODUCTION

A major side effect of modern chemical process technologies and products is the increase of damage to the environment due to industrial releases.

Society, through regulating organizations, has responded to the growing concern over adverse environmental effects caused by major accidents by introducing strict legislative frameworks, such as the directive introduced by the European Commission after the Seveso incident in Northern Italy. At the same time, significant effort has been directed from the industry, environmental agencies, and the scientific community toward analyzing a broad range of aspects and parameters related to environmental degradation. Wahlstrom (1998), commenting on the contribution of industrial releases (routine and accidental) to the possible loss of stability of nature's dynamic equilibrium, Pearce (1991), studying previous industrial practices that led to environmental degradation and suggesting future policies for improvement, and Patton (1998) and Llewellyn (1998), describing the main features required in methods for environmental impact assessment, all stretch the need to build mathematical modeling tools that globally assess the consequences of industrial releases and to design environmentally friendly process systems and products.

Although most available methods dealing with environmental impact and risk are qualitative in nature [such as the utilization of checklists and networks, suggested by HMSO (Department of Environment, 1995), or the hazard identification techniques—HAZOP and FMEA—in Montague (1990)] and focused on postrelease calculations, systematic and quantitative frameworks have started to appear (El-Halwagi and Petrides, 1995).

To address the issue of integrating environmental impact assessment and minimization of routine releases (i.e., process effluent streams) in process design and operation, Pistikopoulos et al. (1995) have introduced a methodology for environmental impact minimization (MEIM), the main steps of which include the following:

- definition of a process system boundary and emissions inventory;
- environmental impact assessment;
- incorporation of environmental impact criteria in process design objectives.

By embedding Life Cycle Analysis (LCA) principles within an optimization framework comprising process specifications and environmental considerations, MEIM provides the best design alternative identifying the existing trade-offs between design economics and environmental impact.

Environmental risk, on the other hand, is associated with unplanned events through their probability and consequences, and hence, to minimize environmental risk, both the probability and the consequences have to be minimized. Consequences relate to the physicochemical characteristics of process design and operation. On the other hand, the probability of unplanned events (e.g., component failure) depends on the impact of reliability and operability upon process design and operation. As shown in the hypothetical risk frequency graph presented in Figure 13.1, the nonroutine's releases, which are related to unexpected plant operation, have significant

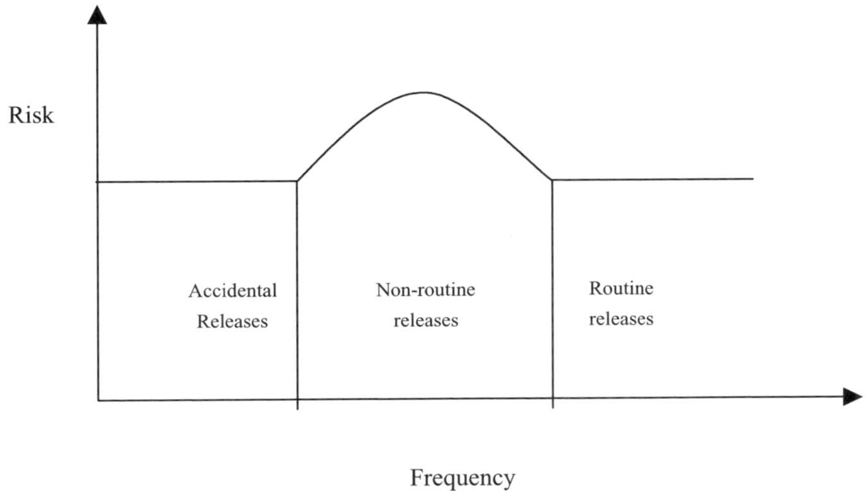

Figure 13.1: Risk frequency graph.

impact on the environmental damage. Unlike extreme cases of major accidents that occur at very low frequencies but with serious consequences and routine releases that are common but cause minor environmental damage, nonroutine releases are placed within this frequency range, often posing moderate adverse effects and therefore resulting in considerable risk levels.

Integrating environmental impact assessment techniques to incorporate risk-related events and nonroutine releases, Aelion et al. (1995) have introduced the frequency/environmental load curve according to which they distinguish release scenarios, depending on whether they result from intended or unintended plant operation, and an estimate of accidental releases load is expressed as a function of the frequency of unexpected events (expected number of accidents per year) and the environmental load during each accident. Although an assessment framework, no explicit information is given on appropriate changes in process design and operation.

The fact, however, that nonroutine releases are mainly attributed to deviations from planned operation and unexpected events, such as equipment breakdown, leaks, etc., suggests that in order to be kept to an acceptable level, high *availability* must be achieved for the environmentally critical process units. Highlighting the importance of availability to the performance of the plant in terms of safety and environmental impact, new and tighter regulations are imposed on industry, demanding for specific availability requirements of equipment, and significant research work in the literature has been concerned with estimating the reliability levels of safety and regulator systems in process plants (Papazoglou and Aneziris, 1990). An analysis of contemporary techniques and future trends in the use of reliability engineering methodologies to facilitate safer operation of process plants can be found in Michelsen (1998). Plant and unit availability is, by definition, a function of process and equipment *reliability* and *maintainability*, the key elements of which (e.g., failure rate characteristics, storage tanks, redundancy and safety systems, system configuration and accessibility to components, etc.) are uniquely determined during the design phase of the process (Kapur and Lamberson, 1977).

During the operating stage of the process, the way to achieve high equipment availability is through the derivation and execution of effective preventive maintenance strategies, such as inspection and replacement of critical units. Toward this direction we have witnessed an increasing interest in including environmental considerations and prioritizing maintenance activities with the objective to reduce environmental impact and risk. The benefits from employing a qualitative framework that considers environmental criteria to prioritize maintenance tasks in one of Exxon's chemical plants are reported in Harnly (1998). Furthermore, Stefanis and Pistikopoulos (1997) have recently extended the environmental impact assessment principles of MEIM to evaluate a criticality index for maintenance, ranking equipment components by the means of the effect their failure would have on process environmental performance. On the other hand, because maintenance in the chemical process industry is very expensive, approaching 30% of all operating costs (Tan and Kramer, 1997), maintenance optimization is traditionally defined as identifying the maintenance policy that optimizes the balance between maintenance costs and benefits (Dekker, 1996), the latter concerning both the minimization of environmental risk and the maximization of process profitability. Considering also the fact that both environmental impact and process profitability are influenced by the existing uncertainty in a large number of process parameters and models (Patton, 1998; Pistikopoulos, 1995), the need for a rigorous integrated framework to account for environmental impact considerations in conjunction with the complex interactions between different operability characteristics, such as flexibility, reliability, and maintenance, at the very early design stage of the process has been strongly emphasized (Van Rijn, 1987; Grievink, 1993). In this chapter a rigorous mathematical framework for addressing these issues is presented. In the next section, an overview of MEIM is given. Then, the various components of a general framework are defined and presented in detail to identify optimal design, operating, and maintenance policies in a simultaneous manner. The trade-offs between maintenance costs, environmental risk, and process revenue are quantified and illustrated by applying the proposed framework to a process example.

2 METHODOLOGY FOR ENVIRONMENTAL IMPACT MINIMIZATION: AN OVERVIEW

Most process design techniques for waste minimization aim at reducing emissions waste, not taking into consideration wastes associated with input to the process (raw materials, energy consumption, etc.). To overcome these drawbacks, the key element behind the development of MEIM is the incorporation of Life Cycle Analysis and Risk Assessment principles within a process optimization framework. The basic steps are discussed next.

2.1 DEFINITION OF SYSTEM BOUNDARY

This step involves expansion of the conventional process system boundary to include all processes associated with raw materials extraction and energy generation. This requires backtracking from the conventional process system all the way to the natural state of pure raw materials, which are available at no environmental penalty. Different technological routes for the production of the same set of raw materials are included in this expanded boundary. The advantage of defining such an extended process system boundary is that input wastes, together with their routes, can also be accounted for together with output emissions forming an aggregated waste vector. Note that although this definition is consistent with the one used in Life Cycle Analysis, it does not include the routes and stages of the product after leaving the process, as the main focus is on optimizing the damage related to a chemical manufacturing route.

2.2 ENVIRONMENTAL IMPACT ASSESSMENT

Having defined a "global" process system boundary, an assessment of the environmental impact of the various wastes is performed in this step. This involves (i) defining an emissions inventory comprising all wastes generated in any stage of the processing network within the global process systems boundary and (ii) grouping these wastes together according to their impact on the environment; this is termed *environmental impact assessment*.

Environmental impact is commonly assessed by defining appropriate environmental indices, which measure air pollution, water pollution, solid wastes, global warming, photochemical oxidation, and stratospheric ozone depletion. Air pollution is measured by defining a CriTical Air Mass, CTAM [=] kg air/h,

$$CTAM = \frac{\text{Mass of Air Emissions (kg pollutant/h)}}{\text{Standard Limit Value (kg pollutant/kg air)}}, \tag{1}$$

water pollution by a CriTical Water Mass, CTWM [=] kg water/h,

$$CTWM = \frac{\text{Mass of Water Pollutant (kg pollutant/h)}}{\text{Standard Limit Value (kg pollutant/kg water)}}, \tag{2}$$

solid wastes by a Solid Mass Disposal, SMD [=] kg solids/h.

In all of the above metrics, the mass of pollutant discharged is assumed to be measured at the point source of release. Long-term interactions like global warming can be depicted by metrics like Global Warming Impact, GWI [=] kg CO_2/h,

$$GWI = \text{Mass of Pollutant (kg/h)} \times \text{GWP (kg } CO_2\text{/kg pollutant)}, \quad (3)$$

where GWP is the global warming potential of each pollutant (Lashof and Ahuja, 1990). Similarly, photochemical oxidation is defined by Photochemical Oxidation Impact, POI [=] kg ethylene/h,

$$POI = \text{Mass of Pollutant (kg/h)} \times \text{POCP (kg } C_2H_4\text{/kg pollutant)}, \quad (4)$$

where POCP is the photochemical oxidation potential (U.K. Ecolabelling Board, 1993). Stratospheric ozone depletion is measured by Stratospheric Ozone Depletion Impact, SODI [=] kg CFC11/h,

$$SODI = \text{Mass of Pollutant (kg/h)} \times \text{SODP (kg CFC11/kg pollutant)}, \quad (5)$$

where SODP is the Stratospheric Ozone Depletion Potential (U.K. Stratospheric Ozone Review Group, 1988).

As a result, for each pollutant w, a vector EI_w can be obtained denoting its corresponding environmental impact, i.e.,

$$EI_w = [CTAM\ CTWM\ SMD\ GWI\ POI\ \text{SODI}]^{\text{T}}_{\text{w}}. \quad (6)$$

By summing up over all pollutants, a global environmental impact vector can be obtained indicating the environmental impact of the entire processing network,

$$GEI = \sum_{w=1}^{W} EI_w = [CTAM\ CTWM\ SMD\ GWI\ POI\ \text{SODI}]^{\text{T}}_{\text{process}}. \quad (7)$$

There are two advantages of using a global environmental impact vector (GEI):

- The vector of waste emissions typically comprising a large number of wastes can be transformed into an aggregated vector of low dimensionality.
- The information provided is directly linked to impact on the environment rather than to mass flow rates of waste materials, for example.

Note that this systematic aggregation of wastes relative to their environmental impact obviously can be used for both conventional and expanded (global) process system boundaries.

2.3 INCORPORATION OF ENVIRONMENTAL IMPACT CRITERIA IN PROCESS SYNTHESIS AND DESIGN OPTIMIZATION

The third step of MEIM involves the direct incorporation of environmental impact criteria in a conceptual process synthesis formulation. The superstructure of the continuous process to include alternative raw materials (such as air or pure oxygen) and separation techniques (steam stripping, distillation) in the most general case can be modeled as a multiobjective MINLP formulation:

$$\min_{x,y}[c^T y + f(x), GEI] \tag{P1}$$

$$\begin{aligned} &\text{subject to} \\ &h(x) = 0, \\ &g(x) \leq 0, \\ &A \cdot x = a, \\ &B \cdot y + C \cdot x \leq d, \\ &x \in X, y \in Y, \\ &GEI(x, y) = [CTAM\ CTWM\ SMD\ GWI\ POI\ \mathrm{SODI}]^{\mathrm{T}}_{\mathrm{process}}. \end{aligned}$$

The continuous variables x represent flows, operating conditions, and design variables. The binary variables y denote the potential existence of process unit blocks and streams. These variables typically appear linearly, as they are included in the objective to represent fixed charges in the purchase of process equipment (in the term $c^T y$) and in the constraints to enforce logical conditions ($B \cdot y + C \cdot x \leq d$). The term $f(x)$ is often a linear term involving purchase costs for process equipment (cost coefficients multiplying equipment capacities or sizes), raw materials purchase costs, utility costs, etc. The sizing equations correspond to $h(x) = 0$, and the inequality constraints $g(x) \leq 0$ include design specifications, which are typically linear inequalities.

Algorithms for solving such multiobjective optimization problems can be found in Acevedo and Pistikopoulos (1996). The solution of the above problem for a fixed structure (fixed y) yields the Pareto curve of noninferior solutions, as shown in Figure 13.2a. If structural alternatives are also included, the solution may be discontinuous, as shown in Figure 13.2b, where different segments correspond to different optimal structural arrangements.

2.4 NUMERICAL EXAMPLE

To gain some more insight, consider the simplified reaction/separation chemical process in Figure 13.3. The process involves two reactors, R-1 and

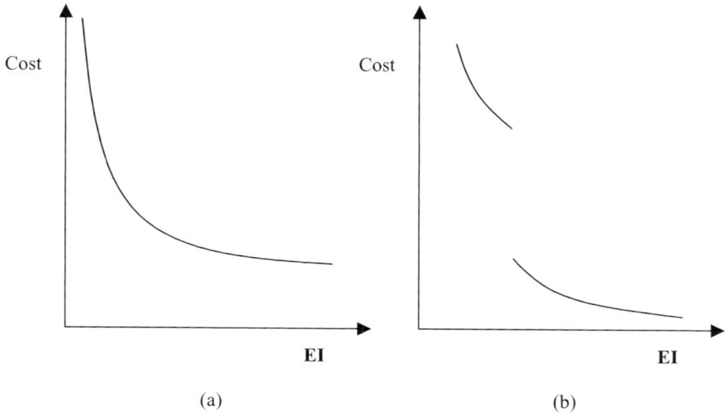

Figure 13.2: Pareto curve for fixed structure.

R-2 in parallel, where two isothermal reactions take place for the production of chemical C from reactant gases A, B, and D, according to a reaction scheme. The system also comprises a flash drum for the separation and two compressors [the main compressor CR-1 for the feed stream (A,B,D) and the recycle compressor CR-2 for the gas mixture (A,B,C,D)]. The chemicals that react in reactors R-1 and R-2 can be both from the feed stream and the stream that is recycled with compressor CR-2. The amount involved in each reaction is determined with a splitting decision (degree

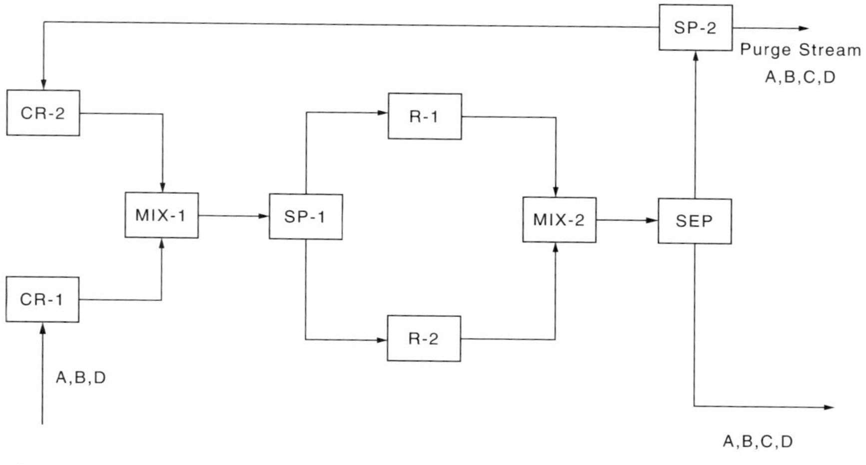

Figure 13.3: Reaction-separation process.

TABLE 13.1
Reaction/separation process specifications

Production rate (kg mols/h)	58
Minimum concentration of C in the liquid product (% mol)	82
Minimum product temperature (K)	400
Maximum reactor volumes (lt)	2000
$0.01 \leq$ purge molar fraction $\leq .01$	

of freedom) at splitter SP-1. The process specifications and environmental legislation limits for pollutants released in the purge stream are given in Tables 13.1 and 13.2, respectively. The releases from the process correspond to the amount of gases A, B, C, and D contained in the purge stream, as this is determined by the process model and operation, and hence the environmental impact of the process is determined as described in Section 2.3.

Application of MEIM in this case results in the following:

- The process system boundary is defined based on Life Cycle Analysis principles. The emissions inventory comprises chemicals A, B, C, D released in the purge stream.
- The CriTical Air Mass (CTAM, tn air/h) is used as a measure of atmospheric damage caused by releases A, B, C, D.
- The Pareto curve of optimal solutions is obtained. The two extreme solution instances of the multiobjective optimization problem for minimum environmental impact and minimum cost, respectively, are shown in Table 13.3. CTAM values are within the following range [13480, 15397] ([=] tn air/h).

By using multiobjective criteria instead of single cost/profit objectives used in traditional design optimization problems, MEIM facilitates the identification of "compromise" solutions to depict pollution prevention strategies which also result in cost savings. In the following sections we will show how the principles of MEIM can be extended to explicitly account for unavailability of equipment due to failure and maintenance planning.

TABLE 13.2
Environmental legislation limits

SLV_A	SLV_B	SLV_C	SLV_D
.667	.167	.25	.11

TABLE 13.3
Cost/environmental optimization results

	Minimum environmental impact	*Minimum cost*
Annual Cost	194.7	175.2
CTAM (tn air/h)	13480	15397
Volume of Reactor 1	169.3	117.4
Volume of Reactor 2	1244.5	863.3
Feed flow rate	89.9	88.8

3 MAINTENANCE-BASED STRATEGIES FOR THE MINIMIZATION OF ENVIRONMENTAL RISK

The performance of a chemical process is a function of both process design and operation. Steady state operation is described by a mathematical model consisting of equality and inequality constraints of the following form:

$$h(d, z, x) = 0,$$
$$g(d, z, x) \leq 0, \tag{8}$$

where

- d is the vector of design variables for the process (e.g., equipment volumes);
- z is the vector of degrees of freedom (e.g., split ratios, etc.);
- x is the vector of state variables (e.g., flow rates, pressures, temperatures, etc.).

Equality constraints (h) correspond to mass and energy balances, and inequality constraints (g) correspond to process design and operation specifications. According to the design selection (d), the mathematical model in equation (8) forms a feasible operating region in which the degrees of freedom (z) can be manipulated to optimize process performance during operation. The feasible operating region for a specific design is defined as

$$FOR(d) = \left\{(z, x): \begin{Bmatrix} h(d, z, x) = 0 \\ g(d, z, x) \leq 0 \end{Bmatrix}\right\}. \tag{9}$$

Process performance is expressed both in terms of profitability and environmental behavior. Environmental impact depends on the quantity and consequences of process releases, which in terms of the mathematical model in equation (8) are represented by certain state variables whose values are determined by the decisions taken during operation within the feasible operating region provided by the selected process design. Therefore, as discussed in Section 2.3, an n-dimensional environmental impact vector can be defined as a function of the process variables and parameters

$$\mathbf{EI} = f_1(d, z, x), \tag{10}$$

comprising n different indices which measure air pollution, water pollution, solid wastes, global warming, photochemical oxidation, and stratospheric ozone depletion.

Process revenue depends on the sales of the end product, which in terms of the mathematical model is also represented by a state variable (x), whose value again is determined by process operation (z) and design (d). Therefore, process revenue can also be expressed as

$$REV = f_2(d, z, x). \tag{11}$$

3.1 MODELING THE MECHANISM OF UNEXPECTED EVENTS

Nonroutine releases are related to unexpected events, mostly accidents resulting from equipment breakdown. Concepts from discrete event theory, such as that of *state, state-space*, and *transition*, can be used to model the mechanism of unexpected events, the occurrence of which is given by some time variant or invariant probability.

Consider a chemical system in steady-state operation whose components are all functioning properly. According to the operating program, the impact of releases to the environment and the revenue of the process can be evaluated from expressions similar to equations (8)–(11). If a failure occurs, the process model determining the operation changes according to which component has failed, and, therefore, the values of environmental impact and process revenue change as well. The state of the system at time t is defined as (Cassandras, 1993) the information required at t such that the output of the system is uniquely determined from this information and the input. Because the outputs of the system are the values of environmental impact and process revenue and the information required to determine them is what components are working and so forth at time t, the state of the process system is defined as the set containing the operating status of each component. The occurrence of an unexpected event, such as equipment breakdown, is represented by a transition from one state to another and all the possible states form the state-space of the system.

Adopting a state-space representation, this nine-component reaction/separation process discussed in Section 2.4 is described by a total of $2^9 = 612$ states in which the system may reside with possible degradation and failure of equipment. According to which units are functioning and which are failed, the process can be in only one of the states at a time. From the complete state-space, however, only six states are *structurally operable*, meaning that only in six states the system can operate to produce chemical C. These states are depicted in Figure 13.4 and can be formally identified from the structure function of the system or its minimal cuts (see, for example, Hoyland and Rausand, 1994).

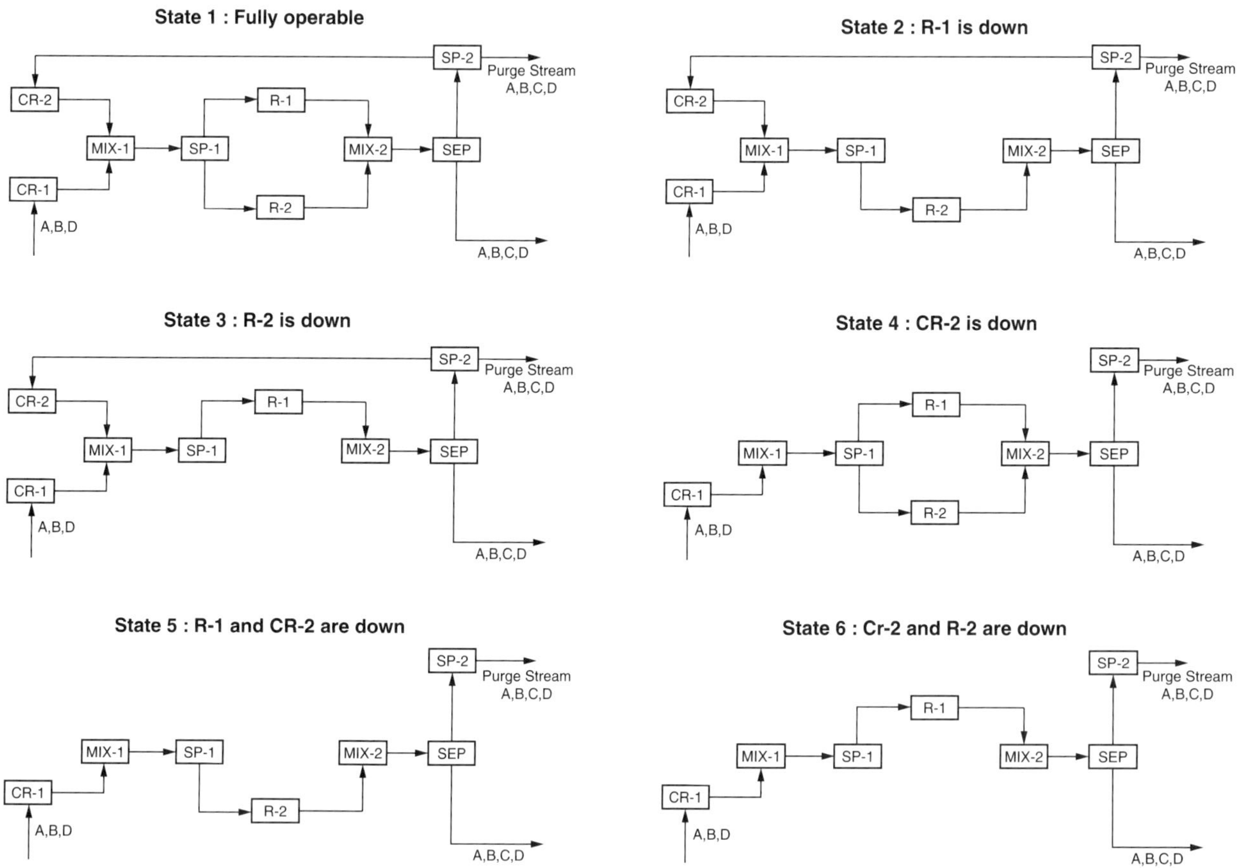

Figure 13.4: Operable system states.

Due to the change in system configuration because of failure, in general, each operable state k is described by a different set of equality and inequality constraints determining a feasible operating region (FOR_k), which is defined as the set of all realizations of the degrees of freedom (z_k) for which feasible operation is guaranteed:

$$FOR_k(d) = \left\{ (z_k, x_k) : \begin{Bmatrix} h_k(d, z_k, x_k) = 0 \\ g_k(d, z_k, x_k) \leq 0 \end{Bmatrix} \right\} \qquad \forall k. \tag{12}$$

Furthermore, because the transitions between the states are related to equipment failure and repair, each state has a probability of occurrence that depends on the time-varying availability of each component determined by the reliability and maintainability characteristics of the equipment.

The new state equations h_k, g_k determine a new operating program which, substituted in equation (10), yields a new value of environmental impact

$$\mathrm{GEI}_k = f_1(d, z_k, x_k) \qquad \forall k. \tag{13}$$

In the case of failure of the recycle compressor ([CR-2]), for example, the process model changes, the feasible operating region becomes smaller, and the releases are expected to be much higher, as nothing is recycled and everything escapes to the environment.

On the other hand, the revenue generated during operation is also a function of the system state and the mathematical model that describes it. Depending on the operation in state k, process revenue in that state is evaluated as

$$REV_k = f_2(d, z_k, x_k) \qquad \forall k. \tag{14}$$

If, for example, one of the reactors fails the capacity of the process to produce the end product, C decreases and the revenue of that state decreases accordingly.

It follows from the above that each state k is characterized by its probability of occurrence $\mathrm{Pr}_k(t)$, its profit-generating capacity and productivity $REV_k(t)$, and the environmental impact caused during operation GEI_k.

3.2 SYSTEM EFFECTIVENESS

Traditional reliability and maintenance practices during the design and the operation of the process aim at preventing the system from entering a failed state and maximizing availability. Not distinguishing, however, between the different levels of performance in different operable states and by treating operation as a binary event in which the system is simply either up or down, they fail to produce maintenance strategies that facilitate efficient process operation. Pyjadas and Chen (1996), for example, underline that in the case of manufacturing systems, traditional reliability and maintenance design methods that focused on simply keeping equipment functioning have produced poor results in terms of the profitability and environmental impact functions of the process. Going a step further, Grievink et al. (1993) and Pistikopoulos (1995) suggest that in order to overcome this drawback of current approaches, detailed process models (fully capturing physicochemical characteristics) should be incorporated into reliability and maintenance design frameworks. In particular, they encourage the use of *system effectiveness* instead of a reliability-related criterion as an optimization objective. System effectiveness depends on the probability of occurrence of each operable state k and the value of a process performance measure at this state (E_k), and is defined as their weighted sum

$$E = \sum_k \mathrm{Pr}_k \cdot E_k. \tag{15}$$

In the case of the reaction-separation process in Figure 13.1, performance is measured both in terms of environmental impact and profitability. Therefore, system effectiveness corresponds to an $(n+1)$–dimensional vector consisting of an expected environmental impact vector and an expected revenue index ($\mathbf{E} = \{\mathbf{EEI}, ER\}$). The expected environmental impact vector over a time horizon H of operation is defined as the weighted sum of the environmental impact generated when the process is operating in state k, using as weight the probability of the system being in that state,

$$\mathbf{EEI} = \int_0^H \sum_k \mathrm{Pr}_k(t) \cdot \mathbf{GEI}_k \, dt, \tag{16}$$

and similarly defined is the expected revenue of the process,

$$ER = \int_0^H \sum_k \mathrm{Pr}_k(t) \cdot REV_k \, dt. \tag{17}$$

Note that the environmental impact index GEI_k and the revenue index REV_k in equations (16) and (17) are determined, according to equations (13) and (14), by the process physicochemical model, process design, and process operation. On the other hand, the state probabilities $\mathrm{Pr}_k(t)$ are a function of the reliability and maintainability characteristics of the process. In particular, the implemented maintenance policy depends on the maintenance policy assumptions (e.g., age replacement, block replacement, etc.) and the maintenance optimization variables (maintenance intervals, number of maintenance activities for each component, sequence of maintenance actions, etc.). Therefore, to optimize the system effectiveness vector, both process design and operation and reliability and maintenance design and implementation should be optimized accordingly.

3.3 MAINTENANCE MODELING ISSUES

Equations (16) and (17) suggest that the system effectiveness vector is closely associated with the implemented maintenance policy, the defining elements of which are the assumptions regarding the type of maintenance that can be performed on the equipment and the maintenance variables regarding the derivation of preventive maintenance strategies.

The selection of the type of maintenance depends on the equipment attributes and specifications, as well as the available maintenance facilities and capabilities. Different types of maintenance are defined at a system level and at a component level. In the case of complex systems, for example, groups of components with similar operating conditions may be identified and treated uniformly during maintenance (e.g., group preventive maintenance policies). Furthermore, at a component level, assumptions are made regarding the effectiveness of maintenance in restoring the component to a good condition. As-Good-As-New (AGAN) policies, for example, restore the component to the original condition it was at the beginning of operation, whereas As-Good-As-Old (AGAO) policies bring it back to where it was immediately before the maintenance task started.

Maintenance optimization variables concern the elements of the maintenance policy that can be treated as decision variables to optimize a maintenance-related objective, commonly system availability. These variables usually regard the number of maintenance actions to be performed, the length of maintenance intervals, etc.

Depending on the nature and complexity of the assumptions and the desired level of detail and depth, different mathematical modeling tools have been employed in maintenance optimization frameworks, such as analytical techniques (e.g. Vatn et al., 1996) and Markov decision processes (Gertsbakh, 1977). In this work, an analytical approach is developed in which the probability of occurrence of each state—appearing in the system effectiveness vector and being the term that is determined by the maintenance strategy—is analytically expressed as a function of the maintenance optimization variables. Therefore, the identification of the reliability and maintenance characteristics that maximize system effectiveness shall be sought through the formulation and solution of an optimization problem.

3.3.1 An Analytical Technique

Consider a system with M components. Let $A_{1,j}(t)$ be the initial (before any maintenance is performed) reliability function of component j as a function of time. Let, also, t_1 and t_N denote the beginning and the end, respectively, of the time horizon of operation. The assumptions regarding the maintenance policy are the following:

- Each time a maintenance action is to be performed, the type of maintenance is determined in the following way:
  corrective, if the unit under maintenance is failed;
  preventive, if it is operable.
  This assumption is common in the literature (e.g. Tseng, 1996) and valid in many real cases.

- All maintenance is of an AGAN type (i.e., the component is restored to its initial condition at the beginning of operation).
- All failure events are independent.
- One maintenance action is performed at a time.

The elements of the maintenance policy to be decided (i.e., the maintenance optimization variables) are the number of maintenance activities required for each component in the time horizon of operation, the sequence in which they will be performed, and the exact maintenance time instants. For that purpose, the following optimization variables are introduced:

- N, which is an integer variable denoting the total number of maintenance activities.
- t_λ, which is a continuous variable denoting the exact maintenance time instant of the $(\lambda - 1)$th maintenance action.
- $u_{\lambda,j}$, which is 0-1 variable depicting whether the $(\lambda - 1)$th maintenance action will be performed on component j or on a different component and is defined as follows:

$$u_{\lambda,j} = \begin{cases} 1, & \text{if the } (\lambda - 1)\text{th maintenance action is performed} \\ & \text{on unit } j; \\ 0, & \text{if the } (\lambda - 1)\text{th maintenance action is performed} \\ & \text{on a unit other than } j. \end{cases}$$

According to the assumptions made for the maintenance policy, analytical expressions are defined for the availability of the equipment and the expected duration of each maintenance task as a function of the maintenance optimization variables (N, t_λ, $u_{\lambda,j}$). From these, analytical expressions can be derived describing the probability of occurrence of each state k and hence the environmental risk and the profitability of the process as a function of the maintenance optimization variables. A detailed analysis on the derivation of such expressions can be found in Pistikopoulos and Vassiliadis (1998).

3.4 OPTIMIZATION FORMULATION

In this section, a multicriteria optimization formulation (P2) is developed to identify the optimal design in terms of process characteristics and reliability and maintenance requirements, minimizing environmental risk and maximizing process profitability in a cost-effective way. For that purpose, two distinct objectives are included, the first being the expected environmental risk and the second being the expected process revenue subtracting investment and maintenance costs expressed analytically as functions of the design and the maintenance variables, respectively:

$$\min_{d, z_k, x_k, N, t_\lambda u_{\lambda,j}} \int_{t_1}^{t_N} \sum_k \Pr_k(N, t_\lambda, u_{\lambda,j}) \cdot f_1(d, z_k, x_k)\, dt, \tag{P2}$$

$$\max_{d, z_k, x_k, N, t_\lambda u_{\lambda,j}} \int_{t_1}^{t_N} \sum_k \Pr_k(N, t_\lambda, u_{\lambda,j}) \cdot f_2(d, z_k, x_k)\, dt,$$

Subject to
design cost (d),
maintenance cost ($N, t_\lambda, u_{\lambda,j}$),
process model (d, z_k, x_k),
maintenance constraints ($N, t_\lambda, u_{\lambda,j}$),

If environmental risk could be translated in terms of an operating cost, then in the context of a cost-benefit analysis, (P2) could be formulated having one objective with the advantage of identifying a unique solution, namely, one optimal design, operating, and maintenance policy; however, because in most cases it is not straightforward how to assign an economic value to environmental damage, we will adopt the use of multiple independent criteria, one of which is going to be environmental risk as quantified using equation (16) and the machinery of MEIM.

In chemical process design, design costs are associated with the sizes of equipment components, normally through nonlinear power law expressions (Biegler et al., 1997). The direct costs of maintenance are a function of the number of maintenance activities, the type of maintenance tasks, and hence the cost of each maintenance task. The indirect costs of maintenance mainly relate to possible loss of production due to down time and repair and are incorporated in the objective function as part of the expected revenue term. The process model represents the set of equality and inequality constraints defining the feasible operating region of each state k of the system. Maintenance constraints correspond to assumptions made when defining a maintenance policy and specific unit requirements. The exact form of (P2) can be found in Pistikopoulos and Vassiliadis (1998).

The solution of (P2) quantifies the trade-offs between the conflicting objectives, and in the form of a pareto surface, it contains the complete set of optimal solutions for different values of design and maintenance costs and different requirements for profitability yielding different values of environmental risk. Therefore, the solution can be used as a road map to identify which are the available combinations of process design and operation strategies as well as reliability and maintenance policies to achieve business targets within the legislation limits imposed for industrial releases.

3.5 NUMERICAL EXAMPLE

Let us revisit the reaction/separation process example discussed in Section 2.4. The volumes of the reactors (design variables) are considered

TABLE 13.4
Reliability and maintenance data.

	$a_j(10^{-4})$	$b_j(10^{-7})$	τ_p	τ_c	$C_p(10^4)$	$C_c(10^4)$
CR-1	3.3	5	5	2	2	8
CR-2	3.3	5	5	2	2	8
MIX-1	1.8	3	4	1	1	5
SP-1	1.8	3	4	1	1	5
R-1	4.5	6	8	3	4	14
R-2	3.5	5.5	6	2.5	3	12
MIX-2	1.8	3	4	1	1	5
SEP	3	5.2	10	4	5	18
SP-2	1.8	3	4	1	1	5

fixed because in this example we want to focus on the interactions of maintenance policies and environmental risk. Formulation (P2), however, allows the identification of optimal process design out of several design alternatives; therefore, we can also study possible process modifications and their impact on environmental risk, process performance, and maintenance costs. For the derivation of the optimal preventive maintenance policies, a time horizon $H = 1000$ is considered. The reliability and maintenance specifications of the units are given in Table 13.4. The failure rate of each unit is assumed to increase linearly with time (equipment is in the wear-out phase) [i.e., $q_j(t) = b_j \cdot t + a_j \; \forall_j$].

According to the amount of chemical C produced, a revenue is generated from the process and can be estimated from a revenue function in a way similar to equation (11). The process also involves a purge stream (A, B, C, D) representing releases to the environment. The environmental impact of these releases can be determined using the Life Cycle Assessment techniques of MEIM. In particular, it is applied in the following steps:

(a) *System boundary and emissions inventory*. The global boundary around the process is defined to allow for input-output waste interactions based on the life cycle analysis principles. The emissions inventory comprises chemicals A, B, C, and D that are emitted through the purge stream in the fully operable state but can also cause significant environmental damage, depending on whether critical equipment fails.

(b) *Assessment of the environmental impact of releases*. The environmental impact of releases in each state k is quantified as

$$GEI_k = CTAM_k = \frac{F_A^k}{SLV_A} + \frac{F_B^k}{SLV_B} + \frac{F_B^k}{SLV_B} + \frac{F_B^k}{SLV_B} \qquad \forall k, \tag{18}$$

where F_w^k denoted the mass flow rate of each emission w at each state k and SLV_w (mol/tn air) its standard legislation value. These values are given in Table 13.2. Note that mass flow rates correspond to state variables (x_k), taking different values in each state according to the values of the degrees of freedom in that state (z_k: e.g., splitting ratio) and the design variables (d: equipment volumes). Therefore, CTAM is equivalent to $f_2(d, z, x)$ in equation (10).

To identify the optimal maintenance (N, t_λ, $u_{\lambda,j}$) and operating schemes (z_k, x_k) and depict the trade-offs between expected environmental impact, process revenue, and maintenance cost (P1) is solved by employing a two-step solution strategy, as described in Pistikopoulos and Vassiliadis (1998).

3.5.1 Trade-offs Between Environmental Risk, Process Revenue, and Maintenance Cost

The complete set of optimal solutions forms the surface (Pareto surface) depicted in Figure 13.5. Each point on the surface represents an optimal maintenance strategy (N, t_λ, $u_{\lambda,j}$), an optimal operating policy (z_k, x_k), and the corresponding values for environmental risk, process revenue, and maintenance cost. Therefore, optimal policies can be selected, satisfying the requirements for the amount of releases allowed, for profitability targets, and

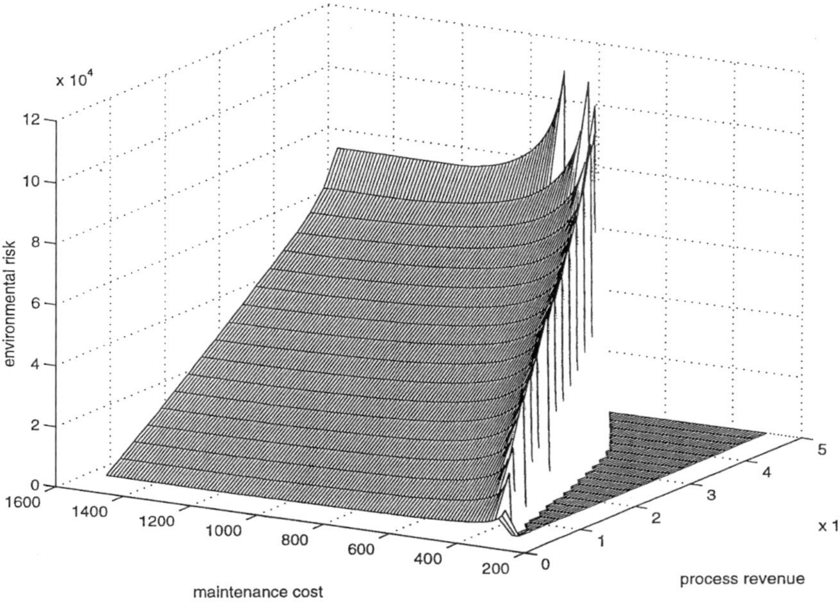

Figure 13.5: Pareto surface of optimal solutions.

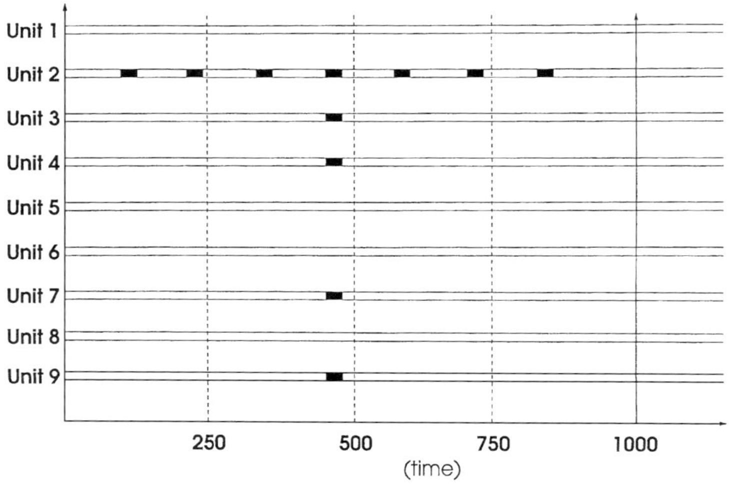

Figure 13.6: Low-risk, low-profit maintenance policy; maintenance cost = 500.

for maintenance expenditures. The following remarks can be made:

- For a fixed amount of available maintenance expenses, trade-offs between environmental risk and process revenue are observed. If the goal is to minimize environmental risk, most of the maintenance actions are devoted to CR-2, which is the environmentally critical unit (Fig. 13.6). On the other hand, if the goal is to maximize profitability, most of the maintenance resources are used to sustain high availability for R-1 and R-2, which are critical to production units (Fig. 13.7).
- For a fixed environmental risk limit, trade-offs are observed between maintenance cost and process revenue. The higher the maintenance resources available, the more the preventive maintenance actions performed to R-1 and R-2 to increase productivity.
- For a fixed profitability target, there exist trade-offs between maintenance cost and environmental risk. The higher the maintenance resources available, the more the preventive maintenance actions performed to CR-2 to decrease the probability of high nonroutine releases.

Note that the quantification of trade-offs between process revenue and environmental risk is not performed to suggest that the maximization of profit may be, in some ways, more favorable than the minimization of environmental risk. It is clear that such an analysis provides guidelines as to what maintenance and operating policies are needed to stay within the environmental legislation limits and maximize profitability.

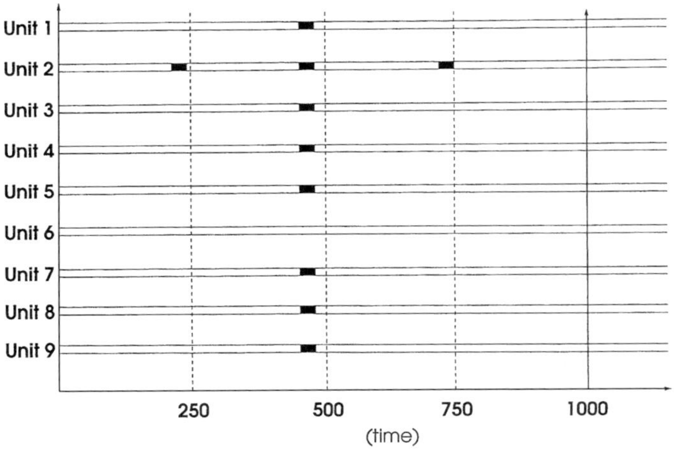

Figure 13.7: High-risk, high-profit maintenance policy; maintenance cost = 500.

REFERENCES

Acevedo, J., and Pistikopoulos, E. N. (1996). A parametric MINLP approach for process synthesis problems under uncertainty. *Industrial and Engineering Chemistry Research* 35:147–158.

Aelion, V., Castells, F., and Veroutis, A. (1995). A life cycle inventory analysis of chemical processes. *Environmental Progress* 14(3):147–158.

Biegler, L. T., Grossmann, I. E., and Westerberg, A. W. (1997). *Systematic Methods of Chemical Process Design*. Prentice-Hall, Englewood Cliffs, NJ.

Cassandras, C. G. (1993). *Discrete Event Systems*. R. D. Irwin, Inc., and Aksen Associates, Inc., Boston.

Dekker, R. (1996). Applications of maintenance optimization models: A review and analysis. *Reliability Engineering and System Safety* 51:229–240.

Department of the Environment (1995). A Guide to Risk Assessment and Risk Management for Environmental Protection. HMSO, London.

El-Halwagi, M. M., and Petrides, D. (1995). *Pollution Prevention through Process and Product Modifications. AIChE Symposium Series*, No. 303, Vol. 90, American Institute of Chemical Engineers, New York.

Gertsbakh, I. B. (1977). *Models of Preventive Maintenance*. Elsevier, New York.

Grievink, J., Smit, K., Dekker, R., and Van Rijn, C. F. H. (1993). Managing reliability and maintenance in the process industry. *Proceedings of the Foundations of Computer Aided Process Operations (FOCAPO)*. Crested Butte, Colorado.

Harnly, J. A. (1998). Risk based prioritization of maintenance repair work. *Process Safety Progress* 17(1):32–38.

Hoyland, A., and Rausand, M. (1994). *System Reliability Theory*. John Wiley & Sons, New York.

Kapur, K. C., and Lamberson, L. R. (1977). *Reliability in Engineering Design*. John Wiley & Sons, New York.

Lashof, D. A., and Ahuja, D. R. (1990). Relative contributions of greenhouse gas emissions to global warming. *Nature* 344:529–531.

Llewellyn, G. (1998). Strategic risk assessment-prioritising environmental protection. *Journal of Hazardous Materials* 61:279–286.

Michelsen, O. (1998). Use of reliability technology in the process industry. *Reliability Engineering and System Safety* 60:179–181.

Montague, D. F. (1990). Process risk evaluation—what method to use? *Reliability Engineering and System Safety* 29:27–53.

Papazoglou, I. A., and Aneziris, O. (1990). Reliability of on-line versus standby safety systems in process plants: A comparative assessment of two systems for controlling accidental gas releases. *J. Loss Prev. Process Ind.* 3:212–221.

Patton, D. E. (1998). Environmental risk assessment: Tasks and obligations. *Human and Ecological Risk Assessment* 4:657–670.

Pearce, D. (1991). Industry and the environment. *Journal of Biological Education* 25:263–269.

Pistikopoulos, E. N. (1995). Uncertainty in process design and operations. *Computers Chem. Eng.* S19:S553–S563.

Pistikopoulos, E. N., and Vassiliadis, C. G. (1998). Reliability and maintenance considerations in process operations under uncertainty. *Proceedings of the Foundations of Computer Aided Process Operations (FOCAPO)*. Snowmass, Utah.

Pistikopoulos, E. N., Stefanis, S. K., and Livingston, A. G. (1995). A methodology for minimum environmental impact analysis. *Volume on Pollution Prevention through Process and Product Modifications*, *AIChE Symposium Series* 90(303):139–151, American Institute of Chemical Engineers, New York.

Pyjadas, W., and Chen, F. F. (1996). A reliability centered maintenance strategy for a discrete part manufacturing facility. *Computers Ind. Eng.* 31(1/2):241–244.

Stefanis, S. K., and Pistikopoulos, E. N. (1997). Methodology for environmental risk assessment of industrial non-routine releases. *Ind. Eng. Chem. Res.* 36:3694–3707.

Tan, J. S., and Kramer, M. A. (1997). A general framework for preventive maintenance optimization in chemical process operations. *Computers Chem. Eng.* 21(12):1451–1469.

Tseng, S. T. (1996). Optimal preventive maintenance policy for deteriorating production systems. *IIE Transactions* 28:687–694.

U.K. Ecolabelling Board (1993). Criteria for Hairspray Ecolabels.

U.K. Stratospheric Ozone Review Group (1988). *Stratospheric Ozone 1988*. Her Majesty's Stationery Office, London.

Van Rijn, C. F. H. (1987). A systems engineering approach to reliability, availability and maintenance. *Proceedings of the Foundations of Computer Aided Process Operations (FOCAPO)*. Park City, Utah.

Vatn, J., Hokstad, P., and Bodsberg, L. (1996). An overall model for maintenance optimization. *Reliability Engineering and System Safety*. 51:241–257.

Wahlstrom, B. (1998). Issues in a world of environmental and societal vulnerability. *Proceedings of the 1998 European Conference on Safety and Reliability (ESREL)*. Trondheim.

CHAPTER 14

Combining Geographic Information Systems, Process, and Environmental Modeling for Macro Level Pollution Prevention

Derya B. Özyurt and Matthew J. Realff
School of Chemical Engineering, Georgia Institute of Technology, Atlanta, GA

Abstract

We define *macro level pollution prevention* as the reduction of waste by interchange of materials from one facility to another, where the distances involved may range from 100 meters to tens of kilometers. In this chapter we present a methodology to address macro scale pollution prevention that combines process engineering tools with geographic and environmental information. A Geographic Information System (GIS) is used to identify the possible sources and sinks of material, provide data for the environmental model, and to store and visualize results. The large number of potential interchanges is first screened using a simple optimization model that utilizes the concept of *feasibility contours*. The optimal interchange system is similarly devised using an optimization model that embeds simple measures of process feasibility and material quality. The objective is to maximize the economic benefits of the interchange, and a multimedia compartment model is used to assess and constrain environmental impact. The analysis and synthesis results can be merged with other pertinent data using GIS.

Keywords: Geographic Information Systems, Industrial Ecosystems, process and environmental models

1 INTRODUCTION

Geographic Information Systems (GIS), computer-based mapping and analysis tools, are widely used as a decision support tool for different applications from environmental modeling to logistics (About GIS, 1998). GIS can combine vast amounts of real world data with a geographic base so that information is connected with its spatial location. Moreover, GIS can be integrated with other tools to increase its efficiency and effect.

In this chapter we examine the use of GIS to aid process engineering in the context of pollution prevention. Our overall objective will be to demonstrate the use of GIS in the design of industrial ecosystems (Graedel and Allenby, 1995). We will also demonstrate that the traditional tools of the chemical engineer, mass and energy balance modeling, synthesis, and design methods, are ideally suited to macro-scale industrial analysis and design in an era of closed-loop manufacturing.

As the name implies, Geographic Information Systems are useful when the locations of the objects are of significance to the analysis or synthesis of a system. This location dependence can stem from two sources. First, it may be necessary to assess the environment local to a given facility. The topography affects the dispersion of airborne discharges and the characteristics of the watershed. The underlying geology affects the dispersion of any discharges to wells and unintended leaks. The location of significant water bodies or streams and rivers may play a role in the acquisition and discharge of water-soluble pollutants and suspended solids and ultimately as the sink for enthalpy discharges. Second, and most significantly for this chapter, the relative location of facilities and the means by which they can be connected can be found from a GIS. This enables us to consider problems of inter-plant/facility material and energy interchange. These types of interchanges will be part of any industrial ecology that attempts to close material and energy loops at a macro as well as plant scale.

This chapter develops an overall framework in which geographic information can be coupled with process and environmental modeling and decision making. In addition, we will give a methodology to aid in the design of ecoindustrial parks as an example of this integration.

2 BACKGROUND

2.1 GEOGRAPHIC INFORMATION SYSTEMS

GIS is, in essence, a computerized database management system; however, it is specifically designed for capture, storage, retrieval, analysis, and display of spatial data (Huxhold, 1991). Its ability to integrate common database operations with unique visualization and geographic analysis capacity of maps makes GIS valuable as a decision support system. In a GIS, the real world is represented by a collection of thematic layers, which can be used separately or together to elicit knowledge about specific locations (Fig. 14.1). Geographic information containing either an explicit geographic reference (e.g., latitude and longitude) or an implicit reference, such as an address, postal code, or census tract name, allows the user to locate features on the earth's surface for analysis. These features can be static events, such as a business or forest stand, or dynamic events, such as an earthquake.

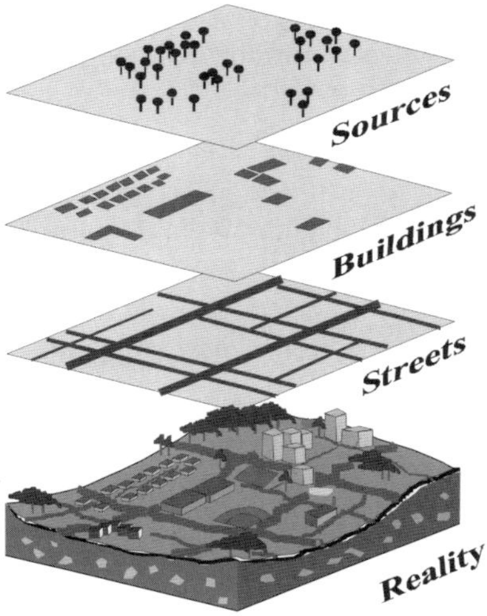

Figure 14.1: Collection of thematic layers to present the reality [based on a figure by Environmental Systems Research Institute, Inc. (ESRI, 1999)].

Using a GIS involves six basic tasks: input, manipulation, management, query, analysis, and visualization (About GIS, 1998). The input task converts analog data to a digitized format that can be used by a GIS system. Scaling, projection transformations, weeding out unnecessary data, and similar processes comprise manipulation of the data and are required to make data compatible with the system used. This is necessary because the data can originate from different data sources with different scales and map projections. Therefore, manipulation is a prerequisite for merging different data to address a given task. The management of large amounts of data that are stored conceptually as a collection of tables (common fields in different tables are used to link them together) is done by a database management system that is one of the components of GIS. Once functioning GIS-containing geographic information has been developed, query tasks can be performed. Questions such as "where is land zoned for industrial use" and "how far is it between two places" can be answered easily. Likewise, analysis task helps the user to find answer for the following types of questions:

- Where are all of the sites that are suitable for building new production facilities?
- If I build a facility here, how will the environment be affected?
- How many facilities lie within 100 m of this water main?

The results of any query and analysis usually are visualized as a graph or map, as maps are very efficient at storing and communicating geographic information. After visualizing the current situation on a map, further manipulation and analysis of the input can be undertaken.

Besides these basic tasks, GIS can also be integrated with other tools. In the following we will give the example of its integration with mathematical models of Process Engineering to increase its capacity beyond being an analysis tool to a system with synthetic capabilities.

2.2 PROCESS ENGINEERING AND POLLUTION PREVENTION

Pollution prevention is a strategy for reducing the net consumption of resources in processes and products by engineering interventions at various scales. At the molecular level, material substitution can lower the inherent toxicity of the process and the energy requirements, and improve the selectivity toward desired products (Allen, 1997). When material substitutions are not feasible due to process requirements and costs, then the level of intervention shifts to the unit operation. Energy and material use can be reduced by manipulation of operating conditions and design changes, such as integrating reaction and distillation to reduce byproduct generation, and hence the energy needed for separation (Agreda et al., 1990). Beyond the scale of the unit operation is the plant and the ability to exchange mass or energy between different units (El-Halwagi, 1997; Gundersen and Naess, 1988). Finally, at the macro-scale, we have the ability to find value in otherwise wasted streams by interchanging them with other plants. Overall, the question is, What combination of technologies at various scales will lead to the lowest cost, minimal discharge, industrial system?

2.3 INDUSTRIAL ECOLOGY AND ECOINDUSTRIAL COMPLEXES

Industrial ecology (IE) is "the means by which humanity can deliberately and rationally approach and maintain a desirable carrying capacity, given continued economic, cultural, and technological evolution" (Graedel and Allenby, 1995). Instead of viewing the industrial system as an entity isolated from its environment, IE enables local decision making with an awareness of broader regional and global impacts. IE embeds approaches which measure, analyze, and simultaneously optimize cost, performance, and environmental impact of activities. One of the examples of the applications of industrial ecological principles is the exchange of wastes, byproducts, production aids, and energy among closely situated firms (Ehrenfeld and Gertler, 1997). These kinds of complexes usually are referred to as an "industrial ecosystem" or "industrial symbiosis." They symbolize the supposed desirable shift from linear to closed-loop material and energy usage. Besides a well-known example of Kalundborg (Ehrenfeld and Gertler, 1997), there are several other evolving ideas and realities of similar complexes (Nemerow, 1995; Ehrenfeld and Gertler, 1997; Bendz, 1999; Saotome et al., 1999).

To motivate the rest of this chapter, one of these extant industrial ecosystems will be described in more detail, paying particular attention to the issues in material and energy interchange.

3 MOTIVATION

The motivation for macro-scale pollution prevention is the growing population of ecoindustrial parks. The archetypal example is that of Kalundborg, which is a seaside industrial town in Denmark. As can be seen in Figure 14.2, this industrial symbiosis consists of 11 physical linkages which were established in an incremental fashion within a 25-year period. The exchanged materials range from steam to molten sulfur, where a 2-mile-long pipeline is needed for the transfer of steam from the power plant to the pharmaceutical plant and the distance between plants is 50 km in the case of sulfur exchange. This gradually evolved symbiosis resulted in wasted energy reductions, raw material substitutions, and reductions in water demands. The symbiotic partners usually established the links when both recognized net cost savings relative to their options one by one. The incremental growth pattern might miss opportunities when synergies exist beyond one specific material or energy link. For example, the wallboard plant may not see the benefit of taking scrubber sludge from the power station without the opportunity to obtain cheap energy from the Statoil refinery. Similarly, the utilization of waste heat from one supplier may be insufficient to justify the

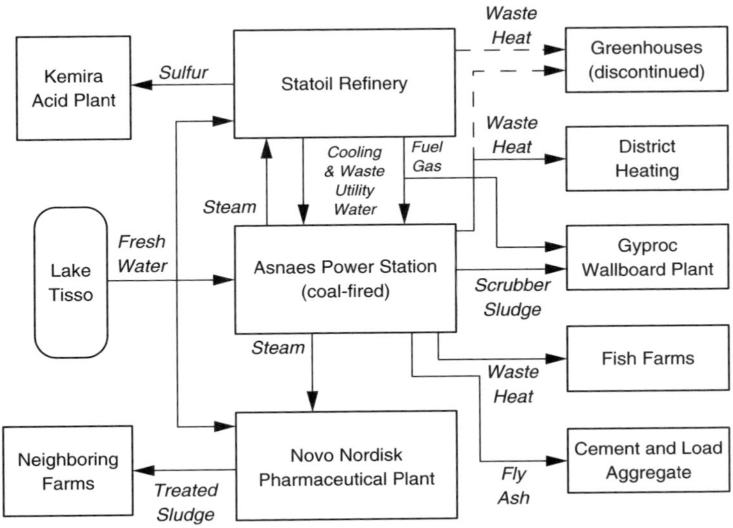

Figure 14.2: Kalundborg Ecoindustrial Park and the linkages between processes.

infrastructure costs, but if several suppliers can be combined, the economics of scale justify the investment.

More recent examples, like Brownsville, Texas/Matamoros, Mexico (Martin et al., 1996), Chattanooga, Tennessee (Ehrenfeld and Gertler, 1997), or Aurora Project (Bendz, 1999), are trying to capitalize on the fortuitous pattern of the Kalundborg example and come up with planning guidelines to address issues and problems for ecoindustrial parks.

The planning of macro-scale material and energy flows is the reproduction of mass and energy integration at the plant scale with interaction distances measured in meters to interactions measured in kilometers. The unique problems that this change of scale engender are

1. *Identification of sources and sinks* for material and energy exchange on the macro-scale.
2. Assessment of the *feasibility and mode of connection* between the sources and sinks.
3. The *design of the optimal system* of linkages that maximizes the global benefit but enables each participant to realize benefits.
4. The inclusion and location of material and energy improvement steps to raise the quality of the interchanges and maximize overall system performance.

In the next section we will define the scope of the problem that will be modeled and solved in Sections 5, 6, 7, and 8. The GIS will play a significant role in addressing issues 1 and 2 above, and process system engineering is applied to address 3 and 4.

4 PROBLEM DEFINITION AND METHODOLOGY

The idea of interplant exchange arises in constructing stable ecoindustrial systems which, like their ecological counterparts, reduce the system's entropic losses by increasing the interdependency of producers and consumers (Ehrenfeld and Gertler, 1997). If a plant has a waste (byproduct in this context) there are several ways to dispose of it. One of them is selling it to another production facility, where it can be used as an alternative or substitute for raw material or as a processing aid. There are several technical factors which can affect the decision of this exchange:

- Infeasibility of other alternatives.
- Transportation costs depending on the mode, volume, and weight of the material.
- Quality of the material.

As a starting point, assume that we have a map depicting several production facilities in a geographic area of interest. For every plant it is assumed that information on major inputs and outputs in amount and quality (one or more of the following: temperature, pressure, concentration, impurity, phase, etc.) is available. For an exchange, these facilities can be partitioned into two classes: possible sinks and sources. Each facility can be both a source and a sink, but not for exactly the same stream. The aim of a source is to sell the byproduct to another company, which will use it as a substitute of virgin raw material or as a process aid. It will seek the interchange with highest profit (i.e., minimizing the labor, transportation, handling, and recall costs) and by maximizing the amount sold. The aim of a sink, on the other hand, is to get the highest quality second hand material available for the minimum price. Its incentive can be constraints on the amount of available virgin raw material and/or processing aids or feasibility of utilizing a lower cost and grade input material without affecting the quality of the end products.

We do not want to represent this system with an inherently antagonistic approach, such as game theory, but rather sublime these differences into the overall system objective of maximizing the net system economic potential for waste interchange. This presumes a willingness of a company to participate, even if the costs of doing so are higher than not participating in the system. We assume that the system benefits will be shared in such a way as to eliminate this local loss of value.

The overall problem can thus be defined as

Given : A geographic region with relevant data about the environment. A database containing major inputs and outputs of facilities and their locations in the region.

Find : The maximum system profit: [Material value—Infrastructure costs].

Subject to : Mass and energy balances;
Quality constraints;
Process performance models;
Environmental models.

The overall methodology for identifying sources and sinks as candidates for possible interplant material exchange and then finding the most economically and environmentally efficient matches among them consists of four main steps (Fig. 14.3).

These steps can be taken sequentially or, as depicted in Figure 14.3, iteratively or simultaneously. The first step of the methodology is finding plants with exchangeable materials. This can be done using the lists of possible materials from the literature and other sources or utilizing case-based reasoning (Kolodner, 1993) to identify the existing or imaginary environmentally balanced industrial complexes. The procedure to find the sources and sinks utilizes a GIS and databases like Manufacturing Directories (GDITT, 1996)

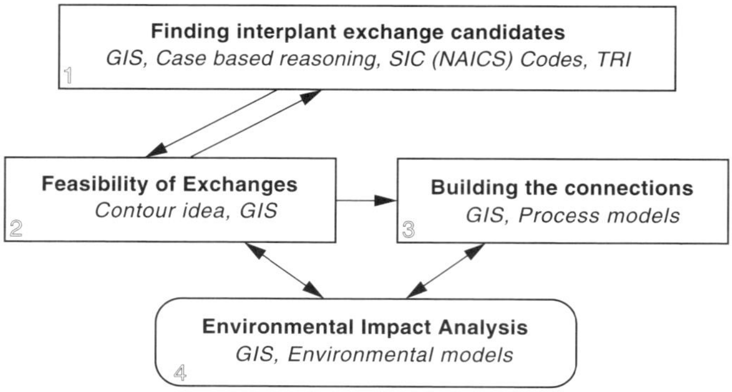

Figure 14.3: Methodology for constructing industrial ecoparks.

or Toxic Release Inventory (TRI) data (Toxic Release Inventory, 1998). From the source's perspective, we can discriminate possible options available for a given material, such as disposing it, reusing it in the plant, or exchanging it with sinks. This leads to the classical process design problem of selecting the optional matching of sources and sinks given a superstructure of possible connections, along with their costs and performance models. A key issue that lies beyond the scope of this chapter is the choice of the mode of interaction between sources and sinks. This can be continuous or batch. The batch mode introduces the additional problems of deciding the batch size and frequency at both ends of the transportation link. In the rest of the chapter we will describe the steps (1–4) of our methodology in detail and give examples for each of them.

5 IDENTIFICATION OF SOURCES AND SINKS

A key step in the synthesis of industrial ecosystems is the identification of the options available for exchange. We propose two alternative routes. The first is *materials driven* and the second is *case based*. In the first method there are two options: source or sink driven. In the former, suppliers of the exchangeable materials are identified in the area of interest, and then the plants nearby that could act as possible sinks are found. In the latter, the material inputs of plants are identified and then sources for those materials, or close analogs, are sought in nearby plants. There are two publicly available data sets which can be used for this purpose. Although they are not intended for designing industrial ecosystems, they enable us to showcase the principles of our methodology. One of these data sets is the Manufacturing Directory, which is a state-specific database available in many states, including Georgia. This data set contains points which represent the locations of all manufacturers listed in the Georgia Manufacturing Directory (GDITT, 1996). Besides the company's name and address, the data contain primary, secondary, tertiary, and quaternary product information and their Standard Industrial Classification (SIC*) codes (*Standard Industrial Classification Manual*, 1987). The second data set is TRI. TRI, in conjunction with the basemap containing the facility locations participating in the inventory, gives amounts of toxic chemicals that are being used, manufactured, treated, transported, or released into the environment. Every item of the inventory is connected to the facility which reports them.

Using these two data sets we can locate the facilities which produce the material as a byproduct for a possible interplant exchange. The materials which are exchanged can be put into categories like acids, alkalis, oils and waxes, solvents, other organic chemicals, other inorganic chemicals, plastics and rubber, textiles and leather, wood and paper, metals and metal sludges, and miscellaneous materials (*Tennessee Material Exchange*, 1998). These categories can then be used to drive interchanges at a higher level of abstraction.

The second, more targeted approach, uses the industry types connected in current ecoindustrial parks. The database can then be searched for the existence of all or some of the elements of the cluster in the area of interest. This differs from the previous approach by identifying candidates using SIC (or NAICS) codes. Some of the current complexes which can be used in this search are summarized in Table 14.1. The feasibility of the candidate cluster of partners can be estimated using the concept of *feasibility contours*, which define the maximum distance for an economically feasible exchange of a certain material from a given facility. This idea will be explained more fully in Section 6.

5.1 EXAMPLES

5.1.1 Finding the Sources According to the Material List

As a specific example, consider methanol as the material which we want to exchange. This material, mainly used as a feedstock for several chemicals like formaldehyde and MTBE, can also be used as a solvent or fuel. Moreover, it may have uses in the agricultural sector or in denitrification of wastewater in treatment plants (Kroschwitz and Howe-Grant, 1995). Utilizing Arcview (*Arcview GIS*, 1998) as our GIS and using basic query tasks, we can easily locate all of the facilities which release methanol within the state of Georgia (Fig. 14.4). Using TRI data for this type of analysis, however, requires some caution, as methanol discharges can be in an extremely dilute form or as an air emission. The purpose of the identification is to make sure that potentially advantageous exchanges are not missed, and thus the

*SIC is a four-digit code which classifies the establishments by type of activity in which they are engaged. It facilitates the collection, tabulation, presentation, and analysis of data relating to establishments. Uniformity and comparability in the presentation of statistical data are promoted by the use of SIC. SIC is replaced by the North American Industry Classification System (NAICS) in 1997.

TABLE 14.1
Some of the existing and possible environmentally balanced industrial complexes (Nemerow, 1995; Ehrenfeld and Gertler, 1997)

Complex	*Involved partners*	*Exchanged materials*
Kalundborg	Refinery, acid plant, wallboard, fish farms, cement and road aggregate, pharmaceutical plant, coal-fired power plant	Sulfur, hot water, steam, water (utility), scrubber sludge, ash, fuel gas, treated sludge
Pulp and papermill	Road binder, vanillin, paperboard, groundwood pulp plants	Sulfite waste liquor, fillers and fines, concentrate sulfate binder, used newspapers, hot water, bark
Tannery	Slaughterhouse and rendering plants; also animal grazing and feed lot facility, biogas and power plant	Wastewater, blood, bone meal, hide and leather trimmings, cattle dung, residential solid wastes
Sugarcane	Sugarcane refinery, boiler-generators, anaerobic digester, agricultural area, alcohol fermenter	Begasse, cachaza, methane, fertilizer
Textile	Cotton growing field, greige goods mill, textile finishing plant, NaOH recovery plant, dye recovery plant	NaOH, wastewater, dye in wastewater, fertilizer
Fertilizer-cement	Phosphate fertilizer, H_2SO_4, cement plants	H_2SO_4, SO_2, phosphogypsum-calcium, sulfate
Fossil fuel power plant	Power plant, cement manufacturing plant, concrete block plant	SO_2, ash waste dust, hot water
Steel mill-fertilizer-cement	Etched glass, steel, hydrofluoric acid, fertilizer, building block plants	Gypsum, SO_2, dust, hydrofluoric acid, pickling liquor ($Fe_2(SO_4)_3$)
Cement-lime-power plant	Cement, lime, power plants	Water, steam, ash, dry limestone
Wood (lumber) mill	Lumber mill, generator, fuel oil, algae cultivation plant	Sawdust and woodchips, oil vapors, CO_2, CO, NO_2
Power plant-agriculture	Fish farming, crop growing	Ash, water, algae, fish waste
Cannary-agriculture	Agricultural growing area, cannery	CH_4, water, hot water

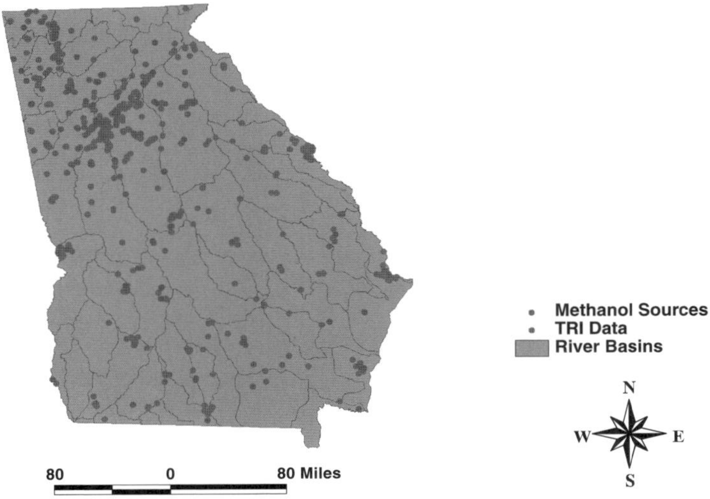

Figure 14.4: The sources of methanol in the state of Georgia.

inclusion of a larger set of sources than could actually provide material is not critical.

5.1.2 Finding the Clusters of Interplant Exchange

To illustrate the use of the case-based approach, we will locate possible steel mill-fertilizer-cement complexes described in Table 14.1. Steel mill, fertilizer, and cement plants can share materials like SO_2, NH_3, gypsum, and slag. As an attempt to generate an ecoindustrial park which contains these three facilities, the Georgia Manufacturing Directory is searched using the SIC codes of these industries. As can be seen on Figure 14.5, two possible clusters are detected. These could be further analyzed with regards to the distances and amounts of material.

6 FEASIBILITY CONTOURS

After identifying sources and sinks using GIS with a relevant database, the economical feasibility of any relation between sources and sinks should be investigated. This can be done by combining a process systems approach with the capabilities of GIS to access and display spatial information. We develop a screening procedure that transforms the value and amount of material at a source into a visual representation of feasible interchanges.

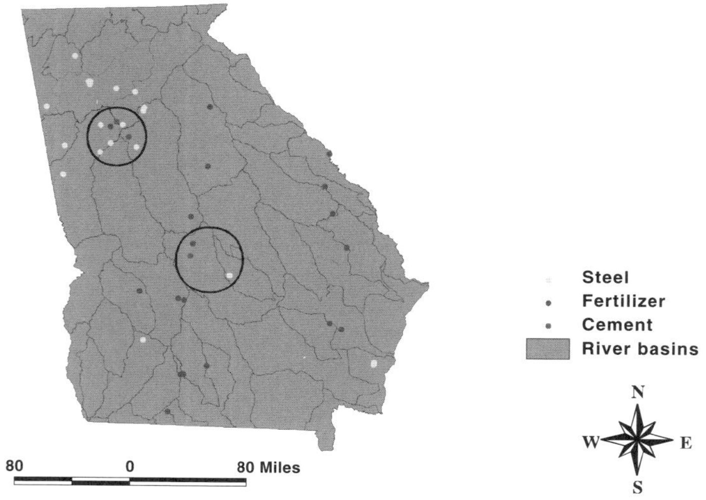

Figure 14.5: Possible steel mill-fertilizer-cement complexes in the state of Georgia.

The problem can be stated as follows:

Given : A plant in a geographic region of interest with several materials which can be disposed, reused, or exchanged.

Find : The maximum distance where a certain amount of a given material can be transported.

Subject to : Mass and energy balances;
Quality constraints;
Break even point for exchange, reuse, and dispose options.

The solutions of these problems for different amounts will be used to construct *contours*. For every source, contours show the maximum possible distance, with a given transportation mode, that a certain amount of the material can be transferred to while still having positive economic value at this distance. The representation of the problem consists of four sets of constraints, namely, splitter balance, constraints on the output temperature and concentrations, treatment efforts, and break even point equation.

6.1 SPLITTER BALANCE

For all materials we have the option of exchanging and disposing/reusing. This can be shown by a simple mass balance around a splitter:

$$f = f_1 + f_2. \tag{1}$$

6.2 CONSTRAINTS ON THE OUTPUT TEMPERATURE AND CONCENTRATIONS

The option to dispose or reuse usually requires that certain specifications should be satisfied. These can be described as constraints on temperature and impurities of the material. Here T^{max} and x_i^{max} are maximum allowable temperature and impurity concentrations for the disposed or reused material, respectively:

$$T^a \leq T^{max}, \tag{2}$$

$$x_i^a \leq x_i^{max}. \tag{3}$$

6.3 TREATMENT EFFORTS

To comply with the quality constraints, treatment techniques which utilizes mass and/or heat exchange can be applied. A simple formulation for the reduction of the impurities in the exchangeable material can be obtained by introducing β_i, fraction reduction in the impurity i:

$$f_1\ x_i\ (1 - \beta_i) = f_1\ x_i^a. \tag{4}$$

We can also use another approach. Assuming that the impurities can be removed sequentially using mass transfer agents and their behavior can be described by linear equilibrium curves independent of the presence of other soluble components in the waste stream (El-Halwagi and Manousiouthakis, 1989), we can write

$$y_i = m_i\ x_i^a + b_i, \tag{5}$$

$$f_1\ \left(x_i - x_i^a\right) = L_i\ \left(y_i - y_i^{in}\right), \tag{6}$$

which contains bilinear terms $f_1\ x_i^a$ (if f_1 is not fixed) and $L_i\ y_i$. m_i is the slope and b_i is the intercept of the equilibrium line for component i. Cost of treatments will then be a function of L_is.

For the excess heat to be removed, heat exchange agents will be used, and this process is modeled as follows:

$$Q = f_{cm}\ c_p^{cm}\ \Delta T, \tag{7}$$

$$f_1\ h_1 = Q + f_1\ h_1^a, \tag{8}$$

where f_{cm} is the amount of cold heat transfer agent, c_p^{cm} is its heat capacity, and ΔT is the temperature difference to achieve the required heat transfer. Obviously, these treatment efforts for a single material can be extended to include other exchangeable materials for HEN and/or MEN synthesis. In this way we can combine material exchange decisions with heat and mass integration within the plant.

6.4 BREAK EVEN POINT

The break even point is achieved when the net profit (benefit) of this exchange-dispose-reuse decision becomes zero. In this chapter we will use cost and revenue to define the net benefit, although other measures of benefit could be devised and used with little change of the overall method. The cost consists of two items. Cost of treatment efforts depends on β_i or L_i and f_{cm}, cost of transportation depends on f_2, T, x_i, distance (d), and mode of transportation (t). Finally, revenues from the exchanged material depend on f_2, T, x_i:

$$r(f_2, T, x_i) - C^{te}(\beta_i, f_{cm}) - C^{tr}(f_2, T, x_i, d, t) = 0. \tag{9}$$

To obtain the contours, we can maximize the distance (d) subject to the constraints 1–4, 7–9, and for a given flowrate f_2. The above optimization problem becomes equivalent to solving a set of (non)linear equations, as the inequalities (2) and (3) become active at the optimum solution provided that cost is monotonically decreasing with temperature. If the cost function for transportation, $C^{tr}(f_2, T, x_i, d, t)$, is separable in d and other variables, the contours can be obtained by solving a single algebraic equation, which is

$$d\left(f_2, T^{max}, x_i^{max}, t\right) = \frac{r\left(f_2, T^{max}, x_i^{max}\right) - C^{te}\left[\beta_i\left(x_i^{max}\right), f_{cm}(T^{max})\right]}{C_s^{tr}\left(f_2, T^{max}, x_i^{max}, t\right)} \tag{10}$$

In this chapter the contours are circular; however, contours can take complex shapes, depending on the accessibility of a given point and the cost of establishing the link. These considerations can be included in the above formulation but at the expense of greater model and computational effort. The purpose of the contours is to help identify the candidates for exchange, and therefore the accuracy of the contours derived by the above method will be considered sufficient. The contours should, in general, try to underestimate costs and overestimate benefits and results in outer approximations to the feasible space. In this way the screening will be certain not to eliminate any candidates that might lead to optimal solutions.

6.5 EXAMPLE

To illustrate the feasibility contours, an example of a power plant which has an excess of 267,000 kg/day low pressure steam is used. Its location (denoted by a star) and its environs are given in Figure 14.6. Instead of condensing the steam and disposing of the water, reuse of the steam is considered as an alternative. The data for the problem can be seen in Table 14.2. Using the above formulation, contours which define the maximum distance where a given amount of steam can be transferred are determined. Besides getting an idea of the maximum distance for this supply to be economically viable, we can also eliminate some of the matches. For instance,

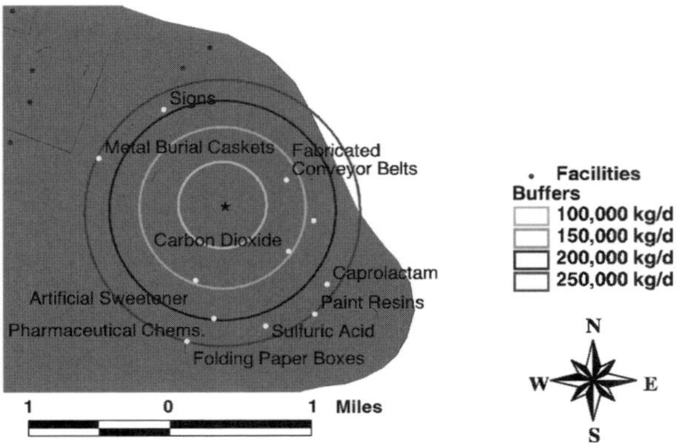

Figure 14.6: Using contour idea for interplant exchange candidates.

if a sink requires less than the amount which the contour closer to the source stands for, we can eliminate this sink from further consideration, unless considering solutions where the excess supply could be shipped to another destination from this sink. In this latter case the contours could be calculated from the intermediate sink now being considered as a source (Fig. 14.7).

7 DESIGN OF THE OPTIMAL SYSTEM OF EXCHANGES

The third step of the methodology is to build the connections between plants which are selected as candidates in the previous two steps. For every plant in this candidate set there are several input and output options of materials. These options are characterized by the material's purity, amount, and transportation costs. The plants have the option to buy their raw materials and/or production aids or use exchanged material. Every input to a plant will be converted to outputs via a production function using amount and purity information. Outputs of plants are defined as the materials which have the opportunity for exchange, such as reusable wastes, production aids, and

TABLE 14.2
Data for the power plant example

Cost of treatment (heat transfer)	$5.284 \times 10^{-5}\ f_{cm}$
Cost of transportation	$9.5 \times 10^{-4}\ f_2^{0.546}$
Revenue	$5.26 \times 10^{-3}\ f_2$
ΔT	20°C

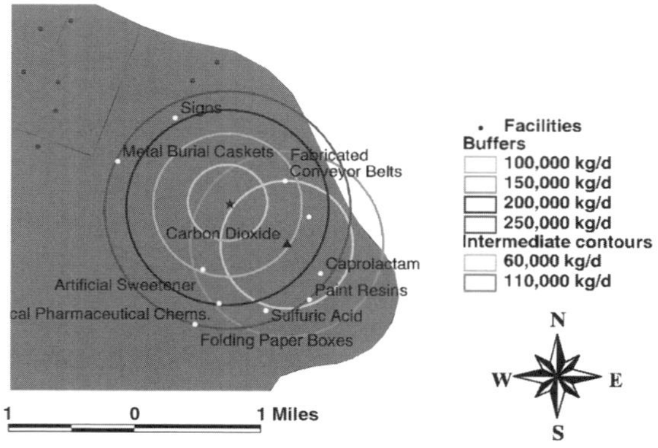

Figure 14.7: Contours from the carbon dioxide producer (intermediate sink with 90,000 kg/day demand for steam).

byproducts. These outputs can be disposed or transported to another facility via available transportation modes, like pipeline, truck, or railroad.

7.1 MATHEMATICAL MODEL

The mathematical formulation consists of nodes describing the source of all-virgin raw materials and production aids (**I**), the sink of all outputs from plants considered (**O**), plants, and transit plants. For example, the objects labeled as I, O, 1, and 1′ in Figure 14.8 are nodes, where 1 is a plant and 1′ is a transit plant. Transit plant nodes are used to describe the material flow to a plant location where it is not used and immediately transferred to another plant location. By introducing these "pseudo" plants, we can get rid of some bilinearities in the model.

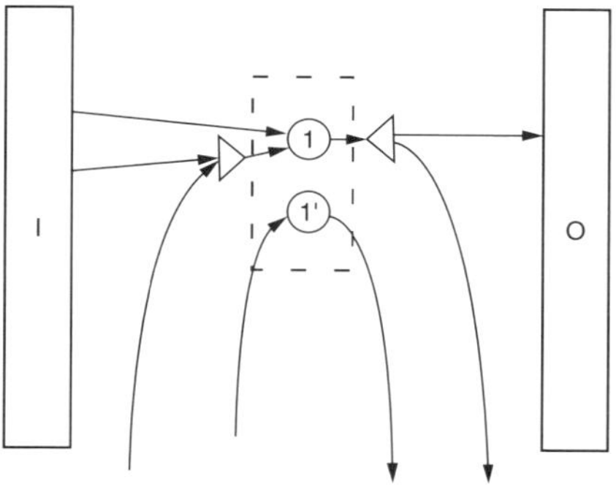

Figure 14.8: Formulation of interplant material exchange.

7.1.1 Mass Balances on the Mixers at the Entrance of the Plants

$$F^i_{n,j} = \sum_{n':\psi(n',n)=1,t} F_{n',n,j,t} \qquad \forall n \in P(n), j \in I_n \cap O_{n'}, \tag{11}$$

$$f^i_{n,j,i} = \sum_{n':\psi(n',n)=1,t} f_{n',n,j,i,t} \qquad \forall n \in P(n), j \in I_n \cap O_{n'}, i \in \theta(j). \tag{12}$$

7.1.2 Heat Balances on the Mixers at the Entrance of the Plants

$$h^i_{n,j}\, F^i_{n,j} = \sum_{n':\psi(n',n)=1,t} h_{n',n,j,t}\, F_{n',n,j,t} \qquad \forall n \in P(n), j \in I_n \cap O_{n'}. \tag{13}$$

7.1.3 Composition and Temperature Specifications for Raw Materials

$$\underline{T}_{n,j} \le T_{n,j} \le \overline{T}_{n,j} \qquad \forall n \in P(n), j \in I_n, \tag{14}$$

$$\underline{x}_{n,j,i} F^i{}_{n,j} \le f^i{}_{n,j,i} \le \overline{x}_{n,j,i}, F^i{}_{n,j} \qquad \forall n \in P(n), j \in I_n, i \in \theta(j). \tag{15}$$

7.1.4 Mass Balances on the Splitters at the Outlet of the Plants

$$F^o_{n,j} = \sum_{n':\psi(n,n')=1,t} F_{n,n',j,t} \qquad \forall n \in P(n), j \in O_n, \tag{16}$$

$$f^o_{n,j,i} = \sum_{n':\psi(n,n')=1,t} f_{n,n',j,i,t} \qquad \forall n \in P(n), j \in O_n, i \in \theta(j), \tag{17}$$

$$\xi_{n,n',j,t}, f^o_{n,j,i} = f_{n,n',j,i,t} \qquad \forall (n,n'):\ \psi(n,n') = 1, j \in O_n, i \in \theta(j), t, \tag{18}$$

$$\xi_{n,n',j,t}, F^o_{n,j,i} = F_{n,n',j,t} \qquad \forall p \in P(n), n':\psi(n,n') = 1, j \in O_n, t, \tag{19}$$

$$\sum_{n':\ \psi(n,n')=1,t} \xi_{n,n',j,t} = 1 \qquad \forall n \in P(n), j \in O_n, \tag{20}$$

$$\xi_{n,n',j,t} \le 1 \qquad \forall n \in P(n), n' \text{ s.t. } \psi(n,n'), j \in O_n, t. \tag{21}$$

7.1.5 Models of Individual Plants

$$F^o_{n,j} = \phi_1\left(F^i_{n,j}\right) \ \forall n \in P(n), j, \tag{22}$$

$$f^o_{n,j,i} = \phi_2\left(f^i_{n,j,i}\right) \ \forall n \in P(n), j, i \in \theta(j), \tag{23}$$

$$h^o_{n,j} = \phi\left(F^i_{n,j}, h^i_{n,j}\right) \ \forall n \in P(n), j, \tag{24}$$

$$F^o_{n,j} \ge 1^K_n \ \forall n \in P(n), j \text{ being the key component of } n. \tag{25}$$

7.1.6 Mass Balances on the Transit Materials

$$F_{n,n',j,t} = F_{n',n'',j,t} \quad \forall n \in P(n),\, n' \in PP(n'),\, n'' \in P(n''):$$
$$\psi(n,n') = 1 \wedge \psi(n',n'') = 1,\ j \in O_n \cap I_{n''},\, t, \qquad (26)$$
$$f_{n,n',j,i,t} = f_{n',n,j,i,t} \quad \forall n \in P(n),\, n' \in PP(n'),\, n'' \in P(n''):$$
$$\psi(n,n') = 1 \wedge \psi(n',n'') = 1,\ j \in O_n \cap I_{n''},\, i \in \theta(j),\, t. \qquad (27)$$

7.1.7 Upper Bounds on Flows

$$F_{n,n',j,t} \leq u_{n,n',j,t} \quad \forall (n,n'):\ \psi(n,n') = 1. \qquad (28)$$

7.1.8 Nonnegativity of Variables

$$F_{n,n',j,t},\, f_{n,n',j,i,t},\, F^{i}_{n,j},\, F^{o}_{n,j},\, f^{i}_{n,j,i},\, f^{o}_{n,j,i},\, \xi_{n,n',j,t} \geq 0. \qquad (29)$$

7.1.9 Objective

$$\min \sum_{(n,n'):\ \psi(n,n')=1,j,t} C_{n,n',j,t}\, F_{n,n',j,t}. \qquad (30)$$

7.2 EXAMPLE

To illustrate the above model we will use an abstract example. Assume there are three plants in the geographic area shown in Figure 14.9.

Plant 1 requires materials A, B, and F, among others. As an output it generates material E. Similarly, Plant 2 gets materials D and E and generates materials A and B, whereas B, A, and E are the inputs to Plant 3, with D being the output. The input-to-output conversions of the plants are given as simple reaction balances with respect to a limiting material, and the impurities in the outputs are described by simple linear models. Because isothermal operation is assumed, no energy balances are included in the model. As an example, we give the equations which describe Plant 1:

$$F^{o}_{1,E} = 2\, \zeta_1\, F^{i}_{1,F}, \qquad (31)$$
$$F^{i}_{1,B} = 2\ 1.25\, \zeta_1\, F^{i}_{1,F}, \qquad (32)$$
$$F^{i}_{1,A} = 0.6\, F^{o}_{1,E}, \qquad (33)$$
$$f^{o}_{1,E,i1} = 0.5\, f^{o}_{1,B,i1} + 0.2\, f^{o}_{1,F,i1} + 2.134, \qquad (34)$$
$$F^{o}_{1,E} \geq 1^{K}_{1}, \qquad (35)$$

where $\zeta_n[n \in P(n)]$ is the extent of the pseudo-reactions in the plant, which is 0.97 for Plant 1. Other relevant data are given in Figure 14.10 and Tables 14.3

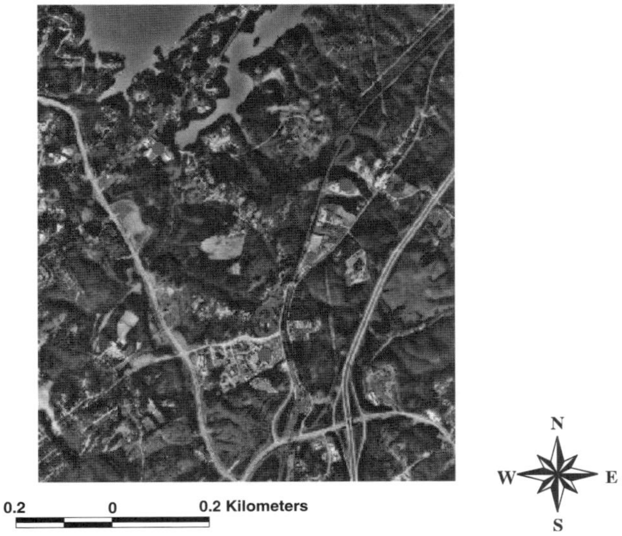

Figure 14.9: Three-plant system for material exchange [the digital orthophoto image depicting NE of Gainesville, GA (Georgia GIS Data Clearinghouse, 1996)].

and 14.4. Only a single mode is used for any transportation of a given material on a specific route.

The formulation for the problem has been implemented using GAMS (Brooke et al., 1992). Since the model includes bilinearities in the constraints that lead to a nonconvex NLP with multiple local minima, standard

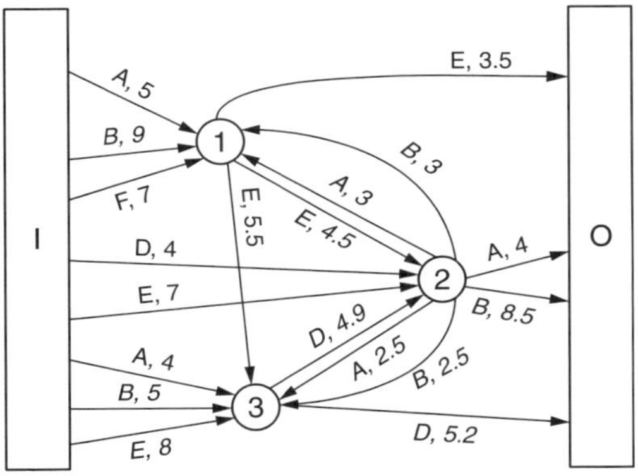

Figure 14.10: Costs (in rmu/kg) of transferring materials between nodes.

TABLE 14.3
Upper limits of mass fractions for impurities in the materials entering the plants

	Plants		
	1	*2*	*3*
1(A)	0.001	—	0.004
2(A)	0.01	—	0.009
1(B)	0.0001	—	0.005
1(D)	—	0.0005	—
1(E)	—	0.02	0.003
1(F)	0.002	—	—

NLP approaches can fail to find the globally optimal solution. Global optimal solutions for this problem class can be found using procedures such as GOP (Floudas and Visweswaran, 1990; Visweswaran and Floudas, 1993) or the branch and bound procedure described in Quesada and Grossmann (1995). In our case we solved the problem using the GAMS/MINOS as the NLP solver with different initial guesses. The global optimality of the best solution found this way was checked by using a lower bound procedure with the LP relaxation of the formulation (McCormick, 1976; Galan and Grossmann, 1998). The overall system cost without any exchange is 8175.56 rmu/day, with the following flow pattern in Figure 14.11. On the other hand, the optimum system configuration is depicted in Figure 14.12 with a total cost of 6483.4 rmu/day.

8 ENVIRONMENTAL IMPACT ANALYSIS

In Sections 5, 6, and 7, a methodology for synthesizing industrial systems that maximize economic benefits of exchanging waste streams was presented. The tacit assumption was that by doing so the overall impact of the system on the environment would be minimized. In this section we

TABLE 14.4
Upper limits (in kg/day) for exchange of materials between plants

	Materials				
Plants	*A*	*B*	*D*	*E*	*F*
1–2	—	—	—	50	—
2–1	25	20	—	—	—

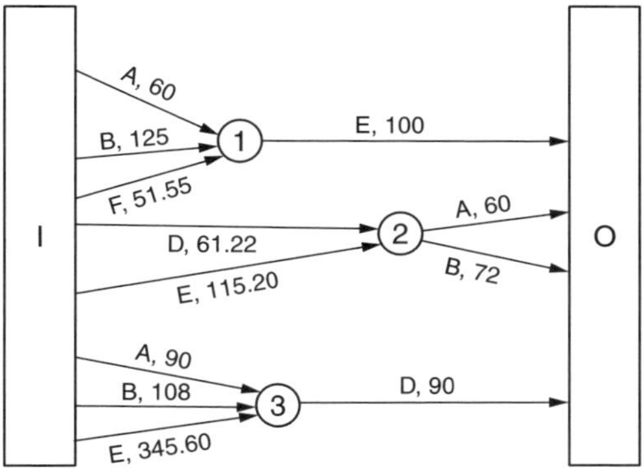

Figure 14.11: Three-plant system without material exchange (flows in kg/day).

will introduce a way to assess this impact on the environment. Because we are dealing with a system of plants, a regional approach is appropriate, as opposed to individual action by the plants. GIS is capable of defining the environment in the required detail, giving most of the parameter information needed for environmental models besides describing the system boundary and eliminating infeasible solutions. Using GIS in conjunction with environmental models to analyze and simulate the fate of discharged materials is a well-established practice (Goodchild et al., 1993; Lahlou et al., 1996). Its use within an optimization framework, combined with process models, has already been explored in Özyurt and Realff (1998).

Here we will extend the approach of Özyurt and Realff (1998) by presenting a multimedia environmental model (Mackay, 1991) incorporated

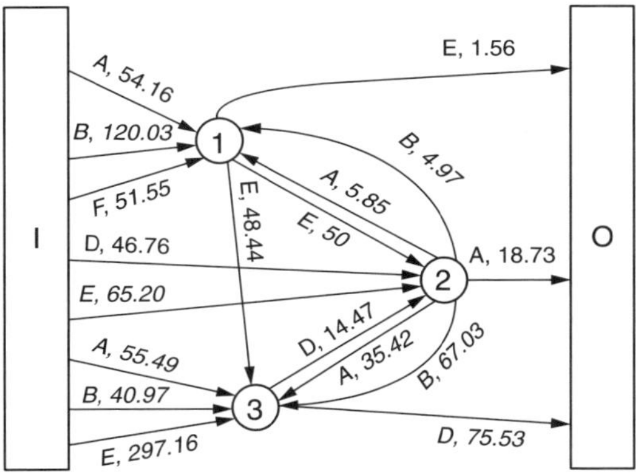

Figure 14.12: Three-plant system with optimum material exchange (flows in kg/day).

into our framework and integrated with GIS. In a multimedia environmental model, the environment is viewed as consisting of a number of compartments which are interacting with each other. This representation of the environment leads to "unit worlds" in which we can study the effects of industrial activities on the surroundings.

Although the models of the compartments in the "unit world" can be detailed using a variety of models of rivers, lakes, air, and sedimentation process, in this presentation we will only use the equilibrium behavior between compartments in addition to advection and exponential decay of the materials entering the environment. Besides being useful in analyzing the impact of the industrial ecosystem we are designing, this multimedia environmental modeling can constrain our construction of the system by placing upper bounds on the concentrations in different compartments.

The idea behind the model we are presenting is that any chemical emitted in the unit world immediately achieves an equilibrium distribution among the compartments; however, this assumption can be omitted when intermedia transport is also considered. On the other hand, the model only treats the emitted compounds, whereas there is a possibility that the products of reactions can be much more harmful than the parent compound which is discharged. Another assumption is that the reactions which degrade the emitted chemical can be written as first-order rate expressions.

The model requires data which are location specific and universal. The location-specific data are average temperature of the environment, the volumes of the compartments and their densities, advection flowrates and chemical concentrations in these flows through the compartments, and, for specific chemicals, reaction rates of decay in compartments. On the other hand, for every chemical emitted, basic thermodynamic information is necessary, which is used to calculate the multimedia partition coefficients. The relationship between fugacity and concentration is denoted by the "Z value" or "fugacity capacity," with units of mol/m^3 Pa. The "Z value" depends on the nature of the chemical and compartment, temperature, pressure, and concentration.

The equations describing the model are as follows:

Calculation of the total emission. The total emission of a given material is the sum of all plant discharges:

$$\sum_{p,t} F_{p,O,j,t} = E_j \quad \forall j. \tag{36}$$

Calculation of the fugacity:

$$fg_j \sum_{c} F_c^{adv} Z_{j,c} + \sum_{c} V_c Z_{j,c} k_{j,c} = E_j + \sum_{c} F_c^{adv} C_{j,c}^{adv} \quad \forall j. \tag{37}$$

Calculation of the concentrations:

$$C_{j,c} = fg_j Z_{j,c} \quad \forall j,c. \tag{38}$$

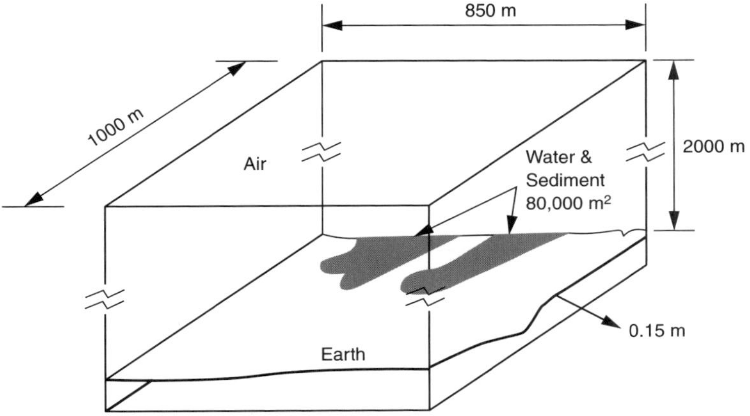

Figure 14.13: Unit world for three-plant material exchange system.

Upper limits for the concentrations:

$$C_{j,c} \le C_{j,c}^{max} \quad \forall j,c. \tag{39}$$

Nonnegativity of variables:

$$E_j, C_{j,c}, fg_j \ge 0. \tag{40}$$

8.1 EXAMPLE

To illustrate the approach, we can use the area shown in Figure 14.9. The unit world, which can be used for this area, is shown in Figure 14.13 with the information required to mathematically define the environment. The material D discharged from Plant 3 is assumed to be the hypothetical chemical described by Mackay (1991). The location-specific data necessary for the solution of the problem are given in Table 14.5.

TABLE 14.5
Compartment properties, Z values, and advection and reaction rates for the area in Figure 14.13

	Compartments			
	Air	*Water*	*Soil*	*Sediment*
Volume (m^3)	$1.7\ 10^9$	880,000	115,500	2,400
Z ($mol/m^3 \cdot Pa$)	4.17×10^{-4}	0.1	12.3	24.6
Advective flow (m^3/h)	$3\ 10^6$	100	0	0
Conc. of D in adv. flow (mol/m^3)	0	0	0	0
k (h^{-1})	6.93×10^{-6}	0.001	0.01	0.0001

TABLE 14.6
Concentration (in mol/m^3) of D in compartments for the three-plant system with optimum material exchange

Compartments			
Air	*Water*	*Soil*	*Sediment*
0	1.0210^{-4}	0.0125	0.0250

The model predicts the concentration of D in the environment, as given in Table 14.6 for the optimum material exchange case (Fig. 14.12). When the model is connected to the above exchange model and the allowed concentration of D in soil is changed to 0.01 mol/m^3, the three-plant material exchange system is required to operate in a different configuration at a higher cost (7346.3 rmu/day) and interaction (Fig. 14.14).

9 SUMMARY AND CONCLUSIONS

Pollution prevention takes place at multiple spatial scales, from the molecular through the plant to a regional or even global scale. This chapter has presented a methodology for addressing *macro-scale pollution prevention* in which facilities or plants are viewed as sources or sinks for materials, and interchanges among them are optimized to reduce total discharges of

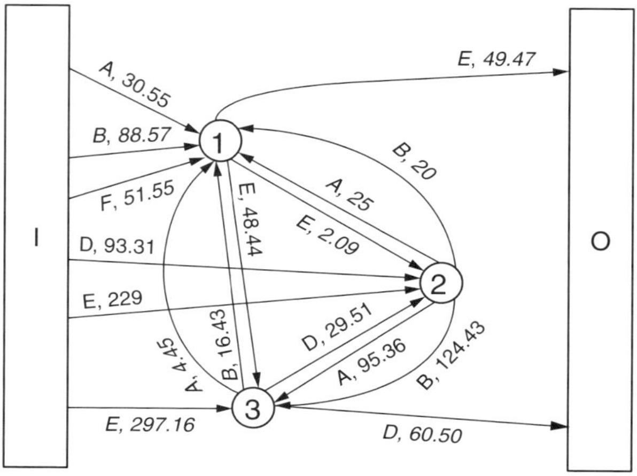

Figure 14.14: Three-plant system with optimum material exchange constrained by environmental concerns.

materials. This abstract problem finds its inspiration in the design of ecoindustrial parks, a paradigm for industrial development that is growing in popularity.

To address macro-scale pollution prevention, we demonstrated solutions to four different but related problems. First, the identification of possible sources and sinks was addressed through the use of reasoning based on the location of the facilities and materials that they generate and consume. Two methods were presented. The first used publicly available databases of manufacturing and emissions reporting, the second was more targeted, using case studies of previous industrial ecosystems. Second, *feasibility contours* were introduced to help prune the search space of possible connections. Third, a model based on the classical process systems formulations of mass and heat exchange was developed to select from the candidate matches. Finally, to assess the impact of the system to the environment, a multimedia model was adopted to facilitate environmentally benign decisions about the possible material exchange. The formulation and solution of these problems relied on the extensive use of GIS. GIS was introduced as a tool to manage the spatial aspects of the problems which are significant when dealing with interplant exchanges as opposed to more spatially local decisions.

We conclude that pollution prevention will involve multiple strategies at multiple spatial and time scales. The representation and coordination of efforts across facilities need to be supported by the appropriate process and environmental engineering tools. GIS, combined with process systems modeling, synthesis, and design, can provide the necessary platform for this important task.

NOMENCLATURE

VARIABLES

f	flowrate of the material (kg/hr)
f_{cm}	flowrate of the cooling medium (kg/hr)
fg_j	fugacity of the system for j
$f^i_{n,j,i}$	amount of the impurity $i \in \theta(j)$ going into $n \in P(n)$ (kg/hr)
$f^o_{n,j,i}$	amount of the impurity $i \in \theta$ (j) going out of $n \in P(n)$ (kg/hr)
f_1	amount of the material treated (kg/hr)
f_2	amount of the material exchanged (kg/hr)
h	enthalpy of the material before any treatment (kJ)
$h^i_{n,j}$	enthalpy of j going into $n \in P(n)$ (kJ)
$h_{n,n',j,t}$	enthalpy of j transferred with t from n to n' (kJ)
$h^o_{n,j}$	enthalpy of j going out of $n \in P(n)$ (kJ)
x_i	mole or mass fraction of the ith component in the material before any treatment
y_i	mole or mass fraction of the ith component in the mass transfer agent

y_i^{in}	mole or mass fraction of the ith component in the mass transfer agent entering the treatment system
$C_{j,c}$	concentration of j in compartment c (mol/m^3)
E_j	amount of j emitted to the environment (mol/hr)
$F_{n,n',j,t}$	transferred amount of j with t from n to n′ (kg/hr)
F_c^{adv}	advective flow (m^3/h)
$F_{n,j}^{i}$	total amount of j going into n ∈ P(n)(kg/hr)
$F_{n,j}^{o}$	total amount of j going out of n ∈ P(n)(kg/hr)
L_i	flowrate of the mass exchange agent for impurity i (kg/hr)
Q	heat removed in the heat treatment (kJ/hr)
T	temperature of the material before any heat treatment (K)
$T_{n,j}$	temperature of j going into n ∈ P(n)(K)
β_i	fractional reduction in the impurity i
$\xi_{n,n',j,t}$	split fraction for the flow out of n ∈ P(n)

PARAMETERS

b_i	intercept of the equilibrium line for component i
c_p^{cm}	heat capacity of the cooling medium for heat transfer (kJ/kg-K)
$k_{j,c}$	rate constants for reaction of j in c (h^{-1})
l_n^k	the minimum daily requirement of key product of the plant n ∈ P(n)(kg)
m_i	slope of the equilibrium line for component i
$u_{\underline{n},n',j,t}$	maximum flow on the arc n-n′ for j using t
$x_{n,j,i}$	upper bound on the concentration of impurity i ∈ θ(j) going into n ∈ P(n)
$\underline{x}_{n,j,i}$	lower bound on the concentration of impurity i ∈ θ(j) going into n ∈ P(n)
$C_{n,n',j,t}$	cost of transport of j using t from n to n′ (rmu)
$C_{j,c}^{adv}$	concentration of advective inputs (mol/m^3)
$C_{j,c}^{max}$	concentration limit for j in compartment c (mol/m^3)
T^{env}	average temperature of the environment (K)
$\overline{T}_{n,j}$	upper bound on temperature of j going into n ∈ P(n)(K)
$\underline{T}_{n,j}$	lower bound on temperature of j going into n ∈ P(n)(K)
V_c	volume of the compartment c (m^3)
$Z_{j,c}$	fugacity capacity of compartment c (mol/m^3 Pa)
ζ_n	extent of the pseudo reactions in the plant n ∈ P(n)
ρ_c	density of compartment c (kg/m^3)
ΔT	temperature difference for the cooling medium to achieve the required heat transfer (K)

SUPERSCRIPTS

a	superscript denoting the quantity after treatment
adv	superscript denoting the quantity of advective terms
max	superscript denoting the maximum allowable quantity

SUBSCRIPTS, INDICES

c	subscript denoting the different compartments of the multimedia model
i	impurities
j	materials
n	nodes
t	mode of transport

FUNCTIONS AND SETS

r	revenues from exchanged material(s)
Cte	treatment costs for dispose and/or reuse options (rmu)
C^{tr}	infrastructure cost for the interplant material exchange (rmu)
I_n	materials going into node n
O_n	materials going out of node n
$P(n)$	plants
$PP(n)$	transit plants
$\theta(j)$	set of impurities (i) in material j
$\psi(n, n')$	boolean function which returns 1 if n and n′ is connected, and 0 otherwise

REFERENCES

About GIS (1998). URL: http://www.esri.com/library/gis/index.html.

Agreda, V. H., Partin, L. R., and Heise, W. H. (1990). High purity methyl acetate via reactive distillation. *Chemical Engineering Progress* 86:40–46.

Allen, D. T., and Rosselot, K. S. (1997). *Pollution Prevention for Chemical Processes*, ed. 1. John Wiley & Sons, New York.

Arcview GIS (1998). URL:http://www.esri.com/software/arcview/index.html.

Bendz, D. J. (1999). Industrial ecology in motion through enterprise integration. *First International Symposium on Environmentally Conscious Design and Inverse Manufacturing*, IEEE Computer Society, pp. 20–26.

Brooke, A., et al. (1992). *Release 2.25: GAMS: A User's Guide*. Boyd and Fraser Publishing Co., Damers, MA.

Ehrenfeld, J., and Gertler, N. (1997). Industrial ecology in practice: The evolution of interdependence at Kalundborg. *Journal of Industrial Ecology* 1:67–79.

El-Halwagi, M. M. (1997). *Pollution Prevention through Process Integration: Systematic Design Tools*. Academic Press, San Diego.

El-Halwagi, M. M., and Manousiouthakis, V. (1989). Synthesis of mass exchange networks. *AIChE Journal* 35:1233–1244.

ESRI (1999). URL: http://www.esri.com/.

Floudas, C., and Visweswaran, V. (1990). A global optimization algorithm (GOP) for certain classes of nonconvex NLPs. I: Theory. *Computers and Chemical Engineering* 14:1397–1417.

Galan, B., and Grossmann, I. E. (1998). Optimal design of distributed wastewater treatment networks. *Industrial and Engineering Chemistry Research* 37:4036–4048.

GDITT (1996). Georgia manufacturing directory, Georgia Department of Industry, Trade and Tourism (GDITT).

Georgia GIS Data Clearinghouse (1996). URL: http://www.gis.state.ga.us.

Goodchild, M. F., Parks, B. O., and Stegaert, L. (1993). *Environmental Modeling with GIS*. Oxford University Press, Oxford.

Graedel, T. E., and Allenby, B. R. (1995). *Industrial Ecology*. Prentice-Hall, Englewood Cliffs, NJ.

Gundersen, T., and Naess, L. (1988). The synthesis of cost optimal heat exchanger networks: An industrial review of the state of the art. *Computers and Chemical Engineering* 12:503–530.

Huxhold, W. E. (1991). *An Introduction to Urban Geographic Information Systems*. Oxford University Press, New York.

Kolodner, J. (1993). *Case-based Reasoning*. Morgan Kaufmann Publishers, San Mateo, CA.

Kroschwitz, J. I., and Howe-Grant, M. (1995). *Kirk-Othmer Encyclopedia of Chemical Technology*, ed. 4. John Wiley & Sons., New York.

Lahlou, M., Shoemaker, L., Paquette, M., Bo, J., Choudhury, S., Elmer, R., and Xia, F. (1996). *BASINS: Version 1.0 User's Manual*. EPA, Washington, DC.

Mackay, D. (1991). *Multimedia Environmental Models: The Fugacity Approach*. Lewis Publishers, Chelsea, MI.

Martin, S., Weitz, K. A., Cushman, R. A., and Sharma, A. (1996). Eco-industrial parks: A case study and analysis of economic, environmental, technical, and regulatory issues. *Technical report*, Research Triangle Institute, Research Triangle Park, NC.

McCormick, G. P. (1976). Computability of global solutions to factorable nonconvex programs: Part I—Convex underestimating problems. *Mathematical Programming* 10:147–175.

Nemerow, N. L. (1995). *Zero Pollution Industry: Waste Minimization Through Industrial Complexes*. John Wiley & Sons, New York.

Özyurt, B. D., and Realff, M. J. (1998). Operational planning for chemical processes using geographic information and environmental modeling. *Proceedings of Foundations in Computer-Aided Operations (FOCAPO)*, CACHE, Snowbird, UT.

Quesada, I., and Grossmann, I. E. (1995). Global optimization of bilinear process networks with multicomponent flows. *Computers and Chemical Engineering* 19:1219–1242.

Saotome, Y., Nakazawa, H., Yoshizawa, H., and Aoki, T. (1999). Development of ecoindustrial park in Gunma. *First International Symposium on Environmentally Conscious Design and Inverse Manufacturing*. IEEE Computer Society. Los Alamitos, CA.

Standard Industrial Classification Manual (1987). Executive Office of the President, Office of Management and Budget.

Tennessee Materials Exchange (1998). URL: http://funnelweb.utcc.utk.edu/~cis/tme_frm.htm.

Toxic Release Inventory (1998). URL: http://www.epa.gov/opptintr/tri/index.htm.

Visweswaran, V., and Floudas, C. (1993). New properties and computational improvement of the GOP algorithm for problems with quadratic objective function and constraints. *Journal of Global Optimization* 3:439–462.

Part Six

MOLECULAR DESIGN AND SELECTION

CHAPTER 15

Computer-Aided Tools for Design/Selection of Environmentally Friendly Substances

Peter M. Harper and Rafiqul Gani

Computer Aided Process Engineering Center (CAPEC), Department of Chemical Engineering, Technical University of Denmark, DK-2800 Lyngby, Denmark

Abstract

In this chapter, computer-aided tools that can be used in the search for environmentally friendly substances are reviewed. The review is presented in the form of a theoretical background with respect to important issues related to computer-aided molecular design/selection, such as compound molecular structure representation, properties prediction, search space management and chemical feasibility, and the Computer-Aided Molecular Design/Selection (CAMD/S) algorithms. A general CAMD/S framework, together with a brief overview of a well-known CAMD algorithm, is presented, illustrating the state of the art in this area. Finally, a tutorial example, highlighting the solution steps of the general CAMD/S framework, is presented. The tutorial example involves the design/selection of solvents for the separation of acetic acid from water and highlights the environmental aspects to solvent design/selection.

INTRODUCTION

In the design and operation of many processes and chemical processes, in particular, external substances are often required in order to perform specified tasks. These substances usually are not reactants or products (hence the term external) but perform specific tasks for the process. The tasks that the external substances perform may be the functions of a solvent, a cooling/heating agent, a miscibility promoter, a viscosity reducer, etc., where a process fluid is required. An important issue with respect to using external substances is the question of spillage or loss (emission) through effluent streams and their consequence on the surroundings. If the spillage or loss exceeds a permitted upper limit, operation of the process may become economically infeasible and/or environmentally unacceptable. Therefore, the design/selection of external substances needs to consider environmental and economic factors in addition to the process functions that they must perform.

Low environmental impact can be obtained in various ways. Essentially, the environmental effect of a compound can be expressed as the product of the discharge to the environment and the environmental effect of the compound per unit mass. Naturally the problem is more complicated than outlined above, as there can be several parallel impacts of a compound, and the effects are dependent on discharge pathway and media as well as local factors; however, by applying the above principle, it can be seen that there are two possible extremes of an environmentally friendly substance: a compound having a high impact but low discharge or a compound having a low impact but a high discharge. Optimally, a compound having a low impact as well as low loss to the surroundings under normal operational conditions can be found. In any case it is imperative that the chosen compound fulfills the operational requirements of the task it is intended for.

Various techniques, such as database search/selection and computer-aided design/selection, exist for design/selection of external substances. A search of existing databases and/or selection based on past experience is often applied when selecting external substances. With advances in computer technology and information technology, use of computer-aided techniques for design/selection of external substances is receiving increased attention. The technique, commonly known as CAMD (Computer-Aided Molecular Design), is particularly suitable for design of substances with specified properties. The specified properties in this case represent the properties related to the function of the substance as well as environmental factors. Considering the specified properties as target values, the CAMD technique can design (generate) molecules that match the target values. In the case of a database search approach (CAMS), generation of molecular structures is not needed. For the design problem, the algorithm needs to generate chemically feasible molecules. For both problems, it needs to examine the properties of the candidate substances. The properties may be retrieved from a database and/or predicted with suitable property estimation methods. The main features of a CAMD technique, therefore, include structure (molecule) representation, generation of structures, property prediction, and the design/selection algorithm. Also, to apply any CAMD technique, it is necessary to formulate the problem (preanalysis of the problem) and evaluate/verify the results (post-analysis of results).

In this text, only CAMD/S solution methodologies will be treated in detail. For further information on CAMS/S, the reader is referred to others (Joback, 1994; Hermansen, 1993; Modi and Stephanopoulos, 1996; Zhao et al., 1996;

Cabezas et al., 1996), who have reported database search–based algorithms for CAMS.

The objective of this chapter is to provide the reader with a theoretical background of the important topics related to CAMD techniques and an illustration of the main steps in the solution of problems related to design/selection of substances with a general CAMD framework. In the theoretical background, the important topics in CAMD techniques are reviewed with respect to the current state of the art. Under CAMD framework, the main solution steps are reviewed through a CAMD technique of Harper et al. (1999). The application of the CAMD technique is illustrated through a tutorial example involving the extraction of acetic acid from an aqueous mixture with a suitable solvent.

THEORETICAL BACKGROUND

The important issues in CAMD/S are representation of the molecular structure of the designed substances, estimation of pure component and mixture properties, generation of chemically-feasible molecular structures, and the solution of the CAMD/S problem with a suitable solution engine. In this section, the above-mentioned topics are reviewed with respect to the current state of the art within the framework of CAMD techniques.

COMPOUND REPRESENTATION

All CAMD techniques need to employ some form of compound representation in order to use molecular structure information for property estimation. In general, the estimation methods used for predicting properties of the designed compounds decide the level of detail needed for the molecular structural information and the chemical representation method to use. Other considerations are compatibility with external programs and databases.

The simplest form of a compound is an atomic representation based on chemical formula. Here, a compound is simply represented by the types of atoms it contains and the number of occurrences of each atom type (Fig. 15.1a). A single representation can describe a large number of compounds of very different types. No direct information regarding the bonds in the compound can be extracted from the representation, although if assumptions of the valency of the different atom types are made, it is possible to calculate bond configurations. A related representation form is the representation of a compound as a collection (or vector) of *groups*. A group is a molecular fragment or substructure defined by the number and types of atoms in the fragment, how the atoms are connected, how many free connections the group has, and where (on which atom) they are located. Figure 15.1b shows an example of a fragment, and Figure 15.1c shows an example of a group vector. A group vector contains some information about the connectivity of the structure of the molecule but does not define it

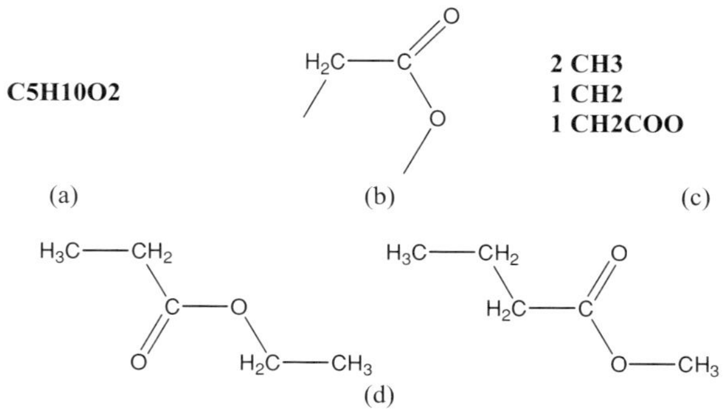

Figure 15.1(a)–(d): Compound representations.

completely. As a result, a group vector can represent more than one possible molecule (isomers); Figure 15.1d illustrates the different compounds possible to construct using the group vector in Figure 15.1c.

The compounds depicted in Figure 15.1d have the connectivity defined. For computerized representation of connectivity, various methods exist (Raman and Maranas, 1998; Churi and Achenie, 1996). One of the most versatile and manageable methods is the connectivity matrix. A connectivity matrix is a square symmetrical matrix with rows and columns representing the atom (or fragments) in the molecule and containing zeroes and nonzeroes indicating bonds or absence of bonds. A connectivity matrix can be on fragment level or atomic level. Conversion from a fragment-based matrix to an atomic-based matrix is achieved by substituting the entry for each fragment with that of the atomic connectivity matrix representing the fragment. Figures 15.1e and 15.1f are the fragment-based and atom-based connectivity matrices, respectively, for the first compound in Figure 15.1d.

Although the connectivity matrix defines the two-dimensional relations between atoms in a compound, it does not contain the steric information needed to distinguish R/S, L/D, and cis/trans isomers. To distinguish between such isomers, it is necessary to have three-dimensional information about the placement of the atoms. For three-dimensional representation, two methods are widely used. The first is the combination of a connectivity matrix with a list of x, y, z Cartesian coordinates for the atoms. The second is the so-called internal coordinate system, where an atom's position is defined by a length, a bond angle, and a torsion angle (Maranas and Floudas, 1994). Choice of the type of representation depends on the computations that are to be performed with the three-dimensional representation.

The generation of molecular representations containing high levels of detail is considerably more computationally intensive than generation of simple group vector representations. Although many methods have been proposed for the generation of group vectors (Pretel et al., 1994; Naser and Fournier, 1991; Odele and Macchietto, 1993; Joback and Stephanopoulos, 1995; Marcoulaki and Kokossis, 1998; Duvedi and Achenie, 1996; Gani et al.,

	CH3	CH3	CH2	CH2COO
CH3	0	0	0	1
CH3	0	0	1	0
CH2	0	1	0	1
CH2COO	1	0	1	0

(e)

	H	H	H	H	H	H	H	H	H	H	C	C	C	C	C	O	O
H	0										1						
H		0									1						
H			0								1						
H				0								1					
H					0							1					
H						0						1					
H							0						1				
H								0					1				
H									0					1			
H										0				1			
C	1	1	1								0		1				
C				1	1	1						0		1			
C							1	1			1		0		1		
C									1	1		1		0		1	
C													1		0	1	2
O														1	1	0	
O															2		0

(f)

Figure 15.1(e)–(f): Compound representations in matrix form.

1991; Vaiddyanathan and El-Halwagi, 1994; Constantinou et al., 1995), the generation of representations of higher detail has not attracted as much attention. Graph reconstruction algorithms have been proposed (Kier et al., 1993 and Gordeeva et al., 1990). Raman and Maranas (1998) and Ourique and Silva Telles (1998) have employed direct generation of connectivity matrices; Venkatasubramanian et al. (1992, 1994, 1995) have employed generation of molecular backbones with sidechains; Harper et al. (1999) have employed multilevel generation leading to three-dimensional structures; and Meniai et al. (1998) have employed direct generation of three-dimensional structure from fragments.

PREDICTION OF PROPERTIES

Success of CAMD techniques depends, to a large extent, on the ability to predict the necessary pure component and mixture properties. Considerable work has been done in this field, yielding methods for the estimation of a wide range of properties. Estimation techniques used in CAMD usually are based on molecular structural information (purely predictive) or based on other properties (empirical). Predictive methods typically are based on property/structure relationships where, given the molecular structure of a compound, the needed properties are calculated using a suitable mathematical relation. Most property estimation methods used in CAMD techniques are based on the *Group Contribution Approach*, or GCA (Franklin, 1949), where the properties of a compound are expressed in terms of functions of number of occurrences of predefined fragments (groups) in the molecule. GCA-based methods (e.g., Joback, 1984; Constantinou and Gani, 1994; van Krevelen, 1990; Lydersen, 1955; Fedors, 1982; Ambrose, 1978, 1979; Franklin, 1949), plus many others, have been applied regularly for pure component property prediction in CAMD techniques. The most widely used GCA-based method for prediction of mixture properties in CAMD techniques is UNIFAC (Fredenslund et al., 1977). Note that because molecular structures are generated, it is not feasible to use molecular methods such as NRTL (Renon and Prausnitz, 1968) or Wilson (Wilson, 1964), which need compound-specific and predetermined molecular interaction parameters for estimation of mixture properties.

Apart from GCA-based methods and empirical correlations, other *Quantitative Structure Property Relationships* (QSPR) and *Quantitative Structure Activity Relationships* (QSAR) (Kier and Hall, 1986; de Waterbeemd, 1995) have also been employed in CAMD techniques. Raman and Maranas (1998) used topological indices correlated to physical and functional properties. Duvedi and Achenie (1996) developed a hybrid estimation technique based on group contribution methods combined with a structural classification system. Several compilations of estimation methods (e.g., Horvath, 1992; Perry and Green, 1984; Reid et al., 1987; Lyman et al., 1990), containing model parameters and recommendations for their use, also exist. As an alternative to the above-mentioned methods, use of molecular modeling tools for the estimation of properties in a CAMD framework has been reported (Meniai and Newsham, 1996; Meniai et al., 1998; Harper et al., 1999).

Estimation of Environmental Properties

Environmental properties need special attention, as they usually do not belong to the standard databases for properties of chemical compounds. For the estimation of environmental properties, such as toxicity, biodegradability, ozone depletion potential, biological oxygen demand, global warming potential, and soil adsorption potential, very few general methods covering a wide range of compounds have been developed; however, a number of methods valid for specific molecular types, such as alcohols, acids, and benzene derivatives, are available (Lyman et al., 1990). These methods are capable of predicting properties of the above-mentioned type. Often, methods for environmental properties rely on the octanol/water partition coefficient (log P) as a known property value.

An in-depth review of the available methods for estimation of properties regarding environment, health, and safety (EH&S) is outside the scope of this work. Instead, the reader is referred to other publications (e.g., Lyman et al., 1991; Hermens and Opperhuizen, 1991; Kaiser, 1987; Hansch and Leo, 1979; van Leeuwen and Hermens, 1995). Sources for experimental and regulatory EH&S values include databases such as CHRIS (Chem-Bank,

1998a), HSDB (Chem-Bank, 1998b), and RTECS (Chem-Bank, 1998c), which store environmental data and properties for a large number of substances.

SEARCH SPACE MANAGEMENT AND CHEMICAL FEASIBILITY

The main motivation for applying computers in molecular design is the size of the search space. Manual examination and selection in molecular design are only possible for very simple problems. If an unrestricted exhaustive enumeration of all possible combinations of groups is performed, the number of alternatives is given by the widely cited formula (Joback and Stephanopoulos, 1995):

$$M_{N,K_{\max}} = \sum_{K=K_{\min}}^{K_{\max}} \frac{(N+K-1)!}{K!(N-1)!}. \tag{1}$$

Equation (1) can be derived from

$$M_{N,K_{\max}} = \sum_{K=K_{\min}}^{K_{\max}} M'_{N,K}, \tag{2}$$

where

$$M'_{N,K} = \binom{N+K-1}{K}. \tag{3}$$

The binomial formula, equation (3), calculates the number of possible combinations when selecting K elements from a population of size N into unordered sets, allowing each element to appear more than once. Applied to molecular design, K is the number of groups in the compound description (allowed to vary between $K_{\min}$ and $K_{\max}$), and N is the number of groups to choose from. For the case of $K_{\min} = 1$, $K_{\max} = 10$, and $N = 40$, the formula reveals that 10,272,278,169 different combinations are possible. This number is for the simplest case of a compound representation using only a group vector. If additional detail were to be included (e.g., connectivity information), the number of alternatives would be even greater due to the increase in the degrees of freedom; however, it is not desirable or necessary for any method to actually search the space mentioned above, allowing free combination of all groups, as it clearly leads to infeasible compounds that do not fulfill simple valency rules. An example of an infeasible compound representation is shown below:

4 CH3

1 OH

By introducing chemical feasibility rules that do not permit the generation of "illegal" compounds, the size of the combinatorial problem can be reduced by several orders of magnitude. Introduction of the simplest chemical feasibility rule, the rule of "complete" compounds (i.e., no free connections in the finished molecules), results in a reduction factor of 2.04 (5,042,724,997 alternative combinations). Note that the $N = 40$ selectable groups are equally divided between having one, two, three, or four free connections.

The effective size reduction of the search space can be easily calculated by applying equation (3) and valency constraints with the procedure described below. The valency constraints are expressed as

$$K = n_S + n_D + n_T + n_Q, \tag{4}$$

$$n_S = n_T + 2n_Q - 2(A - 1), \tag{5}$$

n_s = number of groups having one free connection.

n_D = number of groups having two free connections.

n_T = number of groups having three free connections.

n_Q = number of groups having four free connections.

A = number of rings in molecule.

Note that this formulation allows a group to form a ring with itself. This is a very loose formulation and, in reality, is only valid if all groups have an internal chain length larger than one and have free connections at each end. Additional constraints, such as maximum allowable number of rings with number of groups selected, can also be specified if desired. For any solution ($V_{i,K,N}$) to the valency constraint equations,

$$V_{i,K,N} = \{n_{S,i}, n_{D,i}, n_{T,i}, n_{Q,i}, A\} \text{ ith solution to equations (4) and (5)}$$

with a valency distribution of the available groups given by

$$N = N_S + N_D + N_T + N_Q.$$

The number of feasible alternative combinations is given by

$$\overline{M}'_{i,N,K} = \binom{N_S + n_{S,i} - 1}{n_{S,i}} \cdot \binom{N_D + n_{D,i} - 1}{n_{D,i}} \cdot \binom{N_T + n_{T,i} - 1}{n_{T,i}} \cdot \binom{N_Q + n_{Q,i} - 1}{n_{Q,i}}, \tag{6}$$

$$\overline{M}_{N,K_{\max}} = \sum_{K=K_{\min}}^{K_{\max}} \sum_{i=1}^{I} \overline{M}'_{i,N,K}, \tag{7}$$

where $\overline{M}'_{i,N,K}$ is the number of possible combinations when selecting K groups from a group set having N components for the ith solution to the feasibility constraints, I is the number of solutions to the feasibility constraints when selecting K groups, and $\overline{M}_{N,K_{\max}}$ is the number of possible combinations when selecting from 1 to $K_{\max}$ groups under the restrictions of feasibility constraints.

The combinatorial reduction factor H is then given by

$$H_{N,K_{\max}} = \frac{M_{N,K_{\max}}}{\overline{M}_{N,K_{\max}}} = \frac{\sum_{K=K_{\min}}^{K_{\max}} \frac{(N+K-1)!}{K!(N-1)!}}{\sum_{K=K_{\min}}^{K_{\max}} \sum_{i=1}^{I} \overline{M}'_{i,N,K}}. \tag{8}$$

This quantity is the factor by which the size of search space is diminished when introducing feasibility rules. A block diagram illustrating the solution procedure for calculation of the reduction of search space is given in Figure 15.2.

The procedure described above is the simplest form of introduction of *metagroups* and *chemical feasibility* rules as described by Joback and Stephanopoulos (1995). The reduction in problem size is achieved without disqualifying any valid compound by applying chemical knowledge about the properties of "real" solutions. Table 15.1 lists the achievable combinatorial

TABLE 15.1
Calculated combinatorial reduction factors H at different specifications and group valency distributions

						Results	
Specifications	N_S	N_D	N_T	N_Q	A	$\overline{M}$	H
No restrictions	10	10	10	10	–	5042724997	2.04
	20	10	5	5	–	3436474740	2.99
	5	5	10	20	–	5136044906	2.00
	5	15	15	5	–	5118464479	2.01
	21	11	6	2	–	2638350448	3.89
Only acyclic compounds allowed	10	10	10	10	0	225505280	45.55
	20	10	5	5	0	1122022770	9.16
	5	5	10	20	0	6167475	1665.56
	5	15	15	5	0	96353040	106.61
	21	11	6	2	0	1225436784	8.38
Only monocyclic compounds	10	10	10	10	1	491951449	20.88
	20	10	5	5	1	989184724	10.38
	5	5	10	20	1	24136699	425.59
	5	15	15	5	1	363598822	28.25
	21	11	6	2	1	808160292	12.71
Only bicyclic compunds	10	10	10	10	2	807241425	12.73
	20	10	5	5	2	681802545	15.07
	5	5	10	20	2	76426925	134.41
	5	15	15	5	2	867800075	11.84
	21	11	6	2	2	395814218	25.95
Bi- and monocyclic compounds	10	10	10	10	1,2	1299192874	7.91
	20	10	5	5	1,2	1670987269	6.15
	5	5	10	20	1,2	100563624	102.15
	5	15	15	5	1,2	1231398897	8.34
	21	11	6	2	1,2	1203974510	8.53

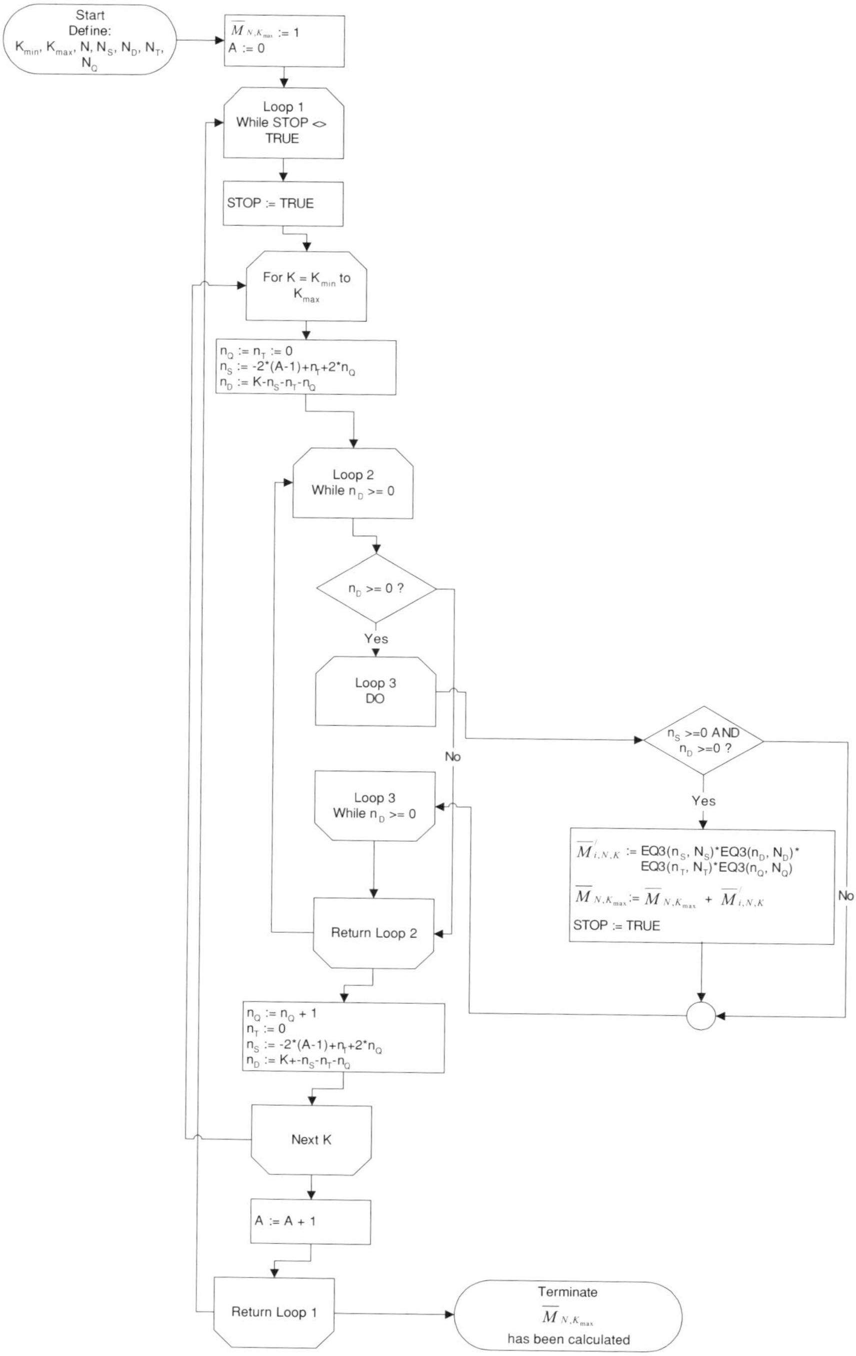

Figure 15.2: Flow diagram of calculation procedure for finding number of combinations when valency constraints have been imposed.

reduction factors for different allowed valency distributions and numbers of rings. An additional and substantial reduction of the size of the search space is achieved by controlling the number and types of functional groups allowed in a molecule. This can be done by subdividing the groups into clusters based on the number of free connections and rules defining the use of groups from each cluster (Gani et al., 1991).

The simplest form of functional group control is to limit the total number of functional groups allowed in a compound. The limitation can serve two purposes: to limit the generated compounds to a complexity the employed property estimation methods can handle and/or to limit the range of generated compounds to types more likely to be commercially available (i.e., monofunctional compounds). Although the combinatorial problem is very large, it can be controlled if the group set is chosen with the combinatorial aspect in mind and carefully selected feasibility criteria are applied.

CAMD ALGORITHMS (SOLUTION ENGINE)

Different approaches have been proposed for solving CAMD problems. The approaches can be grouped into three categories:

1. Mathematical programming (a mathematical representation of the problem is solved with a numerical optimization method).
2. Stochastic optimization (a mathematical representation of the problem is solved by numerical stochastic methods).
3. Enumeration techniques (a combined mathematical and qualitative representation of the problem is solved by hybrid solution approaches).

Common to all of the techniques is that the objective is to find a compound or compounds fulfilling the requirements set forth in the constraints and goals.

Mathematical Programming

Mathematical programming involves formulation of the CAMD problem into an optimization problem where the essential property constraints are treated as mathematical constraints while the desirable and undesirable properties are combined into an objective function weighting the property values into a single score. The objective function is subsequently minimized using an appropriate optimization algorithm. The formulation of the optimization problem—how the constraints, feasibility requirements, objective function, and compound representation are expressed—controls the range of solution algorithms applicable for solving the CAMD problem that can be solved by this approach.

Traditionally, the optimization variables of a mathematical programming–based CAMD algorithm are descriptors used to identify a compound (in its simplest form, a vector giving the number of occurrences of the different groups). The descriptors are by nature discrete and can only hold integer values (number of groups), and the methods used in predicting design properties often are nonlinear relations relating the properties of a compound with the structure. Also, inclusion of chemical feasibility rules means that there will be additional discrete variable–based constraints. The resulting optimization problems are typically solved by methods commonly known as *Mixed Integer NonLinear Programming* (MINLP) (Duran and Grossmann, 1986; Nemhauser et al., 1989). Solution of CAMD problems using MINLP methods have been reported (Duvedi and Achenie, 1996; Odele and Macchietto, 1993; Vaidyanathan and El-Halwagi, 1994). MINLP methods have also been used for mixture design (Churi and Achenie, 1997; Duvedi and Achenie, 1997; Vaidyanathan and El-Halwagi, 1996), whereas Vaidyanathan et al. (1998) have developed an MINLP-based method for the design of composite materials.

The drawbacks of applying MINLP methods lie in the computational cost of the calculation, as well as finding the global optimum, findings of which, in most cases, cannot be guaranteed unless special solution techniques are employed. Vaidyanathan and El-Halwagi (1996) have used a global optimization technique for the design of polymers and polymer blends.

Moreover, the nature of CAMD problems implies the need, in many cases, for a list of candidates that can be further investigated and not for an optimal candidate. This is because it is difficult to represent all candidate selection–related decisions in a suitable mathematical representation. Maranas (1996) and Raman and Maranas (1998) have proposed alternative problem formulations resulting in the development of CAMD algorithms using solution methods based on *Mixed Integer Linear Programming* (MILP). This problem formulation can, of course, find the globally optimal solution (candidate substance); however, it requires that all properties used as design variables can be estimated using linear models. This can be achieved by using models that are indeed linear or by using a linear approximation to a nonlinear model if possible.

Stochastic Optimization

Stochastic optimization relies on the successive pseudo-random generation of solution alternatives. Based on the solution after an iterative step, the next solution is generated by introducing random permutations subject to probability functions. While the stochastic methods are related to the methods used in the mathematical programming approach, in trying to reach an optimum value of an objective function, the underlying theory and practical implementations are very different. One important aspect of the stochastic optimization methods is that they do not require any gradient information, allowing the easy specification of discontinuous properties as design goals.

Two forms of stochastic optimization–based CAMD algorithms have been developed. In the first form of stochastic optimization, Marcoulaki and Kokossis (1998) and Ourique and Silva Telles (1998) have proposed the use of a *Simulated Annealing* method (Kirkpatric et al., 1983) based on Markov probability theory. In the second form of stochastic optimization, Venkatasubramanian et al. (1992, 1994, 1995) have successfully applied a *Genetic Algorithm* (Holland, 1975) based on Darwinian evolutionary theory, where populations of potential solutions are created from the previous population based on "survival of the fittest" and recombination and transmission of attributes from parent to offspring. In both stochastic optimization approaches there are a number of adjustable parameters controlling the optimization sequence. The values of the parameters cannot be calculated a priori, but must be based on experimentation and experience.

Enumeration Techniques

Solution of CAMD problems using enumeration techniques aims at the identification of compounds fulfilling the feasibility and property constraints by generating a series of molecules using a combinatorial approach and testing the generated molecules against the specifications, screening out the molecules that do not meet the requirements. In the enumeration techniques, often termed as the "Generate & Test" approach, the generation and screening are performed separately. First, all molecules are generated and subsequently tested against the design specifications. In the recent CAMD algorithm proposed by Friedler et al. (1998) the generation of molecules is driven by the property constraints, and only one step is needed. Also, Harper et al. (1999) have proposed a multilevel algorithm with generation and test algorithms for each level.

The major challenge in solving practical CAMD problems using enumeration techniques lies in controlling the size of the combinatorial problem. If a simple exhaustive enumeration is performed on CAMD problems of even moderate size, the combinatorial size of the problem may become so large that, even for high-end computers, excessive computational time will be necessary. For a generate and test approach not to suffer from the so-called combinatorial explosion, it is necessary to introduce a control for the generation step without discarding chemically feasible substances. Various control strategies have been proposed, including interval analysis (Joback and Stephanopoulos, 1995; Friedler et al., 1998), group set simplifications based on thermodynamic model insights (Pretel et al., 1994), and hybrid approaches introducing extended feasibility considerations in the generation step (Gani et al., 1991; Constantinou et al., 1996; Harper et al., 1999). The hybrid approach is discussed in more detail in this chapter in the subsection "Generate and Test Approach." As with the stochastic optimization methods, no gradient information is needed for most enumeration methods, and it is possible to test for properties not having an explicit mathematical formulation. For the interval analysis approaches, this is not always possible and can cause an additional testing step to be needed.

GENERAL CAMD FRAMEWORK

A general CAMD framework is illustrated in Figure 15.3. It can be noted from Figure 15.3 that the design/selection of chemicals is an iterative process involving five main solution steps. The first two steps are related to preanalysis of the CAMD problem, whereas the last two steps are related to postanalysis of the results from the CAMD algorithm. Very few CAMD algorithms include all five steps, although, for a solution of the CAMD problem, all steps are needed.

CAMD PROBLEM IDENTIFICATION AND FORMULATION (STEP 1)

The first step in solving a CAMD problem (such as a search for an environmentally friendly substance) is to formulate the ultimate goal. The type of goal determines many of the decisions taken in later steps (such as steps 2 and 4) of the CAMD framework. The goal should describe the function of the desired compound (i.e., what it should be capable of—the essential properties), the environment/equipment where the function should be performed (desirable properties), and properties that are not desirable (such as properties having a negative environmental impact).

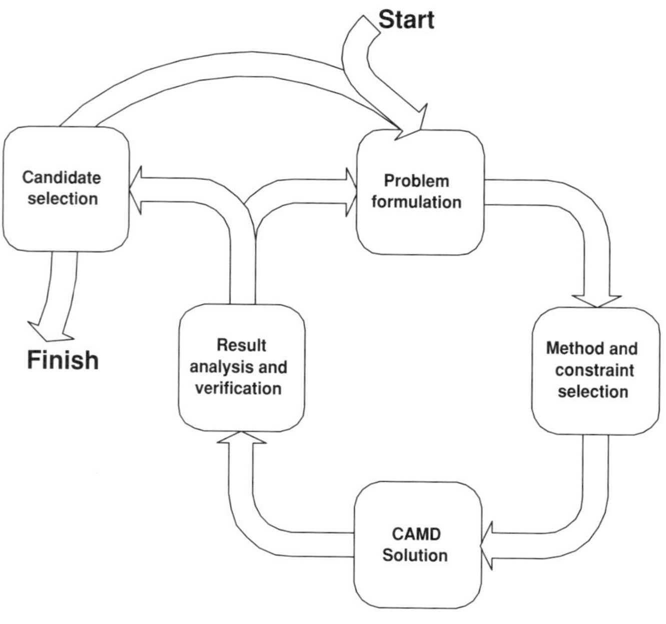

Figure 15.3: CAMD solution steps.

Essential Properties

For a particular use of a substance, usually there is a set of essential properties that a substance must have in order to be used in the intended operation (e.g., if a solvent is to be used at ambient temperatures, it must be in a liquid state at these temperatures). Furthermore, there can be legislative considerations requiring certain properties to fall within predefined values (such as flammability limits allowed for equipment safety certification).

Desirable Properties

Desirable properties are properties that make a particular compound more suited for the intended operation when they have a certain (relative) value. Such properties (e.g., solvent selectivity) can be related to purely operational considerations as well as environmental considerations. Specifications for the desirable properties are not strictly numerical, but rather qualitative, such as "close to a reference value", "as high as possible" or "as low as possible" as the objective is to find the very best alternatives. Because of the qualitative nature of the specifications, such properties are often decision variables rather than direct design variables.

Undesirable Properties

The undesirable properties represent characteristics not wanted in the substance. As for the desirable properties, the specifications for the undesirable properties are typically qualitative. For example, the environmentally friendly substance must not have a high ozone depletion potential.

METHOD AND CONSTRAINT SELECTION (STEP 2)

Method and constraint selection consists of the selection of methods for estimation of the sets of properties identified in the problem formulation step, selection of values for the set of properties identified as the set of constraints, and selection of the particular CAMD solution algorithm. Selection of methods for estimation of properties depends on the availability of a suitable model with the necessary model parameters. Computational accuracy also plays an important role in the selection step.

The starting point here is the set properties identified in the CAMD problem formulation. From this, a subset is selected as constraints with defined goal values plus upper and/or lower bounds. Environmental consideration can be formulated implicitly or explicitly. Explicit considerations relate physical properties to environmental considerations (e.g., ozone depletion potential), whereas implicit considerations are effectuated in the selection of the types of compounds considered in the search/design phase (e.g., the exclusion of aromatic compounds).

The following questions help to define the constraints:

Is the compound a replacement for another compound? If yes, the designed compound should match a set of essential properties, match or surpass a set of desirable properties of the original compound, and avoid a set of undesirable properties.

What should be the operational limits (temperature, pressure, and phase) for the desired compound? The operational limits help define the upper and lower limits of the constraints on the phase and the phase transition–related properties. For example, if the upper and lower bounds of the operational temperature for a compound to be used as a liquid solvent are 273.15 K and 360 K, respectively, the melting point must be lower than 273.15 K and the boiling point higher than 360 K.

What criteria should be used to evaluate the performance of the designed compound? The performance criteria are related to the function of the designed compound in the process operation for which it is designed. As a solvent in solvent-based separations, these criteria usually are specified as; lower bound on selectivity, lower bound on distribution coefficient, upper bound on solvent loss, and many more. A very good source for information regarding key properties for selecting solvents is the review by Zhao and Cabezas (1998).

Are there any downstream processing considerations? The role of the designed compounds in downstream processing, such as solvent recovery, wastewater treatment, and disposal, needs to be considered before application of the CAMD solution step. In cases where downstream processing are considered, the driving force and operational limits need to be extended to cover the additional operations, and consequently, other properties may be considered as design criteria (Pistikopoulos and Stefanis, 1998).

To evaluate the goals rigorously (e.g., the calculation of phase equilibria), a simpler formulation based on the available methods must be used. Table 15.2 lists properties (simple and rigorous) used for the evaluation of compounds in solvent design problems, depending on the type of operation the solvent is intended for. Table 15.3 gives examples of relevant properties with respect to health, safety, and the environment and whether these are handled as explicit or implicit considerations.

Having the estimation methods for property prediction selected and the property constraints defined, an appropriate algorithm needs to be selected before the CAMD solution step can be applied.

CAMD SOLUTION (STEP 3)

The CAMD solution step involves finding the molecules of the desired type having the desired properties. Here, a difference is made between those problems that involve only selection and those that involve selection

TABLE 15.2
Properties used for the design of solvents

		Solvent design									
		L-L Extraction		*Extractive distillation*		*Azeotropic distillation*		*Solid separation*		*Gas absorption*	
	Properties	***Simple***	***General***	***Simple***	***General***	***Simple***	***General***	***Simple***	***General***	***Simple***	***General***
Pure	Solubility parameter	*		*		*		*			
	Surface tension		*								*
	Viscosity		*								
	Boiling point	*	*	*	*	*	*				
	Melting point	*	*	*	*	*	*	*	*	*	*
	Density		*								
	Vapor pressure			*	*	*	*			*	*
	Heat of fusion								*		
Mixture	Selectivity		*		*		*		*		*
	Solvent loss		*								
	Solvent power		*		*		*		*		*
	Distribution coefficient		*								
	Phase split	*	*			*	*				
	Azeotropes		*		*		*				
	Mixture viscosity		*								
	Henry's law constant									*	
Environmental		*	*	*	*	*	*	*	*	*	*

TABLE 15.3
Properties used for addressing health, safety, and environmental considerations

		Environmental concern		
	Properties	*Health*	*Safety*	*Environment*
Implicit	Toxicity	*		*
	Biological persistence			*
	Chemical stability		*	
	Reactivity	*	*	
Explicit	Biodegradability			*
	Vapor pressure	*	*	*
	Henry's law constant in water			*
	log *P*	*		*
	Water solubility			*
	Flash point		*	
	Biological oxygen demand			*
	Vapor density	*	*	
	Evaporation rate	*	*	
	LD_{50}	*		*
	Ozone depletion potential			*

plus design. If the problem is of the selection type (i.e., finding candidates from a database of known compounds—CAMS), the solution step involves one or more database lookup operations in order to identify the subset (if any) satisfying the property and molecule type constraints. For pure component properties-based selection, the search engine is commonly known as pattern matching (Nielsen et al., 1991); that is, find the specified pattern in a database. If mixture properties are also considered, the search is more difficult. Zhao et al. (1996) and Cabezas et al. (1996) have developed tools for efficiently solving CAMS problems. If the problem is defined as a regular CAMD problem, an algorithm is used to design molecules having the properties and of the types defined in the constraint selection and definition. Although different algorithms have been proposed for design of molecules, nearly all algorithms to some degree rely on the creation of molecules from fragments. In the CAMD solution step, the property constraints are evaluated using methods selected in the earlier steps (step 2). It is imperative, however, that the needed model parameters for the selected estimation methods are available in the database. Failure to satisfy this requirement will lead to potentially promising compounds being discarded due to the lack of information.

In this chapter, a CAMD algorithm employing a hybrid generate and test approach is described for purposes of illustration of fundamental aspects of the algorithm—the efficient generation of a wide range of molecular structures and how the molecular structural information is used for other purposes (such as properties prediction, database analysis, and connection to external program packages).

Hybrid Generate and Test CAMD Algorithm

The hybrid generate and test approach–based search engine (CAMD algorithm) of Harper et al. (1999) has four levels, and each level has its own generate and test algorithms. Higher levels use additional molecular structural information compared to lower levels. Levels 1–2 are group contribution based (thereby employing macroscopic representation of the molecule), whereas levels 3–4 are based on atomic (microscopic) representation of the molecule. Switch from level 1–2 to 3–4 needs a conversion of macroscopic representation to microscopic representation. Note that for a solution of a material substitution problem, it may not be necessary to use all levels.

Level 1 This level generates sets of building blocks (*fragments*) by combining first-order functional groups (Constantinou and Gani, 1994). These sets are capable of forming at least 1 feasible molecular structure. Simultaneous calculation of related properties (that are dependent only on first-order groups) and screening of the generated structures are performed to control the problem size and execution time. The algorithm here is based on a modified set of rules (Harper et al., 1999; Constantinou et al., 1996). Building blocks are classified according to type. Feasibility rules are based on the number of groups from a specific class a compound may contain. Valency rules are used to determine the number of groups with 1, 2, 3, and 4 connections that are to be used in molecule structure generation. The main steps of the level 1 algorithm are illustrated in Figure 15.4.

Level 2 This level generates molecular structures by combining elements of the individual fragment sets from level 1 to form molecular structures. First- and second-order groups (Constantinou and Gani, 1994) are considered. The main features of this algorithm are that it is pseudo-recursive, all allowed combinations are considered, and efficiency is maintained by continuous removal of duplicate structures. Also, the combination rules satisfy conditions of chemical feasibility. Use of second-order groups allows the estimation method to differentiate between some isomers. A more detailed version of this algorithm is given by Harper et al. (1999).

Level 3 In this level, the selected candidates from level 2 are given an atomic representation by the algorithm of Harper et al. (1999). Note that the atomic representation also defines the connectivity of the molecules. Therefore, property prediction methods based on connectivity indices can be employed to predict properties that could not be predicted earlier (due to unavailable group contributions) or verifying previously estimated values.

Level 4 In this level, generation and testing enter an interactive mode. For any selected candidate from level 3, it is possible to use molecular modeling programs such as MOPAC or Chem3D (CambridgeSoft Corp., 1997). A three-dimensional graph (or molecular model) is created by applying a set of standard or default bond lengths and angles for the various types of

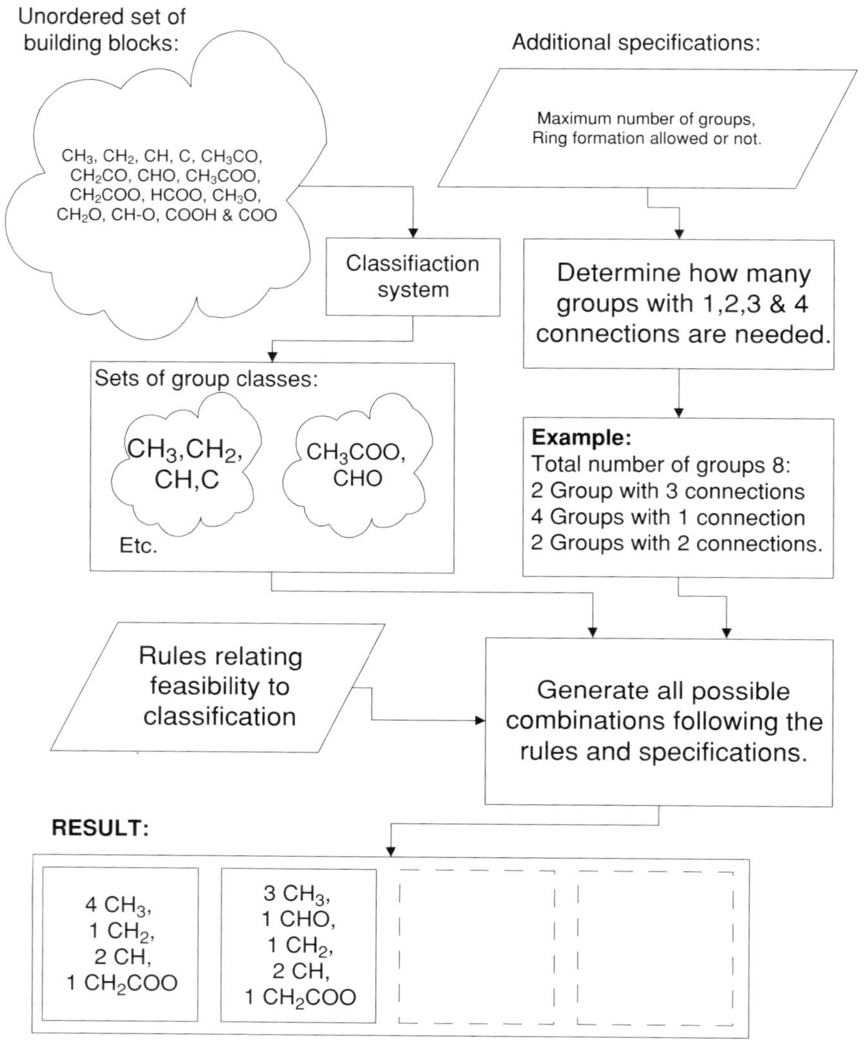

Figure 15.4: Main steps of level 1 algorithm.

connections. As a result, the true molecular model of a compound that can be further analyzed in terms of conformers, stability, properties, etc., is obtained.

RESULT ANALYSIS AND VERIFICATION (STEP 4)

After the CAMD/S solution step, the results need to be analyzed and verified in order to confirm that the compounds are acceptable solutions to the CAMD/S problem. This step involves using other prediction methods, database sources, engineering insight, and, if possible, simulation in order to get an overview of the suitability and capability of the designed compounds for the particular purpose. Which tools to use depend on the original problem formulation. Other tools are mixture analysis tools, property estimation and molecular modeling tools for validation of predicted properties or post-generation screening for properties not handled by the CAMD algorithm, databases for examination of environmental or legislative requirements and data, and reaction pathways or supplier catalogues for price and availability information. Based on the analysis of the candidates a number of prime candidates are selected.

Because estimation of environmental properties is considerably more complex than estimation of physicochemical properties and often requires knowledge and judgment in order to apply the best estimation technique, the consideration of environmental properties is often not part of the main CAMD algorithm. Instead, the environmental properties are assessed during a postanalysis and verification step. Here the advantage of applying CAMD is that all the candidate chemicals fulfill the operational requirements.

CANDIDATE SELECTION (STEP 5)

After validation of the obtained results, the final candidates are selected and final testing is performed. The final testing involves experimental determination of key properties and experimental trials of the compounds used in the intended application. If either the analysis or the final testing steps reveals that none of the selected candidates can fulfill the requirements due to operational, safety, or economic considerations, the problem has to be reformulated (as indicated by the arrows leading back to "Problem formulation" on Figure 15.3). The reformulation can be minor, such as relaxing performance criteria, or major, such as selecting other operational parameters (e.g., different unit operations). After reformulation, the various steps are repeated until a satisfactory solution is found.

TUTORIAL EXAMPLE

In this section, an application of CAMD for the design/selection of environmentally benign solvents for the separation of acetic acid from water is presented as a tutorial example. The purpose of this tutorial is to illustrate the main points of the five-step CAMD technique for environmental problems. Other environmental problems that have been solved with various CAMD techniques include

- CFC Replacement (Joback, 1990; Gani et al., 1991; Duvedi and Achenie, 1996; Ourique and Silva Telles, 1996; Venkatasubramanian et al., 1995; Marcoulaki and Kokossis, 1998).

- Refrigerant mixture design (Churi and Achenie, 1997; Duvedi and Achenie, 1997).
- Solvent substitution (Gani and Harper, 1998; Harper et al., 1999).
- Gas solubility/integrated synthesis/design (Pistikopoulos and Stefanis, 1998).

SOLVENT FOR LIQUID-LIQUID EXTRACTION OF ACETIC ACID FROM WATER

Problem Identification and Solution (Step 1)

From a mixture of water and acetic acid, the acid is to be recovered. Mixture analysis shows that the mixture is close to boiling and distillation is not an economically feasible option. An alternative separation is liquid-liquid extraction of the aqueous phase with a solvent that selectively removes acetic acid. For this operation, Kirk-Othmer (1993) reported isoamyl acetate as a solvent used in industrial processes; however, this compound is reported as being a tumorigen (Chem-Bank, 1998c), and is therefore undesirable in a working environment unless special measures are taken. If the extraction process is an existing installation using isoamyl acetate, the CAMD problem is formulated as a solvent substitution problem where the desirable properties of isoamyl acetate are matched and the undesirable properties are avoided. If the extraction operation is a new process, the parts of the flowsheet surrounding the extraction module are not yet fixed and the CAMD problem is defined with the objective of finding the best candidates based on performance and environmental considerations. Previous applications of CAMD found dipropyl pyridine, 1-nonanol, and decanoic acid as potential solvents (Pretel et al., 1994); however, because dipropyl pyridine contains a heterogeneous aromatic ring, it is not a desirable solvent due to implicit environmental considerations.

In this tutorial example, the CAMD problem solution corresponding to a new process is highlighted. The results, however, are compared with results found in the literature. Assuming that no previous solvent exist, the goal of the CAMD problem is defined as

> Find a nonaromatic solvent for liquid-liquid extraction of acetic acid from water at ambient temperatures. The solvent should be regenerated using distillation. Environmental and workplace health and safety considerations should be addressed.

Method and Constraint Selection (Step 2)

The essential properties for a solvent for liquid-liquid extraction at ambient temperatures are

1. The compound should be liquid at ambient temperatures: $T_m < 280$ K, $T_b > 340$ K.

2. The boiling point of the solvent should be substantially higher than that of the solute: $T_b > 420$ K.
3. The solvent should not form an azeotrope with the solute.
4. The ternary mixture (solvent plus solutes) must show two-liquid phase behavior at 298 K (for a range of solvent amounts at a fixed solute composition).

For the identification of liquid-liquid solvents, the desirable properties are

1. The selectivity (β), distribution coefficient (m), and equilibrium distribution (B) with respect to acetic acid should be as high as possible. The definitions of these terms are

$$\beta = \frac{\gamma_{B,S}^{\infty}}{\gamma_{A,S}^{\infty}} \tag{9}$$

$$m = \frac{\gamma_{A,B}^{\infty} MW_B}{\gamma_{A,S}^{\infty} MW_S} \tag{10}$$

$$B = \left(\frac{x_{A,S}}{x_{B,S}}\right)_{2phase-equilibria}, \tag{11}$$

   with A, B, and S representing acetic acid, water, and solvent, respectively.
2. The solvent loss (Sl) should be as low as possible:

$$Sl = \frac{1}{\gamma_{A,S}^{\infty}}. \tag{12}$$

3. The flash point of the solvent should be higher than ambient temperatures.

Due to implicit environmental considerations, only nonaromatic compounds containing C, H, and O should be considered. Because of stability considerations, compound-containing double and triple bonds are less desirable (Gani et al., 1991).

All of the above requirements can be expressed as

Compound type: Acyclic alkanes, ethers, esters, aldehydes, ketones, and acids.

$T_{flash(open\ cup)} > 310$ K; $T_b > 420$ K; $T_m < 280$ K; $Sl <$ 5E-4; $m > 0.1$; $\beta > 5$; $B > 1$.

Solvent must not form azeotrope with acetic acid.

The ternary mixture (solvent plus solutes) must show two-liquid phase behavior at 298 K (for a range of solvent amounts at a fixed feed composition).

The numerical values of the desirable properties are set at arbitrary nonzero positive values in order to exclude the lowest ranking alternatives.

CAMD Solution (Step 3)

For the solution of the formulated CAMD problem, the PC-based program *ProCAMD* (Harper et al., 1999), based on the previously described generate and test enumeration approach, is used.

The groups used for generation are

CH_3 CH_2 CH C CH_3CO CH_2CO CHO CH_3COO CH_2COO HCOO

CH_3O CH_2O CH-O COOH COO

Apart from the feasibility constraints, the search is limited to finding compounds consisting of 2 to 10 groups and having only one functional group.

In level 1 of the algorithm, 7620 group vectors were generated. After screening the generated group vectors, 122 remained. The selected 122 different group vectors all fulfilled the requirements with respect to size, number of functional groups, and design constraints. In level 2 the 122 group vectors from level 1 were combined into real molecular structures yielding 2385 different compounds. After screening the generated compounds, 2326 alternatives satisfying the design specifications remained. The total time spent in computation for levels 1 and 2 was approximately 5 minutes on a desktop PC. In level 3 the generated molecules were given an atomic representation and a structure search was performed in databases of chemicals to identify the CAS registry numbers for the generated alternatives. Among these, seven candidates were selected using B and $m^*\beta$ as selection criteria. High values of B and/or $m^*\beta$ indicate that the compound would be an effective solvent. The compounds and the predicted values of the specification properties are given in Table 15.4 along with the values for compounds previously reported as potential solvents (isoamyl acetate, 1-nonanol, and *n*-decanoic acid).

Results Analysis and Verification (Step 4)

To address workplace health and environmental considerations, the toxicological and environmental effects of the candidates were examined using the CHRIS (Chem-Bank, 1998a), HSDB (Chem-Bank, 1998b), and RTECS (Chem-Bank, 1998c) databases. The results of this examination are given in the last three columns of Table 15.4. Based on performance as well as environmental considerations, diisobutyl ketone and heptyl acetate were selected for further investigation using mixture analysis tools. Figures 15.5a and 15.5b show the predicted ternary phase diagrams for two ternary systems (acetic acid, water, and the solvents). The predicted phase diagrams confirm the potential of the two selected solvents as promising solvents for the removal of acetic acid from water.

TABLE 15.4
Properties of designed solvents for acetic acid

	CAS-NO	T_b	T_m	S1	β	m	B	Compounds class (RTECS)	Assessment	Comments
5-Nonanone	502-56-7	476	248	7.8e-5	20.3	0.28	1.52	D	Highly toxic	
Ethylhexaldehyde	123-05-7	447	228	2.4e-4	19.8	0.23	1.51	S	Hazardous	Causes skin burns on contact
Diisobutyl ketone	108-83-8	462	227	7.8e-5	20.3	0.28	1.52	S	Low hazard	
Nonyl acetate	143-13-5	506	254	6.2e-6	7.9	0.13	1.56	—	Limited data	
2-Ethylhexyl acetate	103-09-3	483	236	1.8e-5	7.5	0.14	1.45	S	Some hazard	
Heptyl acetate	112-06-1	472	237	5.4e-5	7.0	0.16	1.34	S	Low hazard	
Methyl decanoate	110-42-9	501	245	3.6e-6	9.3	0.11	1.81	—	Limited data	
1-Nonanol	143-08-8	499	250	1.1e-4	7.1	0.50	0.77	T	Some hazard	High aquatic toxicity
Decanoic Acid	334-48-5	541	304	7.9e-5	4.1	0.29	0.72	M,S	Toxic	Causes skin burns on contact
Isoamyl Acetate	123-92-2	421	202	4.7e-4	6.0	0.20	1.07	C	Hazardous	

D = Drug; S = Primary Irritant; T = Reproductive-Effector; M = Mutagen; C = Tumorigen.

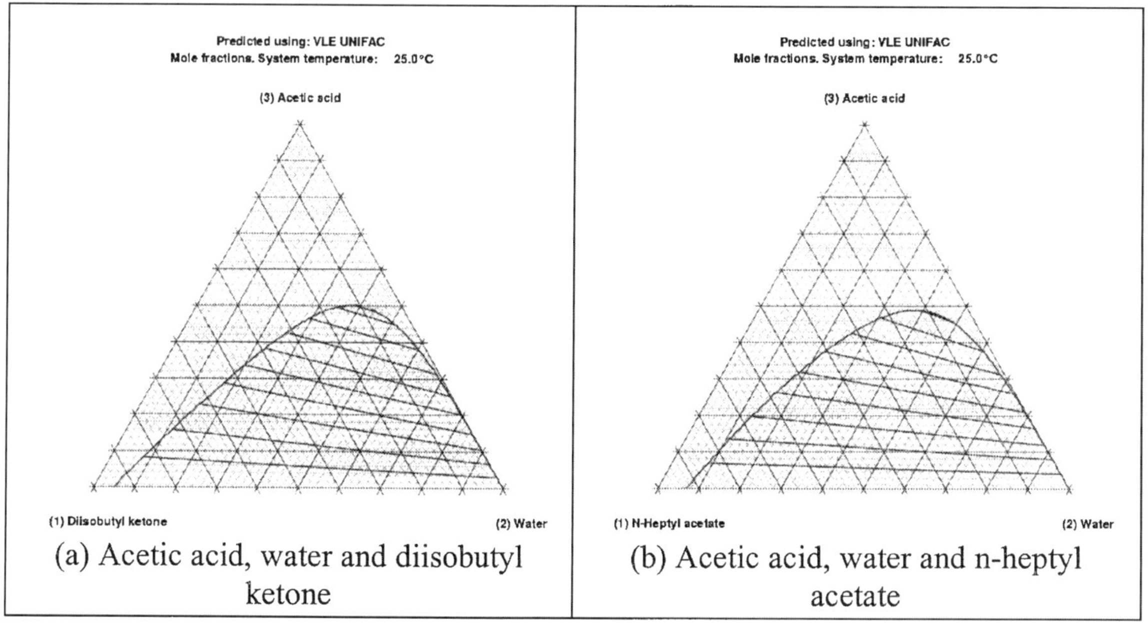

Figure 15.5: LLE diagrams of acetic acid-water-solvent mixtures.

Candidate Selection (Step 5)

For the final selection of a solvent for the extraction, operation experimental validation should be performed and price and availability should be examined for the two candidates. This is beyond the scope of the current example.

ISSUES NEEDING FURTHER ATTENTION

Although the development and application of CAMD techniques have progressed considerably, a number of areas still need further attention. In this section two such issues are highlighted.

PREDICTION METHODS

As mentioned above, the development of reliable general estimation methods for environmental properties is an area facing considerable challenges. Presently, most explicit environmental considerations must be evaluated in the validation and analysis step (postanalysis of results) by using specialized prediction methods or external databases.

Another area is the distinction of property differences in structural isomers. Whereas most structural isomers (e.g., the ortho-, meta-, and para-isomers of substituted aromatics) exhibit differences in physical properties, the majority of available methods (the group contribution methods) cannot distinguish between isomers. This is due to the dimensionality of the molecular structural information used in the property estimation methods. Most group contribution methods do not use information on how the groups are interconnected. To overcome this limitation, Constantinou and Gani (1994) proposed a method using second-order group contributions capable of distinguishing between a restricted set of isomer types.

The alternative to GCA-based property estimation methods is to use methods based on molecular graphs and connectivity matrices; however, the generation of molecules having high structural detail further enlarges the size of the combinatorial problem and poses problems regarding generation of representations without encountering graph isomorphism (Raman and Maranas, 1998).

Missing parameters for some compound classes is a common problem that has been addressed (Meniai and Newsham, 1992, 1996; Meniai et al., 1998) using molecular modeling techniques.

RULES FOR STRUCTURE GENERATION

The danger of introducing rules not founded in a fundamental theoretical basis may result in the exclusion of valid compounds due to the formulation of overly restrictive feasibility constraints. It is necessary, however, to apply

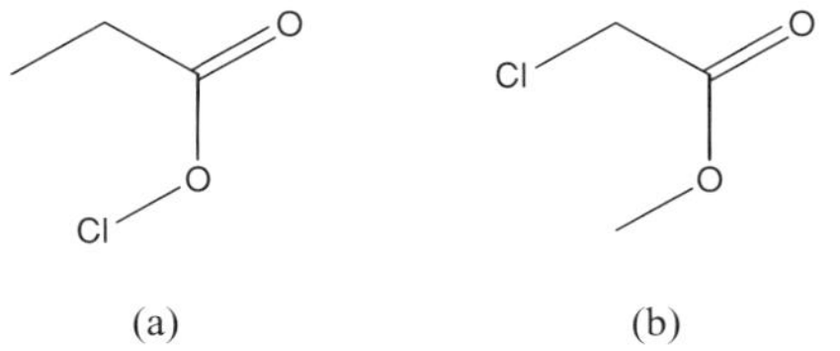

Figure 15.6: Example of an infeasible (a) and a feasible (b) compound.

some rules when designing compound-containing functional groups (defined as groups containing atoms other than C and H and/or groups having free connections of an order higher than 1) in order to avoid compounds that are physically unrealistic (Joback and Stephanopoulos, 1995). An example of a physically unrealistic compound is shown in Figure 15.6.

DISCUSSION/CONCLUSIONS

The use of computer-aided tools for the identification and selection of compounds with specified properties has been addressed by a number of researchers using different approaches for the CAMD solution step. Regardless of the method used, the preanalysis and postanalysis steps of the general CAMD framework remain the same, although, in some cases, not all the postanalysis steps are performed.

For the successful solution of a design/selection problem that includes environmental considerations, certain key issues need to be addressed. The objective of the design problem must be defined in terms of the functionalities a compound should have with respect to the task the compound is being designed for. The defined goals must be expressed as explicit or implicit design constraints in order to formulate a problem that is solvable with the currently available CAMD techniques. The preanalysis steps help to guide the user in this direction. For successful application of the CAMD algorithms, compatible property estimation methods must be available for evaluation of the explicit constraints and performance criteria.

External validation and analysis tools, preferably based on all available experimental data, should be used for the selection of the final candidates. This is a postanalysis of CAMD results and is important for validating the results obtained by the CAMD algorithm. This is necessary because of the predictive nature of the property estimation methods. Finally, for any real application of the results from a CAMD technique, experimental testing of the ability of the designed compounds to fulfill the original criteria should be performed before a final selection is made. It should be noted that the aim of CAMD techniques is not to replace laboratory experimentation, but to assist in the optimal allocation of cost-intensive laboratory resources by rapid identification of the most promising candidates.

NOMENCLATURE

A	number of rings in a compound
B	equilibrium distribution coefficient [see equation (11)]
$H_{N,K_{\max}}$	reduction factor obtained by the introduction of feasibility constraints
K	number of groups in a compound
$K_{\min}$	minimum number of groups considered
$K_{\max}$	maximum number of groups considered
m	distribution coefficient
$M_{N,K_{\max}}$	number of possible combinations when selecting from 1 to $K_{\max}$ groups from a groups set having N components
$M'_{N,K}$	number of possible combinations when selecting K groups from a group set having N components
$\overline{M}'_{i,N,K}$	number of possible combinations when selecting K groups from a group set having N components for the ith solution to the feasibility constraints
$\overline{M}_{N,K_{\max}}$	number of possible combinations when selecting from 1 to $K_{\max}$ groups from a group set having N components under the restrictions of feasibility constraints
MW_B	molecular weight of component B
MW_S	molecular weight of component S
n_S	number of groups in a compound having one free connection
n_D	number of groups in a compound having two free connections
n_T	number of groups in a compound having three free connections
n_Q	number of groups in a compound having four free connections
N_S	number of groups to choose from having one free connection
N_D	number of groups to choose from having two free connections
N_T	number of groups to choose from having three free connections
N_Q	number of groups to choose from having four free connections
N	total number of groups to choose from ($N = N_S + N_D + N_T + N_Q$)
Sl	solvent power
T_b	normal boiling point
$T_{flash(open\ cup)}$	open cup flash point
T_m	normal melting point
$V_{i,K,N}$	ith solution to the imposed feasibility constraints when selecting K groups from N possible
$x_{A,S}$	mole fraction of component A in solvent phase
$x_{B,S}$	mole fraction of component B in solvent phase

β	selectivity
γAB^{∞}	infinite dilution activity coefficient for component pair *A-B*
γAS^{∞}	infinite dilution activity coefficient for component pair *A-S*
γBS^{∞}	infinite dilution activity coefficient for component pair *B-S*

REFERENCES

Ambrose, D. (1978). Correlation and Estimation of Vapour-Liquid Critical Properties. I: Critical Temperatures of Organic Compounds. National Physics Laboratory, Teddington, NPL Rep. Chem. 92, corrected March 1980.

Ambrose, D. (1979). Correlation and Estimation of Vapour-Liquid Critical Properties. II: Critical Pressures and Volumes of Organic Compounds. National Physics Laboratory, Teddington, NPL Rep. Chem. 98, 1979.

Cabezas, H., Zhao, R., Bare, J. C., and Nishtala, S. R. (1996). A Computer Program for Assisting the Design and Replacement of Environmentally Objectionable Solvents. AIChE Annual Meeting, Chicago, Paper 46h.

CambridgeSoft Corp. (1997). Chem3D Pro Users Guide. CambridgeSoft Corp., Cambridge, MA.

Chem-Bank (1998a). Chemical Hazards Response Information System (CHRIS) Database, Silver Platter Information Inc, Norwood, MA.

Chem-Bank (1998b). The Hazardous Substances Data Bank (HSDB), Silver Platter Information Inc, Norwood, MA.

Chem-Bank (1998c). The Registry of Toxic Effects of Chemical Substances (RTECS), Silver Platter Information Inc, Norwood, MA.

Churi, N., and Achenie, L. E. K. (1996). Novel mathematical programming model for computer aided molecular design. *Ind. Eng. Chem. Res.* 53:3788–3794.

Churi, N., and Achenie, L. E. K. (1997). The optimal design of refrigerant mixtures for a two-evaporator refrigeration system. *Comp. Chem. Eng.* 21(suppl.): S349–S354.

Constantinou, L., and Gani, R. (1994). New group-contribution method for the estimation of properties of pure compounds. *AIChE J.* 10:1697–1710.

Constantinou, L., Bagherpour, K. , Gani, R., Klein, J. A., and Wu, D. T. (1996). Computer aided product design: Problem formulations, methodology and applications. *Comp. Chem. Eng.* 20:685–702.

de Waterbeemd, H. V., (1995). *Chemometric Methods in Molecular Design*. VCH, Weinheim, Germany.

Duran, M. A., and Grossmann, I. E. (1986). Simultaneous optimization and heat integration of chemical processes. *AIChE J.* 32:123.

Duvedi, A., and Achenie, L. E. K. (1997). On the design of environmentally benign refrigerant mixtures: A mathematical programming approach. *Comp. Chem. Eng.* 21:915–923.

Duvedi, A. P., and Achenie, L. E. K. (1996). Designing environmentally safe refrigerants using mathematical programming. *Chem. Eng. Sci.* 51:3727–3739.

Fedors, R. F. (1982). Relationship between chemical structure and the critical tempearatures. *Chem. Eng. Commun.* 16:149–151.

Franklin, J. L. (1949). Prediction of heat and free energies of organic compounds. *Ind. Eng. Chem.* 41:1070.

Fredenslund, Aa., Gmehling, J., and Rasmussen, P. (1977). *Vapour-Liquid Equilibria Using UNIFAC*. Elsevier, Amsterdam.

Friedler, F., Fan, L. T., Kalotai, L., and Dallos, A. (1998). A combinatorial approach for generating candidates molecules with desired properties based on group contribution. *Comp. Chem. Eng.* 22:809–817.

Gani, R., and Harper, P. M. (1999). Computer aided product design and pollution prevention through substitution, *Tools and Methods for Pollution Prevention* (S. Sikdar and U. Dwekar). *NATO Science Series* 62:299–315, Kluwer Academic Publishers, The Netherlands.

Gani, R., Nielsen, B., and Fredenslund, Aa. (1991). A group contribution approach to computer aided molecular design. *AIChE J.* 37:1318–1332.

Gordeeva, E. V., Molchanova, M. S., and Zefirov, N. S. (1990). General methodology and computer program for the exhaustive restoring of chemical structures by molecular connectivity idexes. Solution of the Inverse Problem in QSAR/QSPR. *Tetrahedron Comp. Methodol.* 3:389.

Hansch, C., and Leo, A. J. (1979). *Substituent Constants for Correlation Analysis in Chemistry and Biology*. John Wiley, New York.

Harper, P. M., Gani, R., Ishikawa, T., and Kolar, P. (1999). Computer aided molecular design with combined molecular modeling and group contribution. *Fluid Phase Equilibria* 158:337–347.

Hermansen, R. D. (1993). Development of a Solvent Database Software Program, Pollution Technology Review, U.S. Department of Energy, U. S. Air Force, Noyes Data Corp., Park Ridge, NJ, 212:159.

Hermens, J. L. M., and Opperhuizen, A. (1991). *QSAR in Environmental Toxicology - IV*. Elsevier, Amsterdam.

Horvath, A. L. (1992). *Molecular Design: Chemical Structure Generation from the Properties of Pure Organic Compounds*. Elsevier, Amsterdam.

Joback, K. G., and Stephanopoulos, G. (1989). Designing Molecules Possessing Desired Physical Property Values. FOCAPD '89, Snowmass, CO, p. 363.

Joback, K. G., and Stephanopoulos, G. (1990). Designing Molecules Possessing Desired Physical Property Values. Foundations of computer-aided process design. Proceedings of the Third International Conference, pp. 363–387.

Joback, K. G., and Stephanopoulos, G. (1995). Searching spaces of discrete solutions: The design of molecules possessing desired physical properties. *Adv. Chem. Eng.* 21:257–311.

Joback, K. G., and Reid, R. C. (1987). Estimation of pure-component properties from group contributions. *Chem. Eng. Commun.* 57:233–243.

Joback, K. G. (1984). A Unified Approach to Physical Property Estimation Using Multivariate Statistical Techniques. M.Sc. Thesis, MIT, Cambridge, MA.

Joback, K. G. (1994). Solvent substitution for pollution prevention. *AIChE Symp. Ser.* 40:98.

Kaiser, K. L. E. (1987). *QSAR in Environmental Toxicology – II*. D Reidel, Dordrecht.

Kier, L. B., and Hall, L. H. (1986). *Molecular Connectivity in Structure Activity Analysis*. John Wiley & Sons, New York.

Kier, L. B., Lowell, H. H., and Frazer, J. F. (1993). Design of molecules from quantitative structure-activity relationship models. 1: Information transfer between path and vertex degree counts. *J. Chem. Inf. Comput. Sci.* 33:142.

Kirk-Othmer, *Encyclopedia of Chemical Technology*, Vol. 10, Ed. 4 (1993). (J.I. Kroschwitz & M. Howe-Grant). John Wiley & Sons, New York.

Kirkpatrick, S., Gelatt, C. D., Jr., and Vecchi, M.P. (1983). Optimization by simulated annealing. *Science* 220:671–680.

Lydersen, A. L. (1955). Estimation of Critical Properties of Organic Compounds. Univ. Wisconsin Coll. Eng., Eng. Exp. Stn. Rept. 3, Madison, WI.

Lyman, L. J., Reehl, W. F., and Rosenblatt, D. H. (1990). *Handbook of Chemical Property Estimation Methods, Environmental Behavior of Organic Compounds*. American Chemical Society, Washington DC.

Macchietto, S., Odele, O., and Omatsone, O. (1990). Design of optimal solvents for liquid-liquid extraction and gas absorption processes. *Trans. IChemE* 68:429–433.

Maranas, C. D., and Floudas, C. A. (1994). A deterministic global optimization approach for molecular structure determination. *J. Chem. Phys.* 100:1247–1261.

Maranas, C. D. (1996). Optimal computer-aided molecular design: A polymer design case study. *Ind. Eng. Chem. Res.* 35:3403–3414.

Marcoulaki, E. C., and Kokossis, A. C. (1998). Molecular design synthesis using stochastic optimisation as a tool for scoping and screening. *Comp. Chem. Eng.* 22:S11–S18.

Meniai, A.-H., and Newsham, D. M. T. (1992). The selection of solvents for liquid-liquid extraction. *Trans. IChemE.* 70:78–87.

Meniai, A.-H., and Newsham, D. M. T. (1996). A computer-aided molecular design of solvents for liquid-liquid extraction. *Trans. IChemE* 74:695–702.

Meniai, A.-H., Newsham, D. M. T., and Khalfaoui, B. (1998). Solvent design for liquid extraction using calculated molecular interaction parameters. *Trans. IChemE* 76:942–950.

Modi, A., Aumond, J. P., Mavrovouniotis, M., and Stephanopoulos, G. (1996). Rapid plant-wide screening of solvents for batch processes. *Comp. Chem. Eng.* 20:S375–S380.

Naser, S. F., and Fournier, R. L. (1991). A system for the design of an optimum liquid-liquid extractant molecule. *Comp. Chem. Eng.* 15:397–414.

Nemhauser, G. L., Rinnoy Kan, A. H. G., and Todd, M. J. (1989). *Optimization: Handbook in Operations Research and Management Science*, Vol. 1. Elsevier, Amsterdam.

Nielsen, J. M., Gani, R., and O'Connell, J. P. (1991). TMS: A knowledge based expert system for thermodynamic model selection and application. *Computer-Oriented Process Engineering* (L. Puigjaner and A Espuna), 10:29–34. Elsevier, New York.

Odele, O., and Macchietto, S. (1993). Computer aided molecular design: A novel method for optimal solvent selection. *Fluid Phase Equilibria* 82:47–54.

Ourique, J. E., and Silva Telles, A. (1996). Computer-aided molecular design with simulated annealing and molecular graphs. *Comp. Chem. Eng.* 22(Suppl.):S615–S618.

Perry, R. H., and Green, D. W. (1984). *Perry's Chemical Engineers' Handbook*, ed. 6. McGraw-Hill, New York.

Pistikopoulos, E. N., and Stefanis, S. K. (1998). Optimal solvent design for environmental impact minimization. *Comp. Chem. Eng.* 22:717–733.

Pretel, E. J., López, P. A., Bottini, S. B., and Brignole, E. A. (1994). Computer-aided molecular design of solvents for separation processes. *AIChE J.* 40:1349–1360.

Raman, V. S., and Maranas, C. D. (1998). Optimization in product design with properties correlated with topological indices. *Comp. Chem. Eng.* 22:747–763.

Reid, R. C., Prausnitz, J. M., and Poling, B. E. (1987). *The Properties of Gasses and Liquids*, ed. 4. McGraw-Hill, New York.

Renon, H., and Prausnitz, J. M. (1968). Local compositions in thermodynamic excess functions for liquid mixtures. *AIChE J.* 14:135–144.

Vaidyanathan, R., and El-Halwagi, M. (1994).Computer-aided design of high performance polymers. *J. Elastomers Plastics* 26:277–293.

Vaidyanathan, R., and El-Halwagi, M. (1996). Computer-aided synthesis of polymers and blends with target properties. *Ind. Eng. Chem. Res.* 35:627–634.

Vaidyanathan, R., Gowayed, Y., and El-Halwagi, M. (1998). Computer-aided design of fiber reinforced polymer composite products. *Comp. Chem. Eng.* 22:801–808.

van Krevelen, D. W. (1990). *Properties of Polymers: Their Correlation with Chemcial Structure; Their Numerical Estimation and Prediction from Additive Group Contributions*, ed. 3. Elsevier, Amsterdam.

van Leeuwen, C. J., and Hermens, J. L. M. (1995). *Risk Assessment of Chemicals: An Introduction*. Kluwer Academic Publishers, The Netherlands.

Venkatasubramanian, V., Chan, K., and Caruthers, J. M. (1994). Computer-aided molecular design using genetic algorithms. *Comp. Chem. Eng.* 18:833–844.

Venkatasubramanian, V., Chan, K., and Caruthers, J. M. (1995). Genetic algorithmic approach for computer-aided molecular design. *ACS Symp. Ser.* 589:396–414.

Venkatasubramanian, V., Chan, K., Sundaram, A., and Caruthers, J. M. (1992). Designing molecules with genetic algorithms. *Trans. IChemE* 72:695–702.

Wilson, G. M. (1964). A new expression for excess energy of mixing. *J. Am. Chem. Soc.* 86:127.

Zhao, R., and Cabezas, H. (1998). Molecular thermodynamics in the design of substitute solvents. *Ind. Eng. Chem. Res.* 37:3268–3280.

Zhao, R., Cabezas, H., Govind, R., and Fang, Y. (1996). Computer aided solvent substitution for pollution prevention. AIChE Spring Meeting, New Orleans, Paper 40c.

CHAPTER 16

Environmentally Benign Synthesis of Refrigerants and Solvents

Luke E. K. Achenie and Manish Sinha

Computer Aided Process and Product Design Laboratory, Department of Chemical Engineering, University of Connecticut, Storrs, CT 06269

Abstract

The search for new solvents and refrigerants is driven by (a) the needs of new applications and processing requirements, (b) the need to replace solvents or refrigerants whose continued use pose a threat to environmental health and safety, (c) a response to changing environmental regulations, and (d) a response to market demands. Many traditional solvents and refrigerants are on the environmental "hit list" (Krishner, 1995) and are to be phased out within the next few years. In particular, cleaning solvents, which are used as blanket washes in the lithographic printing industry, and chlorofluorocarbon (CFC)-based refrigerants present a number of environmental, health, and safety concerns.

To respond to these concerns and, more importantly, to respond to rapidly changing operational requirements and market forces, there is a need for efficient and systematic strategies for "just-in-time" optimal design of product alternatives. A good strategy has to consider economics, performance, and environmental restrictions simultaneously.

In this chapter we discuss product design research in our laboratory. The framework we have developed considers performance objectives, system interactions, and environmental, health, and safety constraints concurrently. We have presented a number of case studies on design of refrigerants and solvents.

Keywords: computer-aided product design, MINLP optimization, molecular design

1 INTRODUCTION

The chemical process industry (CPI) employs materials such as heat transfer fluids, solvents, coolants, and refrigerants used in manufacturing processes.

Ultimately these chemicals are released to the environment. Solvents are widely used in the CPI as mass separating agents (e.g., solvent extraction), cleaning agents, or as suspension media (e.g., during reaction). Many organic solvents are being phased out in response to directives from federal government agencies, such as the United States Environmental Protection Agency (EPA) (Krishner, 1995). For example, the EPA's TRI (Toxic Release Inventory) and 33–50 program calls for voluntary emission reduction of 17 chemicals, including 6 commonly used solvents (Gessner, 1998). Environmental regulations place strict limits on the concentrations of pollutants in emission streams. Many industries that use high environmental impact organic solvents (e.g., 1,1,1-trichloroethane) are actively searching for solvent alternatives that have similar performance characteristics but have a lower environmental impact (Hairston, 1997).

Closely related to solvents are refrigerants that are also being regulated. Fully halogenated CFC refrigerants have extremely adverse environmental effects in that they deplete atmospheric ozone. The accepted mechanism is one in which chlorine-free radicals catalyze the conversion of stratospheric ozone to atomic oxygen (Gow et al., 1996). The Montreal Protocol (1987) and its London Amendments (1990) have resulted in legislation restricting future production of CFCs. This has stimulated design and discovery of suitable environmentally safe substitutes for these fluids. Several hydrohalomethanes and hydrohaloethanes (e.g., HCFC124, HFC125, and HFC134a) have emerged as potential short- and intermediate-range CFC replacements (Gow et al., 1996; Spauschus, 1991). Most environmental regulations target solvent and refrigerant properties that include global warming potential, ozone-depleting potential (ODP), bioconcentration factor (BCF), and flammability. In addition to environmental, health, and safety issues, global economic competition has also greatly increased the pressure for bringing new compounds to market in a timely fashion.

In response to the above pressures, the chemical industry has been exploring ways to synthesize compounds that seamlessly incorporate cost considerations, and environmental health and safety factors, in addition to desirable physical property attributes. Traditionally, the search for substitute solvents has been done through experience, database searches, and bench scale synthesis of solvent alternatives. All of these approaches have severe limitations and often are unable to effectively find the optimal solvent substitute. It is reasonable to expect that systematic identification of products that possess desired thermophysical properties, satisfy environmental constraints, and have favorable economics is key to competing and adapting to an ever-changing global market.

With this in mind, several researchers have approached the product design problem at a macroscopic level via computer-aided product design (CAPD). CAPD has been applied to the design of new products, such as solvents, polymers, biomaterials, and refrigerants. In CAPD, models of certain attributes of the products are *inverted* in a systematic fashion (e.g., via mathematical programming) to design products with the desired levels of these attributes (Fig. 16.1). Usually the product attributes are physical properties that are modeled by structure-property relations.

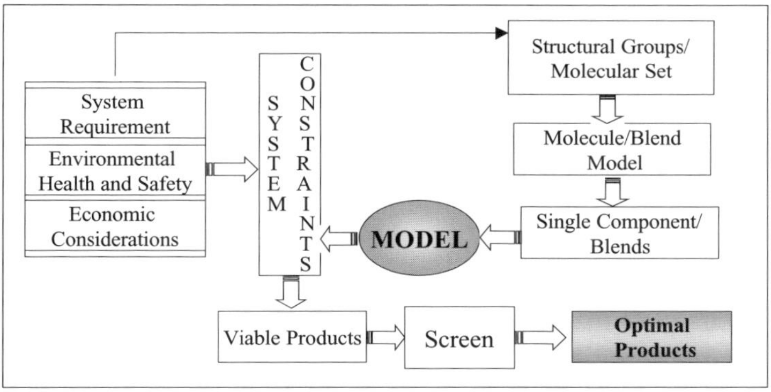

Figure 16.1: Schematic of computer-aided product design.

In a recent article, Constantinou et al. (1996) discuss the general CAPD problem formulation and its applications. Gani and Brignole (1983) and Brignole et al. (1986) have presented methods of solvent design related to liquid-liquid extraction. Gani et al. (1991) have developed a knowledge-based system for selection of solvents for liquid-liquid extraction and solvents for gas absorption. Joback and Stephanopoulos (1989) have examined the design of replacement refrigerants for automotive air conditioners. The CAPD methodology has also been used for liquid-liquid extraction and gas absorption (Macchietto et al., 1990; Odele and Machietto, 1993). Pistikopoulos and Stefanis (1998) have proposed a CAPD model for multicomponent gas absorption processes, and Klein et al. (1992) have considered solvent blends for paint formulation. Industrial successes of CAPD include the design of fuel additives, herbicides, and fungicide, catalysts for manufacture of polyethylene foam, and substitutes for CFC-based refrigerants. A recent article by Hairston (1998) discusses the impact of CAPD in designing new compounds.

In this chapter we discuss product design approaches developed in our research laboratory and their application to the design of refrigerants and solvents. In the product design formulation, we have accounted for environmental considerations. The chapter is organized as follows. We start with a discussion of environmental health and safety models, followed by other pertinent physical property models. Next, we describe a mathematical programming framework for solvent and refrigerant design, followed by solution approaches. Some case studies are presented and future directions in this field are discussed.

2 ENVIRONMENTAL, HEALTH, AND SAFETY MODELS

As discussed earlier, environmental and health concerns need to be incorporated to assist the design of environmentally benign solvents and refrigerants. Fortunately, the current understanding of the interaction of compounds with the environment, although limited, has resulted in reliable environmental-fate prediction models.

Many risk assessment models have also been developed. Compartmental models that predict the partitioning of chemicals in the environment have been reported (van Leeuwen and Hermens, 1995). The distribution of chemicals released to the environment (air, soil, water, and biota) is an important measure for determining environmental impact. Examples of distribution measures are bioconcentration factor (BCF), soil sorption (K_{ow}), and bioaccumulation.

Property models that predict the impact of chemicals on environmental factors, such as global warming potential (GWP) and ozone depletion potential (ODP), are available in the literature. LC_{50} and LD_{50} (concentration or dosage above which 50% of test-laboratory animals die) are indicators of chemical toxicity. Measures such as flammability, ignitability, and corrosivity are important operational safety measures. Some of the health, environmental, and safety measures are available either through the open literature or Material Safety Data Sheets (MSDS). Although the theoretical basis of the models discussed below is limited, they provide practical tools for risk assessment. In addition, we note that all correlations, including the ones discussed below, have a limited region of validity.

2.1 OZONE DEPLETION

The ODP of a compound is a measure of the environmental impact of volatile compounds. Compounds with a minimum of one chlorine, bromine, or iodine group are known to have a nonzero ODP; however, ODP data are scarce and limited to a few halogenated compounds. For hydrochlorofluorocarbons with one or two carbon atoms, the ODP can be quantitatively expressed as (Nimitz and Skaggs, 1992)

$$ODP = 0.05013 n_{Cl}^{1.510} exp(-3.858/\tau), \tag{2.1}$$

where the tropospheric lifetime τ is given by

$$\tau = 0.787(M/nH)\, exp(-2.060 n_{\alpha\text{-}Cl} - 4.282 n_{2C} + 1.359 n_{\beta\text{-}F} + 0.926 n_{\beta\text{-}Cl}). \tag{2.2}$$

The α designation implies that a substituent is attached to the same carbon atom as the hydrogen atom being considered; β means the substituent is attached to the carbon atom adjacent to the carbon atom bonded to the hydrogen; the subscript *2C* indicates a term to be added only if two carbon atoms are present; $n_{\alpha\text{-}Cl}$ represents the number of α chlorine atoms present; the other symbols are similarly defined; nH represents the number of abstractable hydrogen atoms.

The ODP values for a number of CFCs with one or two carbon atoms have been reported (Soloman et al., 1992; Stamm, 1989). Using these values, we formulate the following general expressions for the ODP of CFCs with one or two carbon atoms:

$$ODP = 0.585602 n_{Cl}^{-0.0035} exp(M/238.563)(\text{one carbon}), \tag{2.3}$$

$$ODP = 0.0949956 n_{Cl}^{-0.0404477} exp(M/83.7953)(\text{two carbon}). \tag{2.4}$$

In addition to the above compounds, the ODP has been reported for other compounds (Duvedi and Achenie, 1996). Although, their ODPs are not known, $C_1 - C_2$ bromides, iodides, and higher carbon halides are believed to have high ODPs.

2.2 METRICS BASED ON K_{OW}

For solvent selection, properties targeted by environmental regulations include BCF, K_{oc}, and hydrophile-lipophile balance (HLB). Most of these metrics are correlated to the octanol-water partition coefficient (K_{ow}) by many researchers (Horvath, 1992; Lyman et al., 1981; van Leeuwen and Hermens, 1995). K_{ow} is a measure of hydrophobicity and relates the equilibrium partition of solvent between water and a water-immiscible liquid phase. For most chemicals, $\log K_{ow}$ ranges from -3 to 7 and may be estimated by Hansch and Leo's approach (Hansch and Leo, 1979) as follows:

$$\log K_{ow} = \textit{sum of fragments}(\chi^0) + \textit{factors}(\chi^1). \tag{2.5}$$

Here, empirically derived atomic or group fragment constants (χ^0) and structural factors (χ^1) are used.

A solvent with a high bioconcentration factor has a greater likelihood of accumulating in and harming living tissue. One of several correlations for *BCF* [equation (2.6) from Veith and Konasewich, 1975] is

$$\log BCF = 0.76(\log K_{ow}) - 0.23. \tag{2.6}$$

The *soil sorption coefficient* ($\log K_{oc}$) provides a measure of the extent of chemical partitioning between solid and solution phase of water-saturated or unsaturated soil. K_{oc} can be thought of as the ratio of chemical adsorbed per unit weight of organic carbon in the soil or sediment to the concentration of chemical at equilibrium conditions:

$$K_{oc} = \frac{\mu\text{g adsorbed/g organic carbon}}{\mu\text{g/ml solution}}. \tag{2.7}$$

The higher the K_{oc}, the more likely the chemical will bind to the soil than to remain in water. The typical range for K_{oc} is 1 to 10,000,000. One commonly used correlation for nonionic organics is (Chiou et al., 1979)

$$\log K_{oc} = 0.99(\log K_{ow}) - 0.35. \tag{2.8}$$

The HLB is another parameter that, like octanol-water partition coefficient, measures the partitioning between biota and environment. Archer (1996) has related HLB to the interaction distance R^{ij} [see equation (3.2)] between dodecane and a solvent by

$$HLB = 14.306 - 3.489 \log^{ijR} . \tag{2.9}$$

Note that the *BCF* and K_{oc} depend linearly on K_{ow} on a log-log scale. Thus, a product design criterion that constrains the $\log K_{ow}$ by an upper limit will ensure that the solvent will have a lower bioconcentration and soil sorption. For example, $\log K_{ow}$ less than 4.0 ensures that the log *BCF* is less than 2.81 according to equation (2.6).

2.3 HEALTH AND SAFETY CONSIDERATIONS

The *toxicity* of a chemical is often difficult to predict from molecular structural information because of our lack of understanding of toxicity itself; however, there exist predictive correlations for aquatic toxicity of chemicals. LC_{50} is a parameter that measures the lethal concentration of material. The higher the LC_{50} value, the less toxic the substance is. LC_{50} value has been correlated to K_{ow} (Konemann, 1981) by the following relation:

$$\log LC_{50} = -0.87(\log K_{ow}) - 0.11 . \tag{2.10}$$

Flammability is a measure of the ability of a chemical to form ignitable, potentially explosive vapor mixtures with air, and is often characterized by the flash point. Under the 40 CFR (Code of Federal Regulations) standard, which applies to solvents, a chemical is considered ignitable if it is a "liquid other than aqueous solution containing less than 24 % alcohol by volume and has a flash point less than 60°C." Most of the predictive techniques rely on boiling point data. Butler et al. (1956) proposed a more general correlation for paraffins, aromatics, and cycloparaffins:

$$T_f = 0.683 T_b - 119. \tag{2.11}$$

A product design criterion that limits the flash point to a prespecified upper limit, in turn, ensures that the solvent will be less flammable.

Other environmental measures may not be very easy to incorporate in the CAPD framework; however, environmental models are constantly being developed and can be incorporated in the future. With current CAPD tools, we can generate compounds that satisfy some environmental constraints. Other more complicated environmental and toxicity evaluation can then be performed on this limited set of designed products.

3 OTHER PHYSICAL PROPERTY MODELING

Now we discuss some property requirements used in the case study. For refrigerant design, other performance properties considered are (i) heat of vaporization (h_v), (ii) liquid heat capacity (C_{pl}), and (iii) vapor pressure (P_{vp}). The C_{pl} and primary properties, such as the normal boiling point (T_b), critical temperature (T_c), and critical pressure (P_c), can be evaluated by group contribution. T_b, T_c, and P_c are subsequently used by certain correlations to estimate h_{vp} and P_{vp}. Another important correlated property is the theoretical compressor displacement (CD), which determines the required size of the compressor for a given tonnage of the refrigerant.

For solvent design, capacity and selectivity are important issues. These properties can be modeled by the equilibrium relationships that govern the distribution of a solute in a solvent, which in turn can be modeled by an activity coefficient model, such as UNIFAC (Fredenslund et al., 1977). In addition, other models for characterizing solvent performance are prevalent in the solvent industry (Archer, 1996). Some of these models are based on the solubility parameter.

The solubility parameter has been successfully used in the solvent-based industry. Applications in which the solubility parameter approach have been employed include finding (a) compatible solvents for coating resins, (b) effective mass separating agents, (c) predicting the swelling of cured elastomers by a solvent, and (d) estimating liquid-liquid equilibrium in polymer systems (Archer, 1996; Grulke, 1989). In a recent article, Gessner recommends the use of the solubility parameter for designing solvents for surface cleaning and surface coating applications (Gessner, 1998). In our work (Sinha et al., 1999) we employ the Hansen solubility parameter (δ_T) (Hansen, 1969), which is made up of hydrogen bonding (δ_H), polar (δ_P), and dispersive (δ_D) components, as shown below (also see Fig. 16.4):

$$\delta_T = \sqrt{\delta_D^2 + \delta_P^2 + \delta_H^2}. \tag{3.1}$$

Solute and solvents are represented as points in the three-dimensional Hansen solubility space. The distance, R^{ij}, of a solute (i) from a solvent (j) is a measure of the solute-solvent interaction (see Fig. 16.4). Associated with the solute point is a radius of interaction, R^*. If R^{ij} is less than R^*, then the solvent is likely to dissolve the solute. R^{ij} is given by

$$R^{ij} = 4(\delta_D - \delta_D^*)^2 + (\delta_P - \delta_P^*)^2 + (\delta_H - \delta_H^*)^2 \leq (R^*)^2, \tag{3.2}$$

where

$$\delta_D = \frac{\sum_i n_i F_{Di}}{V_0 + \sum_i n_i V_i},$$

$$\delta_P = \frac{\sqrt{\sum_i n_i (F_{Pi})^2}}{V_0 + \sum_i n_i V_i},$$

$$\delta_H = \sqrt{\frac{\sum_i n_i(-U_{Hi})}{V_0 + \sum_i n_i V_i}}.$$

Here, F_{Di}, F_{Pi}, and U_{Hi} are the group contribution parameters associated with the dispersion, polar, and hydrogen bonding solubility parameters, respectively, for the ith group. n_i is the number of times the ith group is present in a molecule. V_0 is a constant associated with the liquid molar volume prediction model, and V_i is the group contribution associated with the ith group for prediction of liquid molar volume.

The group contribution parameters are taken from van Krevlen and Hoftyzer's work on the dispersion and polar components (van Krevelen and Hoftyzer, 1976), and from Hansen and Beerbower for the hydrogen bonding component (Hansen and Beerbower, 1971). For predicting the liquid molar volume, the model proposed by Constantinou and Gani (1994) is used.

4 SYSTEMS ENGINEERING APPROACH TO PRODUCT DESIGN

In a systems engineering framework, such as the mathematical programming framework we employ in our work, different attributes of a molecule may be modeled by the use of integer variables. Such a framework can be used to find integer solutions that describe the structure of the designed solvent molecule (Churi and Achenie, 1996; Odele and Machietto, 1993; Constantinou et al., 1996). In recent years, the CAMD problem has often been posed as an MINLP formulation in which product specifications are modeled as constraints and/or performance objective functions. We formulate the product design problem as the following mathematical program (Duvedi and Achenie, 1996):

$$\max_{x,y} f(x, y) \tag{4.1}$$

subject to

$$g(x, y) \leq 0 \quad \text{(structural feasibility constraints)}, \tag{4.2}$$

$$c(x, y) \leq 0 \quad \text{(chemical feasibility constraints)}, \tag{4.3}$$

$$p(x, y) \leq 0 \quad \text{(physical property constraints)}, \tag{4.4}$$

$$h(x, y) \leq 0 \quad \text{(process constraints)}, \tag{4.5}$$

$$x^L \leq x \leq x^U \quad \text{(bounds on continuous variables)}, \tag{4.6}$$

$$y \in \{0, 1\} \quad \text{[binary (discrete) variables]}. \tag{4.7}$$

The performance index f is often expressed in terms of target property values, such as the heat of vaporization x_1 and the liquid heat capacity x_2. For example, f may be chosen as

$$f = \frac{x_1}{x_2}. \tag{4.8}$$

A component of the vector variable y indicates whether a particular *basis element* is present in the compound or mixture. For single component product design, the basis elements are structural or functional groups. On the other hand, in mixture design, the basis elements are pure component compounds. The components of the continuous vector variable x may refer to mole fractions, physical properties, and process variables, among other things.

Structural feasibility constraints (Churi and Achenie, 1996; Odele and Machietto, 1993) may be used to ensure that a compound formed from structural or functional groups has no net electronic charge (i.e., the molecule is neutral). For some molecules this may mean the satisfaction of the *octet rule*. In this case, structural feasibility constraints depend only on y [i.e., $g(x, y) = g(y)$] (Churi and Achenie, 1996). In the design of blends, structural feasibility constraints may be used to limit the number of pure component refrigerants that participate in the mixture. The constraints are also used to express the trivial fact that the mole fractions of single component refrigerants in the blend should sum up to one.

Chemical feasibility constraints can be used to prevent CAPD from designing molecules that may be unstable and impossible to synthesize. For example, compounds with a large number of triple bonds may decompose or polymerize. Another example is the presence of three or more ether groups in a relatively small molecule. Some of these rules can be inferred by looking at a large data set of molecules or consulting of an expert synthetic chemist.

Very often, it is important to impose physical property constraints to ensure that the refrigerant (single component or blend) has certain desirable attributes (Duvedi and Achenie, 1996). Examples of pertinent properties are heat of vaporization, liquid heat capacity, vapor pressure, solubility parameters, flash point, ozone depletion potential, and compressor displacement (CD).

CD is not a physical property of the refrigerant; however, it can be calculated directly from the other property values. CD is also an example of a process constraint. Another process consideration in solvent and refrigerant substitution is the desire to use existing equipment and processing technology. This requires that both the old and the substitute compounds should have similar transport and other properties, such as viscosity, diffusivity, density, surface tension, heat capacity (C_p), and vapor pressure (P_{vp}). Zhao and Cabezas (1988) have reviewed molecular property considerations for design of substitute solvents.

5 SOLUTION APPROACHES

Efficient and reliable strategies for inverting solvent design models are critical to optimal solvent design. Two classes of solution approaches, namely enumeration techniques and knowledge-based strategies, have been employed for certain CAMD formulations (Brignole et al., 1986; Gani and Brignole, 1983; Joback and Stephanopoulos, 1989). These methods rely significantly on heuristics and user intervention. Other solution approaches are stochastic approaches, such as genetic algorithms (Venkatasubramanium et al., 1994) and simulated annealing (Marcoulaki and Kokossis, 1998).

Mathematical programming–based solution strategies have been successfully employed for solving the CAMD problem (Duvedi and Achenie, 1996; Macchietto et al., 1990; Maranas, 1996; Odele and Machietto, 1993; Pistikopoulos and Stefanis, 1998; Vaidyanathan et al., 1998). These approaches can handle complex design constraints, and because they have a wide body of literature behind them, there are methods to ascertain the optimality of the solutions. Some of the methods for solving these programs are based on the outer approximation algorithm (Viswanathan and Grossman, 1990). One major weakness of these methods is the convergence to locally optimal solutions when nonconvex models (objective function and constraints) exist in the mathematical programming model. The nonconvexities in the thermodynamic models used in CAPD formulation are often a source of multiextramality (multiple local optima).

Interval analysis (Moore, 1966)–based global optimization (Vaidyanathan and El-Halwagi, 1996) has been employed to solve small CAPD problems. Realistic CAPD problems, however, have several hundred variables and constraints. To address this issue we have developed a reduced space branch and bound global optimization algorithm (Sinha, et al. 1999) that is tailored to large CAPD problems. In developing this algorithm, we have exploited the mathematical structure of the CAPD problems.

6 EXAMPLES

CASE STUDY R1

Problem Statement

This case study was considered by Duvedi and Achenie, and details can be found in Duvedi and Achenie (1996). Here the problem of finding alternative refrigerants to freon-12 (Joback and Stephanopoulos, 1989) is considered. CFC-based refrigerants (such as freon-12) have found wide uses in home refrigerators and automotive air conditioners, primarily due to their nontoxic, nonflammable nature (Masters, 1991), in addition to their overall efficiency; however, as discussed earlier, it is widely believed that CFC refrigerants damage the Earth's protective ozone layer (Stamm, 1989). The refrigeration process is defined by the following temperatures: (a) evaporating temperature (T_e) = 30°F, and (b) condensing temperature (T_{cd}) = 110°F. Saturated conditions are assumed so that the superheat temperature (T_s) is equal to the T_e (Fig. 16.2).

Several factors must be considered in developing feasible CFC replacements. The primary objective is to find fluids that have low ozone depletion potentials with thermodynamic and transport properties similar to CFCs 11, 12, 113, 114, and 115. Mixtures, however, offer much greater flexibility in matching multiple thermodynamic properties, but generally involve more complex cycle design (Gow et al., 1996). Several pure halohydrocarbons have been exhaustively studied (Baehr and Tillner-Roth, 1991; Kubota et al., 1988; McLinden, 1990; McLinden et al., 1989; Sukornick, 1989). Some of

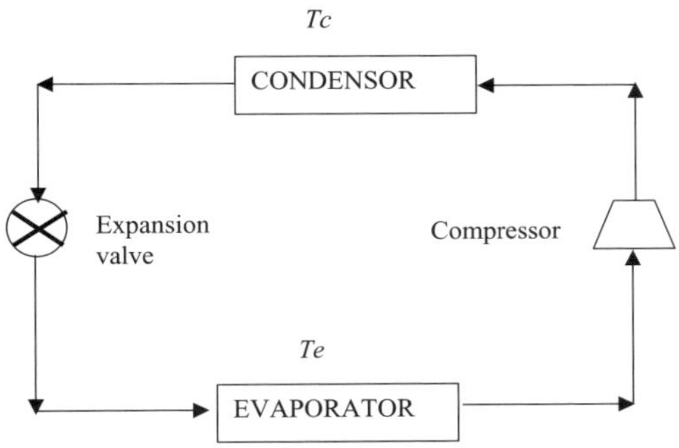

Figure 16.2: Refrigerant cycle schematic.

the refrigerant substitutes in current use are HCFC141b and HFC134a in place of CFC11 and CFC12, respectively.

Problem Setup

The first step is to choose an initial (basis) set of structural groups (Duvedi and Achenie, 1996). We have included an implicit constraint to ensure that the refrigerant would not polymerize over a long period. Thus, double or triple bonds should not exist in the molecule. In addition, most refrigerants used for automotive air conditioning are aliphatic compounds. It is practical, therefore, to concentrate on just aliphatic structural groups. Based on the above criteria, a set of 15 structural groups is identified in Table 16.1.

The objective is to find a refrigerant as good as or better than freon-12; that is, a refrigerant with a larger heat of vaporization (h_v) and with a smaller liquid heat capacity (C_{pl}) than that of freon. Subsequently, the objective function can be expressed as

$$p = x_1/x_2, \tag{6.1}$$

where x_1 and x_2 are continuous variables that are set equal to h_v and C_{pl} in the MINLP formulation. The relevant physical properties for this problem are (a) h_v (T_e), (b) $C_{pl}(T_{mean})$, (c) $P_{vp}(T_e)$, and (d) $P_{vp}(T_{cd})$.

In addition to structural feasibility constraints (Duvedi and Achenie, 1996), the following constraints on physical property values are expressed

TABLE 16.1
Structural groups used in case R1

CH_3	CH_2	CH	C	F	Cl	Br	I
OH	O	NH_2	NH	N	SH	S	

as bounds on the corresponding continuous variables:

$$h_v(T_e) \geq h_{v,freon}(T_e), \tag{6.2}$$

$$C_{pl}(T_{mean}) \leq C_{pl,freon}(T_{mean}), \tag{6.3}$$

$$P_{vp}(T_e) \geq 1.4\,\text{bar}, \tag{6.4}$$

$$P_{vp}(T_{cd}) \leq 14\,\text{bar}. \tag{6.5}$$

Equation (6.2) expresses the fact that a large h_v reduces the amount of refrigerant needed. The value for the enthalpy of vaporization of freon-12 (estimated by the hybrid method) is $h_{v,freon}(30°\text{F}) = 18.74$ kJ/g-mol. Because the objective is to find a refrigerant better than freon-12, this value is used instead of the 18.4 kJ/g-mol reported in the literature (ASHRAE, 1972).

It is desirable to have a low liquid heat capacity. A low value of C_{pl} reduces the amount of refrigerant that flashes upon passage through the expansion valve. The heat capacity constraint in equation (6.3) is evaluated at an average temperature of 70°F. An estimated value of $C_{pl,freon}(70°\text{F}) = 27.12$ cal/g-mol is assumed for the liquid heat capacity of freon-12.

The lowest pressure in the cycle should be greater than the atmospheric pressure (Dossat, 1981) as expressed in equation (6.4). This reduces the possibility of air and moisture leaking into the system. On the other hand, a high system pressure increases the size, weight, and cost of equipment (Dossat, 1981). A pressure ratio of 10 is considered to be the maximum for a refrigeration cycle (Perry and Chilton, 1973); this value is used in equation (6.5) to put a bound on the system pressure.

Results

A solution was obtained by using an implementation (Duvedi and Achenie, 1996) of the outer approximation method (Viswanathan and Grossmann, 1990) together with NPSOL (Gill et al., 1986) and OSL (IBM Corp.). Table 16.2 lists the compounds obtained as the iterations proceed, and their computed physical property values. $F_{1.38}I_{0.62}$ was obtained after

TABLE 16.2
Iteration sequence for objective function = h_v/C_{pl} (h_v[=] kJ/g-mol; C_{pl}[=] cal/g-mol; P_{vp}[=] bar)

Itn no.	*Compound type*	*Obj function*	h_v (−1.1°C)	C_{pl} (21.1°C)	P_{vp} (−1.1°C)	P_{vp} (43.3°C)
0	$F_{1.38}I_{0.62}$	1.9985	21.91	10.96	1.40	6.17
1	CH_3OF	1.0266	21.81	21.24	1.79	8.60
2	**CH_3Cl**	**1.1737**	**20.44**	**17.42**	**2.45**	**9.19**
3	CH_3NH_2	0.9686	22.11	22.82	2.29	8.51

solving the initially relaxed problem. CH_3Cl was eventually identified as the locally optimal compound.

CASE STUDY R1A

Problem Setup

In this variant of Case Study R1, a quantitative measure (the ODP) of the environmental damage caused by the depletion of the earth's ozone layer is incorporated in the refrigerant design objective in addition to the CD.

The CD is directly proportional to the size of the compressor required in a refrigeration cycle, and is thus a measure of the performance or the efficiency of the refrigerant used. For cost reasons, it is desirable to use refrigerants that have a low CD. The incorporation of the ODP and the CD ensures that the resulting refrigerant is environmentally safe and efficient. In the objective function, the ODP and the CD are traded off as follows:

$$p = -Wt_1 xCD - xODP, \tag{6.6}$$

where x_{ODP} and x_{CD} refer to the ODP and CD, respectively, in the MINLP formulation. W_{t1} is a relative weight assigned to the CD term in the objective function. In addition to the variables defined in the previous case study, there exist two continuous variables corresponding to the ODP and the CD.

The performance or the efficiency of the refrigerant is reflected better in the CD than the heat of vaporization. There may exist refrigerants that have a low CD and are therefore efficient despite a low heat of vaporization. Imposing a strict lower bound on the heat of vaporization can exclude such efficient compounds from the search. Thus, the corresponding constraint is relaxed as follows:

$$h_v(T_e) \geq 10.0 \text{ kj/g-mol}. \tag{6.7}$$

The lower bound is arbitrarily chosen as 10.0. The remaining constraints, which include the assignment constraints, the structural feasibility constraints, and the constraints which assign the calculated physical property values to the continuous variables, are the same as those used in the first case study.

Results

The results are summarized in Table 16.3 with CH_3CH_2F as the locally optimal refrigerant using a weight W_{t1} of 10.0. This means that a greater importance is placed on the CD than the ODP in the objective function. The relaxed problem yielded a fractional, nonoptimal solution, which was used to initiate the search.

For the case studies considered, sensitivity and flexibility can be seen qualitatively in the performance of two refrigerants (CH_3Cl and CH_3CHF_2) in terms of the predicted as well as experimental values (Meacock, 1979; Nimitz and Skaggs, 1992) of the target properties (see Table 16.4). Here, the

TABLE 16.3
Iteration sequence: $Wt_1 = 10.0$. $T_e = 30°F$, $T_s = 30°F$, $T_{cd} = 110°F$ [h_v[=] kJ/g-mol; C_{pl}[=] cal/g-mol; P_{vp}[=] bar; CD [=] m^3/(min)(ton)]

Iteration no.	*Compound*	*Obj Fn*	$h_v(T_e)$	$C_{pl}(T_m)$	$P_{vp}(T_e)$	$P_{vp}(T_{cd})$	*CD*	*ODP*
0	$(CH_3)_{1.09}(F)_{0.08}(Cl)_{0.75}(NH_2)_{0.08}$	−0.8919	19.42	17.56	3.19	11.49	0.0884	0.008
1	**CH_3CH_2F**	**−0.806**	**17.78**	**20.14**	**4.21**	**13.11**	**0.0806**	**0.0**
2	CH_3Cl	−1.123	20.44	17.42	2.45	9.19	0.1118	0.005

"experimental value" of the CD is the value that is calculated using experimental enthalpies and specific volumes (Meacock, 1979).

It can be seen that the property prediction is much better for CH_3Cl than CH_3CHF_2; however, the same ranking (CH_3Cl is a better refrigerant for this process) is seen to be retained even when the experimental values are compared.

CASE STUDY R2

This case study was also considered by Duvedi and Achenie, and details can be found in Duvedi and Achenie (1997). In this case study we consider binary mixtures of known single component refrigerants. Altogether, 21 single-component refrigerants have been considered in the basis set. These include the most commonly used refrigerants. One particular refrigeration cycle is considered in which the condensing, evaporating, and superheat temperatures are given by (a) $T_e = 40°F$, (b) $T_s = 40°F$, and (c) $T_{cd} = 100°F$. We consider the performance objective

$$p = Wt_1 xCD - Wt_2 xODP,$$

where Wt_1 and Wt_2 are weights. Equal weights are assigned to the CD and the ODP in the objective function. To initialize the design procedure, the relaxed problem (in which all the binary variables are assumed to be continuous) is solved and a hypothetical mixture containing fractional values for the binary

TABLE 16.4
Comparison of predicted properties to experimental values [$Wt_1 = 1.0$, $T_e = 30°F$, $T_s = 30°F$, $T_{cd} = 110°F$, h_v[=] kJ/g-mol; P_{vp}[=] bar; CD [=] (m^3/(min)(ton))]

	$h_v(T_e)$		$P_{vp}(T_e)$		$P_{vp}(T_{cd})$		*CD*		*ODP*		*Obj func*	
Mol	*Pred*	*Expt*	*Pred*	*Expt*	*Pred*	*Expt*	*Pred*	*Expt*	*Pred*	*Expt*	*Pred*	*Expt*
CH_3Cl	20.44	20.5	2.45	2.46	9.19	9.27	0.1118	0.1077	0.005	0.005	−.1168	−.1127
CH_3CHF_2	20.61	20.20	2.07	2.56	7.85	9.96	0.1404	0.1131	0.000	0.000	−.1404	−.1131

TABLE 16.5
Iteration sequence (Maximization problem) [$Wt_1 = 1.0$, $Wt_2 = 1.0$. $T_e = 40°F$, $T_s = 40°F$, $T_{cd} = 100°F$ (Units: P_{vp}[=] bar; CD [=] m^3/(min)(ton))]

Mixture type Comp1/Comp2	*Composition Comp1/Comp2*	*Obj function*	*CD*	*ODP*	P_{vp} (T_e)	P_{vp} (T_{cd})
**	**	0.04720	0.04563	0.00157	5.86	14.0
$CH_2F_2 + CH_3Cl$	0.344/0.656	−0.05300	0.05170	0.00130	5.45	14.0
$CHF_2CF_3 + CH_3Cl$	0.486/0.514	−0.05769	0.05708	0.00061	5.46	14.0

** Relaxed problem solution: $(F)_{0.164}(R)_{0.836}$.

variables is obtained as the solution. This solution has no physical significance. CH_2F_2/CH_3Cl is eventually identified as the local optimum. Table 16.5 lists the solutions and their property values.

Table 16.6 compares the properties of the designed solution obtained in Table 16.6 to both the estimated and experimental values of the properties of the refrigerant mixture R22/R11 ($CHClF_2/CCl_3F$; Radermacher, 1989) at the same conditions. This mixture has a concentration of 60% by weight (70.44% mole percent) of R22. The experimental values are obtained from the pressure-enthalpy diagram for this refrigerant mixture (Radermacher, 1989). It can be seen that the designed refrigerant mixture (CH_2F_2 [34.4% mole]/ CH_3Cl [65.6% mole]) is better in terms of a lower compressor displacement in addition to a lower ODP.

Finally, the properties of the some locally optimal refrigerant mixtures (Duvedi and Achenie, 1997) are compared to those obtained for a few available single-component refrigerants under similar conditions (Table 16.7).

TABLE 16.6
Comparison of properties of R32 (0.344)/R40(0.656) and R22(0.7044)/R11(0.2956) [$Wt_1 = 1.0$, $Wt_2 = 1.0$. $T_e = 40°F$, $T_s = 40°F$, $T_{cd} = 100°F$ (Units: P_{vp}[=] bar; CD [=] m^3/(min)(ton))]

Property	*CH_2F_2(R32)/ CH_3Cl(R40) 0.344/0.656 (pred)*	*$CHClF_2$(R22)/ CCl_3F(R11) 0.7044/0.2956 (pred)*	*$CHClF_2$(R22)/ CCl_3F(R11) 0.7044/0.2956 (expt)*
Comp Disp CD	0.0517	0.1013	0.0984
Oz Dep Pot ODP	0.0013	0.4448	—
Evap Press $P_{vp}(T_e)$	5.45	2.56	2.70
Cond Press $P_{vp}(T_{cd})$	14.0	7.42	7.45

TABLE 16.7
Refrigerant mixtures vs single component refrigerants [$Wt_1 = 1.0$, $Wt_2 = 1.0$. $T_e = 30°F$, $T_s = 30°F$, $T_{cd} = 110°F$, $P_{vp}(T_e) \geq 1.4$ bar, $P_{vp}(T_{cd}) \leq 14.0$ bar (Units: CD [=] $m^3/(min)(ton)$, P_{vp} [=] bar)]

Ref	*CD pred*	*CD expt*	*ODP pred*	*ODP expt*	*Ob Fn pred*	*Ob Fn expt*	P_{vp} *evap pred*	P_{vp} *evap expt*	P_{vp} *cond pred*	P_{vp} *cond expt*
Mix**	0.0718	—	0.0015	—	−.0733	—	3.92	—	14.0	—
CH_3Cl	0.1078	0.1077	0.002	0.002	−.1098	−.1097	2.47	2.46	9.36	9.27
CH_2F_2	0.0438	0.0423	0.0	0.0	−.0438	−.0423	7.85	7.84	26.73*	26.86*
CH_3CHF_2	0.1187	0.1130	0.0	0.0	−.1187	−.1130	2.61	2.57	9.94	9.96

* Upper bound on the condenser pressure is violated.
** CH_2F_2 (0.227)/CH_3Cl (0.773).

The experimental values of the CD are the values that are calculated using experimental values of enthalpies and specific volumes. The *experimental objective function* is the value obtained from the experimental values of the CD and the ODP. It can be seen that the mixture properties are superior to those of the single component refrigerants for the given process conditions and constraints. CH_2F_2 is a better refrigerant in terms of a low CD and a zero ODP; however, its vapor pressure at the condensing temperature is too high and it violates the upper bound of 14.0 bars. This "undesirable attribute" of CH_2F_2 is overcome in the mixture of CH_3Cl and CH_2F_2.

CASE STUDY S1

Problem Statement

In this case study we consider the problem of finding substitute solvents in the lithographic printing industry. In a lithographic printing process, ink is carried to the impression plate by means of a train of rubber rollers commonly called "blankets" (Fig. 16.3). The cleanliness of the blanket is of primary concern for production of high-quality images. One of the most used solvents in the printing industry is called the "blanket wash," which is specially formulated to clean residue ink from lithographic printing presses. There are more than 52,000 lithographic printers in the United States (Adrian, 1991). Many solvents used in commercial blanket washes are to be phased out in the near future due to environmental regulations. The "Printing Industry of America" (PIA) and the EPA have taken a major initiative to finding environmentally benign solvent substitutes for this application (Design for the Environment Program, 1997).

The detailed model development has been reported in Sinha and Achenie (1998) also to appear in *Advances in Environmental Engineering* (2001). In short, an ideal blanket wash solvent should have the following attributes: (a) ability to rapidly dissolve dried ink, (b) have minimal drying time, (c) should not swell the rubber blanket, (d) be nonflammable, and (e) be environmentally benign.

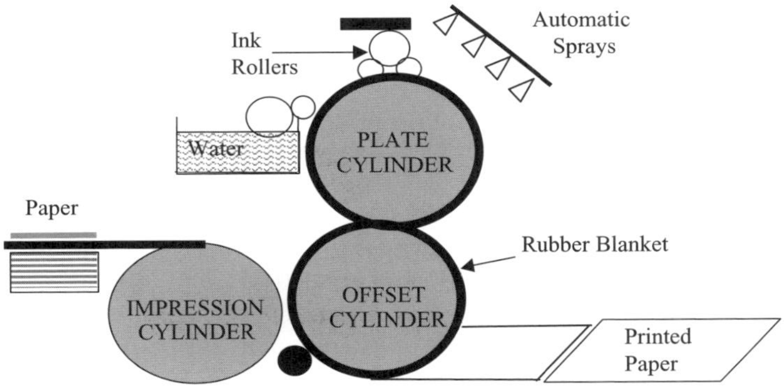

Figure 16.3: Schematic of lithographic printing process with automatic blanket washing.

Problem Setup

The CAPD problem is designed as a single objective optimization problem:

$$\min_{x,y} / \max \quad performance\_objective(\boldsymbol{x}, \boldsymbol{y}, \boldsymbol{\theta})$$

subject to

Structural constraints (as defined in Section 4)

Property constraints:

a: be liquid at operating temperature

b: be able to dissolve ink (6.8)

c: not swell blanket

d: be environmentally benign

Two case studies are considered. In the first case study the objective is to design solvents with minimal drying time. In the second case study the objective is to design solvents with minimal flammability with drying time as a constraint. Drying time is correlated with the heat of vaporization of the solvent. The flash point, which is an indicator of flammability, is linked to operational safety requirements. The use of lower flammability solvents often results in lower insurance costs.

A basis set of 12 groups is used (Table 16.8); the maximum number of groups allowed in a molecule is 7. The phase constraint on boiling and melting point ensures that the solvent will be liquid at the operating conditions. Specifically the following constraints are employed:

$$T_b > 350°K, \tag{6.9}$$

$$T_m < 250°K. \tag{6.10}$$

TABLE 16.8
Structural groups for case study S1

CH_3-	CH_2-	Ar-	-Ar-	-OH	CH_3CO-
-CH_2CO-	-COOH	CH_3COO-	-CH_2COO-	-CH_3O	-CH_2O-

A typical ink formulation is composed of pigments dispersed in phenolic resins. Lithographic ink is made of 10% pigments, 10% resins, and a balance of a hydrophobic solvent, such as hydrocarbon oil. The hydrocarbon oil is a high viscosity oil having a boiling range of 310 to 550°C. Common resins used are gilsonite and phenolic resins. Most of the blankets are made of cured rubber. The ink residue is assumed as phenolic resin, for example Super Bakacite (Barton, 1985), having the nonpolar (δ^d), polar (δ^p), and hydrogen bonding (δ^h) solubility parameters of 23.3, 6.6, and 8.3 MPa$^{1/2}$, respectively (Barton, 1985). The radius of interaction (R^*) is 19.8 MPa$^{1/2}$ (Fig. 16.4). Thus, solvents that can effectively dissolve the ink residue must satisfy the following constraint:

$$^{ij}R = \left[4\left({}^{i}\delta_D - 23.3\right)^2 + \left({}^{i}\delta_P - 6.6\right)^2 + \left({}^{i}\delta_H - 8.3\right)^2\right]^{1/2} < 19.8. \qquad (6.11)$$

Lithographic blanket is generally made of polyisoprene rubber. Polyisoprene rubber has 25% swelling range between nonpolar (δ^d) 14.6–20.1, polar (δ^p) 0–6.3, and hydrogen bonding (δ_h) 0–7.4 MPa$^{1/2}$ (Bhaswarker and Cussler, 1997; Cunningham, 1998). Thus, the following constraints are employed to ensure that solvent will result in minimal blanket swelling:

$$\delta_p > 6.3,\ \delta_h > 7.3,\ \delta_d > 20.1,\ \text{and}\ \delta_d < 14.6. \qquad (6.12)$$

A satisfaction of any of the four constraints will ensure that the solubility parameter will lie outside the swelling box. This, in turn, will ensure that the solvent causes minimal blanket swelling.

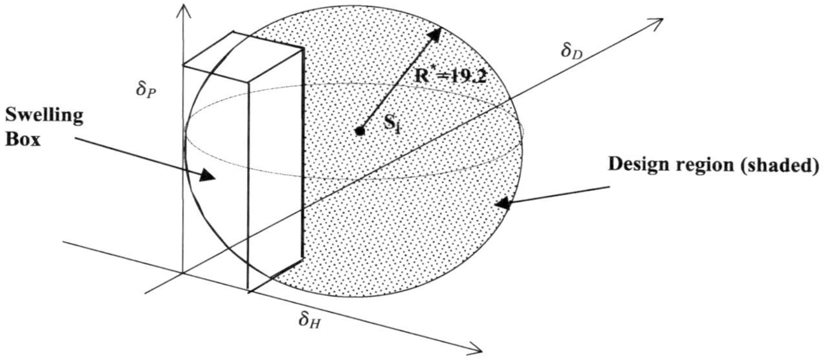

Figure 16.4: Solute-solvent interaction sphere. Solvent having solubility parameter within the sphere of interaction will be able to dissolve solute S_i.

Environmental factors are accounted for by restricting the value of partition coefficient (K_{ow}). This will ensure that the designed solvent will have low bioaccumulation, bioconcentration, and toxicity:

$$\log K_{ow} \leq 4.0. \tag{6.13}$$

The above CAPD model was solved using the MINLP solver DICOPT++ (Brooke et al., 1996) on an IBM RS6000 computer. After a solution is obtained, one may eliminate it from the design space with the use of a functional cut (i.e., constraining the objective function to be inferior to the best found so far). This will result in progressively inferior solutions (from an objective function point of view); however, the danger with using a functional cut is that the global solution may never be reached if the first solution is not a global solution.

Results

Table 16.9 lists the designed solvent with minimization of drying time as the objective function. The objective values for the best and seventh best molecule are 33.913 and 36.145 kJ/mol. For this case study we have tried to verify whether the solution is indeed a global solution. For this verification step, the nonconvex solubility constraint and the blanket interaction constraint [equations (6.11) and (6.12)] were removed from the design problem, and the resulting relaxed convex problem was solved using DICOPT. The globally optimal solution resulted in the same molecule (i.e., molecule 1 with objective value of 33.913 kJ/mol). This molecule satisfies the solubility and blanket interaction nonconvex constraints, and hence is a valid solution to the original problem. Therefore, in this particular case, the solution is indeed globally optimal with respect to the original problem.

The objective of the second case study was to maximize the flash point (T_f). In this case, the heat of vaporization was restricted to 37 kJ/mol,

TABLE 16.9
Solvent molecules designed and ranked in the increasing order of heat of vaporization (ΔH_v) (minimization of the heat of vaporization is the objective function)

	Solvent molecule	ΔH_v	δ_D	δ_P	δ_H	*logK OW*
1	$-CH_2O-CH_2-CH_2O-CH_2O-$*	33.913	14.1	12.5	12.13	−3.4
2	$CH_3CH_2-CH_2O-CH_3O$	33.992	14.3	7.4	9.3	−1.02
3	$CH_3-CH_2O-CH_2-CH_2O-CH_3$	34.667	14.5	6.45	8.7	−.48
4	$CH_3-CH_2CO-CH_3$	35.471	15.12	8.64	4.65	0.87
5	$-CH_2-CH_2-CH_2-CH_2O-CH_2O-$*	35.735	14.41	7.62	9.45	−.940
6	$CH_3O-(CH_2)_3-CH_3$	35.814	14.55	3.41	6.3	1.46
7	$CH_3O-CH_2O-CH_3O$	36.145	14.00	12.20	12.95	−3.50

* Cyclic molecule.

TABLE 16.10
Solvent molecules designed and ranked in decreasing order of flash point (T_f) (maximization of the flash point is the objective function)

	Solvent molecule	$T_f(°K)$	ΔH_v	δ_D	δ_P	δ_H	$logK_{OW}$
1a	-CH_2O-CH_2O-CH_2O-CH_2O-*	275.4	36.74	14.11	15.53	13.48	−5.36
2a	CH_3O-$(CH_2)_3$-CH_3¶	261.9	35.814	14.55	3.41	6.3	1.46
3a	CH_3CO-OCH_3	259.7	36.738	8.96	14.86	9.14	−1.61
4a	-CH_2O-CH_2-CH_2O-CH_2O-*¶	259.1	33.913	14.1	12.5	12.13	−3.4
5a	CH_3CH_2-CH_2O-CH_3O¶	256.5	33.992	14.3	7.4	9.3	−1.02
6a	CH_3-CH_2CO-CH_3¶	256.1	35.471	15.12	8.64	4.65	0.87

* Cyclic molecule.
¶ Molecules common in Tables 16.9 and 16.10.

ensuring a high drying rate. Table 16.10 lists the solvents designed. Four solvents are common to both case studies. The best molecule in the first case study is the fourth best in the second case study.

Molecules 2, 3, 4, and 6 (see Table 16.9) can be synthesized or are available as common solvents from a vendor. Molecules 1 and 5 are cyclic. This can be suppressed by incorporating additional constraints as discussed in Churi and Achenie (1996). Molecule 7 (with three ether groups) cannot be found in most chemical databases. Molecules 2a, 4a, 5a, and 6a (Table 16.10) also appear in Table 16.9.

Molecule 2 (also 5a) is an isomer of 1,2-dimethoxyethane, also known as ethylene glycol dimethyl ether (EGME). This is a commercial solvent. The actual boiling and melting point are 85 and −68.9°C, respectively. The predicted values are 82 and −83°C, respectively. Solvent 3 is an isomer of ethyl propyl ether. This is another commercial solvent. Molecule 4 (also 6a) is 2-butanone, also known as methyl ethyl ketone (MEK). The actual boiling point for MEK is 86°C and melting point is −79°C; these values are close to the predicted values of 81.61 K and −80 K, respectively. MEK is a common component in printing inks.

Solvent 6 is methyl butyl ether. Several commercial blanket wash solvents are ketones and ethers. This is consistent with the results from this case study. It is interesting to note that solvents 2, 3, 4, and 6 are components of commercial blanket washes (United States Environmental Protection Agency, 1997). In particular, MEK is found in many different commercial blanket wash formulations. This reinforces the validity of the framework proposed in this study. Unfortunately, glycol ethers and MEK are on the regulatory list of the Toxic Release Inventory (TRI) and Clean Air Act (Krishner, 1995) for health factors. For example, monomethyl and monoethyl ethers are blamed for miscarriage rates among semiconductor industry workers. Modeling such health factors is difficult. Some researchers are attempting to estimate such factors from structural information via QSAR models (Veith and

Konasewich, 1975). One important issue in CAPD is to verify if a molecule (e.g., molecule 5) can be synthesized. If it can, would such a molecule have the properties as predicted? A synthetic organic chemist may be able to answer these questions. If such molecules (i.e., cyclic with more than one ether group) cannot exist, then such knowledge has to be generalized and incorporated as constraints. This would prevent CAPD from designing such compounds.

7 FUTURE RESEARCH DIRECTIONS

In this chapter we have presented a mathematical programming–based system engineering framework for product design. Through several case studies we have demonstrated that CAPD can be a very effective tool for product design; however, we believe that several issues need to be addressed before CAPD becomes a standard tool for product design in the chemical industry.

One such issue is the presence of model uncertainty. Many researchers are attempting to address this issue by developing more realistic and accurate property prediction models. For example, we have developed a hybrid method for accurate estimation of physical properties (Duvedi and Achenie, 1996). Such models will improve the model reliability but may never eliminate model errors. For CAPD to be industrially relevant, it should be able to design products in the presence of uncertainties in the property prediction models. Tools should be developed that also provide some guarantees on the degree of validity of the products. We are currently incorporating flexibility analysis (Ostrovsky et al., 1999, 2000) into the MINLP framework, thus enabling CAPD to design products under model uncertainty.

Another pertinent issue is stability (thermodynamic and kinetic) of the structures found from a CAPD approach. The "Gibbs free energy" criterion offers a way to enforce thermodynamic stability. Finally, the presence of nonlinear models often results in nonconvex CAPD formulations. These formulations have many local minima. Furthermore, as the problem size increases, the solution of CAPD problems can become computationally intensive. Thus there is a need for robust global optimization algorithms that can solve large MINLP problems. To this end, we have developed a branch and bound deterministic global optimization algorithm that exploits the mathematical structure of CAPD problems (Sinha et al., 1999). Finally, a prototype parallel implementation of this algorithm (on a cluster of personal computers) has been developed (Sinha et al., 1998).

NOTATION

CD	compressor displacement
C_{pl}	liquid heat capacity
h_v	heat of vaporization
$n_{\alpha\text{-}Cl}$	number of α chlorine atoms

$n_{\beta\text{-}Cl}$	number of β chlorine atoms
M	molecular weight
$n_{\beta\text{-}F}$	number of β fluorine atoms
n_{Cl}	number of chlorine atoms
n_H	number of hydrogen atoms
n_j	number of groups of type j
n_j^l	lower limit on n_j
$n_{\max}$	maximum number of groups
n_{NH}	number of NH groups
n_{NH2}	number of NH_2 groups
n_N	number of N groups
p	performance index
P_c	critical pressure
P_{vp}	vapor pressure
P_{vpr}	reduced vapor pressure
τ	tropospheric lifetime
x	continuous variables
v_j	valency of group of type j
δ_d	nonpolar solubility parameter
δ_p	polar solubility parameter
δ_h	hydrogen bonding solubility parameter
W_{t1}	relative weight in the objective function
F_{Di}	group contribution for δ_d
F_P	group contribution for δ_P
U_H	group contribution for δ_H
ΔH_v	heat of vaporization
K_{ow}	octanol water coefficient
K_{oc}	soil sorption coefficient
T_b	normal boiling point
T_{br}	reduced boiling point
T_c	critical temperature
T_{cd}	condensing temperature
T_e	evaporating temperature
T_r	reduced temperature
T_s	superheat temperature
T_f	flash point
R^*	the radius of interaction
ODP	ozone depletion potential
BCF	bioconcentration factor

REFERENCES

Adrian, J. R. (1991). Managing solvents and wipers. *Pollution Prevention Review* 419–425.

Archer, W. L. (1996). *Industrial Solvent Handbook*. Marcel Dekker Inc., New York.

Baehr, H. D., and Tillner-Roth, R. (1991). Measurement and correlation of the vapor pressures of 1,1,1,2-tetrafluoroethane (R134a) and of 1,1-difluoroethane (R152a). *J. Chem. Thermo.* 23:1063.

Barton, A. F. (1985). *CRC Handbook of Solubility Parameters and Other Cohesion Parameters*. CRC Press, Inc., Boca Raton, FL.

Bhaswarker, A. N., and Cussler, E. L. (1997). Pollution-preventing lithographic inks. *Chemical Engineering Science* 52:3227–3231.

Brignole, E. A., Bottini, S., and Gani, R. (1986). A strategy for the design of solvents for separation processes. *Fluid Phase Equilibria* 29:125–132.

Butler, R. M., Cooke, G. M., Lukk, G. G., and Jameson, B. G. (1956). Prediction of flash points of middle distillates. *Industrial and Engineering Chemistry* 48:808–812.

Chiou, C. T., Peters, L. J., and Freed, V. H. (1979). A physical concept of soil-water equilibria for nonionic organic compounds. *Science* 206:831–832.

Churi, N., and Achenie, L. E. K. (1996). Novel mathematical programming model for computer aided molecular design. *Industrial and Engineering Chemistry Research* 35:3788–3794.

Constantinou, L., and Gani, R. (1994). New group contribution method for estimating properties of pure compounds. *AIChE Journal* 40:1697–1710.

Constantinou, L. K., Bagherpour, R., Gani, R., Klein, J. A., and Glen, R. C. (1996). Computer-aided product design: Problem formulations, methodology, and applications. *Computers and Chemical Engineering* 20:685.

Cunningham, H. W. (1998). (National Joy Ink Information Center) *Typical Ink Composition* (Private communication).

Design for the Environment Program. (1997). Cleaner Technologies Substitute Assessment: Lithographic Blanket Washes. EPA 744-R-97-006, USEPA.

Duvedi, A. P., and Achenie, L. E. K. (1995). A mathematical programming based approach to refrigerant design. *Simulators International XII, Simulation Series* 27:254–259.

Duvedi, A. P., and Achenie, L. E. K. (1996). Designing environmentally safe refrigerants using mathematical programming. *Chemical Engineering Science* 51:3727–3739.

Duvedi, A. P., and Achenie, L. E. K. (1997). On the design of environmentally benign refrigerant mixtures: A mathematical programming approach. *Computers and Chemical Engineering* 21:915–923.

Fredenslund, A., Gmehling, J., and Rasmussen, P. (1977). *Vapor-Liquid Equilibria Using UNIFAC*. Elsevier, Amsterdam.

Gani, R., and Brignole, E. A. (1983). Molecular design of solvents for liquid extraction based on UNIFAC. *Fluid Phase Equilibria* 13:331.

Gani, R., Nielsen, B., and Fredenslund, A. (1991). A group contribution approach to computer-aided molecular design. *AIChE Journal* 37:1318–1331.

Gessner, A. W. (1998). Switch materials to prevent pollution. *Chemical Engineering Progress* (December), 94:59–64.

Gow, A. S., Stein, F. P., Xinhou, G., and Lucia, A. (1996). Simulation of binary refrigerant mixture vapor-liquid equilibria using a quasi-regular nonrandom fluid theory and local composition mixing rules. *Fluid Phase Equilibria*. 116: 60–67.

Grulke, E. A. (1989). Solubility parameter values. *Polymer Handbook* (J. Brandrup and E. H. Immergut), pp. 519–559. John Wiley, New York.

Hairston, D. (1998). New molecules get on the fast track. *Chemical Engineering* (September), pp. 30–33.

Hairston, D. W. (1997). Solvent alternatives. *Chemical Engineering* (February), pp. 55–58.

Hansch, C., and Leo, A. J. (1979). *Substituent Constants for Correlation Analysis in Chemistry and Biology*. John Wiley, New York.

Hansen, C. (1969). The universality of the solubility parameter. *Ind. Eng. Chem. Prod. Res. Dev.* 8:2.

Hansen, C. M., and Beerbower, A. (1971). Solubility parameters. *Kirk-Othmer Encyclopedia of Chemical Technology* (A. Standen). Interscience, New York.

Horvath, A. L. (1992). *Molecular Design*. Elsevier, New York.

Joback, K. G., and Stephanopoulos, G. (1989). Designing Molecules Possessing Desired Physical Property Values. *Proceedings FOCAPD 1989*, Snowmass Village, Colorado.

Klein, J. A., Wu, D. T., and Gani, R. (1992). Computer aided mixture design with specified property constraints. *Computers and Chemical Engineering* 16(Suppl): S229.

Konemann, H. (1981). Quantitative structure-activity relationships in fish toxicity studies. 1: Relationship for 50 industrisal chemicals. *Toxicology* 19:209–221.

Krishner, E. M. (1995). Environment, health concerns force shift in use of organic solvents. *Chemical and Engineering News* (June), pp. 13–20.

Kubota, H., Tanaka, Y., Makita, T., Kashiwagi, H., and Noguchi, M. (1988). Thermodynamic properties of 1-chloro-1,2,2,2-tetrafluoroethane (R124). *Int. J. Thermophys.* 9:85.

Lyman, W. J., Reehl, W. F., and Rosenblatt, D. H. (1981). *Handbook of Chemical Property Estimation Methods*. McGraw-Hill Book Company, New York.

Macchietto, S., Odele, O., and Omatsone, O. (1990). Design of optimal solvents for liquid-liquid extraction and gas absorption processes. *Transactions of the Institute of Chemical Engineers* 68:429.

Maranas, C. D. (1996). Optimal computer-aided molecular design: A polymer design case study. *Industrial and Engineering Chemistry Research* 35:3403.

Marcoulaki, E. C., and Kokossis, A. C. (1998). Molecular design synthesis using stochastic optimisation as a tool for scoping and screening. *Computers and Chemical Engineering* 22:S11–S18.

Masters, G. M. (1991). *Introduction to Environmental Engineering and Science*. Prentice-Hall, Englewood Cliffs, NJ.

McLinden, M. O. (1990). Thermodynamic properties of CFC alternatives: A survey of the available data. *Int. J. Refrig* 13:149.

McLinden, M. O., Ward, D., Chae, H. B., Gallagher, J. S., Goodwin, A. R. H., Bruno, T. J., Weber, L. A., Moldover, M. R., Ely, J. F., Morrison, G., Schmidt, J. W., and Huber, M. L. (1989). Measurement and formulation of the thermodynamic properties of refrigerants 134a (1,1,1,2-tetrafluoroethane) and 123 (1,1-dichloro-2,2,2-trifluoroethane). *ASHRAE Transactions 95* 95:263.

Moore, R. E. (1966). *Interval Analysis*. Prentice-Hall, Englewood Cliffs, NJ.

Nimitz, J. S., and Skaggs, S. (1992). Estimating tropospheric lifetimes and oxone depletion potentials of one and two-carbon hydrofluorocarbons. *Environmental Science and Technology* 26:739–744.

Odele, O., and Machietto, S. (1993). Computer aided molecular design: A novel method for optimal solvent selection. *Fluid Phase Equilibria* 82:47–54.

Ostrovsky, G. M., Achenie, L. E. K., and Wang, Y. (1999). Deterministic methods of flexibility analysis. *Computers and Chemical Engineering* 23(Suppl):387–390.

Ostrovsky, G. M., Achenie, L. E. K., Wang, Y., and Volin, Y. (2000) A New Algorithm for Computing Process Flexibility, I&EC Research 39:2368–2377.

Pistikopoulos, E. N., and Stefanis, S. K. (1998). Optimal solvent design for environmental impact minimization. *Computers and Chemical Engineering* 22:717–733.

Sinha, M., and Achenie, L. E. K. (1998). Computer aided solvent design for lithographic blanket wash system. *Topical Conference on Pollution Prevention and Risk Reduction*, Annual AIChE Meeting, Paper 13e, Miami, FL.

Sinha, M., Achenie, L. E. K., and Ostrovsky, G. M. (1998). Product design and global optimization in parallel computing environment. *AIChE Annual Meeting*, Miami, FL.

Sinha, M., Achenie, L. E. K., and Ostrovsky, G. M. (1999). Environmentally benign solvent design by global optimization. *Computers and Chemical Engineering* 23:1381-1394.

Soloman, S., Pollock, W. H., Heidt, L. E., Lueb, R., Vedder, J. F., and Mills, M. J. (1992). On the age of stratospheric air and ozone depletion potentials in polar regions. *Journal of Geophysical Research* 97:12993–12999.

Spauschus, H. O. (1991). Emerging HVAC\&R Issues: Energy, environment, and economics. *ASHRAE J.* 33:23.

Stamm, R. (1989). The CFC problem: Bigger than you think. *Heating Piping and Air Conditioners* 61:51–54.

Sukornick, B. (1989). Potentially acceptable substitutes for chlorofluorocarbons: Properties and performance features of HFC-134a, HFC-123, and HCFC-141b. *Int. J. Thermophys.* 10:553.

United States Environmental Protection Agency. (1997). Cleaner Technologies Substitutes Assessment: Lithographic Blanket Washes. EPA 744-R-97-006, United States Environmental Protection Agency.

Vaidyanathan, R., and El-Halwagi, M. (1996). Computer-aided synthesis of polymers and blends with target properties. *Industrial and Engineering Chemistry Research* 35:627–634.

Vaidyanathan, R., Gowayed, Y., and El-Halwagi, M. (1998). Computer-aided design of fiber reinforced polymer composite products. *Computers and Chemical Engineering* 22:801–808.

van Krevelen, D. W., and Hoftyzer, P. J. (1976). *Properties of Polymers: Their Estimation and Correlation with Chemical Structure*. Elsevier, Amsterdam.

van Leeuwen, C. J., and Hermens, J. L. M. (1995). *Risk Assessment of Chemicals: An Introduction*. Kluwer Academic Publishers, Boston, MA.

Veith, G. D., and Konasewich, D. E. (1975). Structure-activity correlations in studies of toxicity and bioconcentration with aquatic organisms. Proceedings of a Symposium held in Burlington, Ontario, at the Canada Center for Inland Water, Ontario, Canada.

Venkatasubramanium, V., Chan, K., and Caruthers, J. M. (1994). Computer-aided molecular design using genetic algorithms. *Computers and Chemical Engineering* 18:833–844.

Viswanathan, J., and Grossman, I. E. (1990). A combined penalty function and outer-approximation method for MINLP optimization. *Computers and Chemical Engineering* 14:769–782.

Zhao, R., and Cabezas, H. (1988). Molecular themodynamics in the design of substitute solvents. *Industrial and Engineering Chemistry Research* 37:3268–3280.

CHAPTER 17

Computer-Aided Solvent Design for the Environment

Renhong Zhao and Heriberto Cabezas

Systems Analysis Branch,
Sustainable Technology Division,
National Risk Management Research Laboratory,
U.S. Environmental Protection Agency,
Cincinnati, OH 45268

Subba R. Nishtala

Research Triangle Institute
Center for Environmental Analysis,
Research Triangle Park,
NC 27709-2109

Abstract

Solvent substitution is an effective and useful means of eliminating the use of harmful solvents, but finding substitute solvents that are less harmful and as effective as currently used solvents presents significant difficulties. Solvent substitution is a form of reverse engineering where one tries to inversely find the solvents from the required properties rather than estimating property values for given solvents. In some ways it is actually different and more difficult than reverse engineering because there are no general or well-established methods for finding solvents from required properties. Here we present a brief and systematic introduction to computer-aided design of substitute solvents and illustrate these concepts with examples involving the design of substitutes for methyl ethyl ketone under different environmental requirements. In fact, we demonstrate that substitutes can be designed regardless of the environmental concern.

1 INTRODUCTION

Solvents are used in almost every industry for products or processes. In the United States alone, solvent sales represent a market of about $3 billion per year and solvent consumption amounts to about 11 billion pounds per year. With increasing awareness of the consequences of pollution and environmental and health concerns, investigations on solvents have received tremendous impetus over the last 10 to 15 years. Currently, many solvents, including some very common ones, such as toluene, methanol, methyl ethyl ketone, and 1,1,1-trichloroethane, are on the Toxic Release Inventory and other regulatory lists. Thus, elimination of these regulated solvents from current applications has become an urgent and global mission for both pollution regulators and solvent manufacturers and users.

There are two basic strategies for eliminating regulated solvents from use. The first strategy is to develop new technologies that do not use solvents. Unfortunately, these new technologies always require the design and use of new processes, new equipment, and new operations. Laser cleaning is an example of a new cleaning technology that does not use solvents. The second strategy, which is often called solvent substitution, tries to find alternative solvents that can use existing processes, equipment, and operations with no or minor modification. To most solvent users and solvent applications, the second strategy is more attractive in terms of economics and feasibility; however, technically, the second strategy is more complicated than the first, as discussed in Section 2.

Substitute solvents must satisfy specifications with regard to effectiveness, operability, cost, safety, environmental impacts, and health concerns. Finding successful substitute solvents is often very complex and difficult. The complete process of finding substitute solvents includes three consecutive stages: (1) determining the replacement formulations, including components and compositions, (2) performance and evaluation testing, and (3) full-scale validation and operational test. Among all three stages, the first one is the most important one because it is the decision-making stage. The first stage is also the most difficult stage among all three because there are neither generally effective rules nor commonly accepted procedures available for this stage, especially for finding mixtures as substitutes.

Nonetheless, one should note that there could be difficult questions in the second and third stages, particularly for solvents that are used in chemical processes. These questions can be related to cost, process objectives, and integration of new replacement solvents into existing process equipment. There is literature on the subject as it relates to mass integration (Hamad and El-Halwagi, 1998), and there also is literature on the use of solvent recovery and reuse (Ahmad and Barton, 1999). These strategies will ensure the most effective use of solvents in chemical processes, minimize the consumption of fresh solvent, and hopefully minimize the emission of solvents to the environment. These can be effective means of addressing certain classes of solvent-related environmental problems; however, they have a limitation in that they do not eliminate the use of undesirable or regulated solvents. Further, these strategies are more properly in the realm of process design, and therefore outside the solvent design scope of this chapter. For this reason, they will not be further discussed here. The role of computer-aided solvent design in relation to process design is to suggest environmentally better substitute solvents that can be evaluated in terms of cost, process objectives, process integration, etc., by existing process design methodologies.

Traditionally, the determination of replacement solvent formulations has been completely based on experience or information from solvent handbooks, manuals, technical reports, or publications. The use of experience can be limited except when the current solvent substitution problems are very similar to previous problems. Because of the problems in determining replacement formulations, often many more tests than necessary are done in a manual solvent search. Previous experience has shown that it is not unusual to spend hundreds of thousands of dollars and years of work in finding one successful substitute solvent. This expense and time can be greatly reduced by the use of computers.

There are mainly two categories of computer applications for solvent substitution. Essentially, the first category of computer applications is information systems. They provide technical guidance and information related to specific application problems. Many different and advanced search and link functions often are included to facilitate search. The on-line service Enviroene and the solvent alternative guide tool SAGE, developed by the U.S. Environmental Protection Agency (EPA), are examples in this category. The second category of computer applications tries to design substitute solvents from an analysis of solvent properties and behavior. Within this second category there are the following three subcategories: (1) solvent database search, (2) molecular design, and (3) mixture replacements.

Solvent database searches try to find substitutes through screening solvent databases according to some criteria. Typically, one needs to specify the allowed range or extreme values for each required solvent property. Then the included solvent database is searched, looking for those compounds which meet the requirement for all specified properties, and these are taken as substitution candidates for performance and evaluation tests. Some efforts in this subcategory include the Solvent Substitution for Pollution Prevention (Joback, 1994), the Solvent Database Software Program (Hermansen, 1993), and the Tool for Systematic Solvent Screening for Batch-Process Development and Revamping (Modi and Stephanopoulos, 1996). This subcategory can be used for searching for single chemicals but not mixtures as substitutes. The reason is that the solvent database usually only contains property information for a limited number of pure compounds. Database solvent searches do not always find single chemical substitutes because solvent databases only contain limited numbers of chemical compounds or because there may not exist a chemical that can meet the specified property requirements.

Molecular design uses molecular structure analysis to find chemical compounds that have specified required properties. Depending on how it is used, molecular design can either identify already existing chemicals or design entirely new ones. A characteristic of work in this subcategory is that a solvent database is not a necessary requirement because the molecular design algorithms include property prediction. Some efforts in this subcategory include Molecular Design of Solvents for Liquid Extraction Based on UNIFAC (Gani and Brignole, 1983), Designing Molecules Possessing Desired Physical

Property Values (Joback and Stephanopoulos, 1990), Group Contribution Approach to Computer-Aided Molecular Design (Gani et al., 1991), Design of Optimal Solvents for Liquid-Liquid Extraction and Gas Absorption Process (Macchitto et al., 1990), Computer-Aided Molecular Design of Solvents for Separation Processes (Pretel et al., 1994), and Design of Environmentally Safe Refrigerants Using Mathematical Programming (Duvedi and Achenie, 1996).

Mixture replacement tries to design mixtures as substitutes if no single chemicals can be found as substitutes. The design of replacement mixtures includes two parts: (1) selecting solvent components and (2) determining compositions. The early work in this area focused on finding optimal compositions of mixtures formed from specified chemical components. Solvent Selection by Computers (Hansen, 1973) used a computer to determine solvent compositions for selected components. Since the early 1990s, research on selecting solvent components also has achieved some progress. One way is to use a knowledge-based method to narrow the number of solvents to be considered for mixture replacement. Computer Aided Mixture Design with Specified Property Constraints (Klein et al., 1992) is an example of this work. Another way is to extend the molecular design method for designing mixtures where the number of components is specified. Optimal Design of Solvent Blends for Environmental Impacts Minimization (Buxton et al. 1999) is in this category. The Program for Assisting the Replacement of Industrial Solvents, Version 2, or PARIS II presents a new method of forming mixtures as substitutes systematically by recommending candidates from a ranked list that is updated as components are chosen. The work in this area includes Computer Aided Solvent Substitution for Pollution Prevention (Zhao et al., 1996), Computer Program for Assisting the Design and Replacement of Environmentally Objectionable Solvents (Cabezas et al., 1996), The Design of Technologically Effective and Environmentally Benign Solvent Substitutes (Zhao et al., 1998), and Designing Environmentally Benign Solvent Substitutes (Cabezas et al., 1999).

The objective of this chapter is to present the fundamental ideas, methods, and results the authors believe are important in solvent design. Section 2 discusses the technical challenges encountered in solvent design. Section 3 proposes some new methodology for solvent substitution. Section 4 describes the basic structure and characteristics of the PARIS II program and describes the procedures for using PARIS II. Section 5 illustrates the methods by designing replacements for methyl ethyl ketone under different environment and health concerns. Section 6 is the summary.

2 THE SUBSTITUTE SOLVENT DESIGN PROBLEM

For most solvent users and for most solvent applications, solvent substitution is more attractive than developing new technology because of the economic and feasibility advantages previously mentioned. On the other hand, however, solvent substitution also presents three significant technical challenges: (1) the need to find substitutes that are as effective as currently used solvents, (2) the need to use existing processes, equipment, and operations, and (3) the need to simultaneously have less impact on human health and the environment. Meeting the first requirement can be achieved either by the development of new technologies or by designing solvent substitutes, but meeting both the first and the second requirement can be achieved only by designing solvent substitutes. If we only have the first two requirements, then the current solvent is always the best substitute for itself, but it is the third requirement that brings a special and additional difficulty that even the current solvent cannot, by definition, satisfy.

Because no two chemicals are exactly alike, the design of a substitute solvent frequently involves a mixture, but depending on the requirements, the substitute can also be a pure component. Depending on whether the current solvent is a pure component or a mixture, there are four possible substitute solvent design problems: (1) replacing a one-component solvent with another single component, (2) replacing a one-component solvent with a mixture, (3) replacing a mixture solvent with a single component, and (4) replacing a mixture solvent with another mixture. Although it is possible to encounter cases 1 and, very rarely, 3, most practical solvent substitutions involve cases 2 and 4.

3 SOLVENT DESIGN THEORY

3.1 SOLVENT STRUCTURE AND PROPERTIES

In solvent substitution, one wants the substitute to be similar to the current solvent in terms of desirable aspects, such as effectiveness, and different from the current solvent in terms of undesirable aspects, such as toxicity and environmental and health impact. It is, therefore, necessary to predict the solvent behavior from known solvent characteristics, such as chemical or molecular structures and physical and chemical properties. Chemically similar compounds usually have similar behavior. For example, methyl ethyl ketone and methyl isopropyl ketone have similar molecular structure and similar behavior. Unfortunately, using structure similarity to select or design substitute solvents has limitation and difficulties. First, the chemical structure of a compound determines all of its behavior regardless of whether the behavior is desirable or not. This means that two chemicals having similar effectiveness usually have similar toxicity and environmental impacts. This is true for methyl ethyl ketone and methyl isopropyl ketone. Second, although chemical structure similarity may be useful if both the substitute and current solvent are single chemicals, if mixtures are involved, chemical structure similarity is not useful because it does not represent the nonidealities of mixtures.

In addition to chemical structure, solvent properties also represent solvent behavior, and property similarity is, therefore, a better and more feasible way of representing solvent behavior for solvent substitution. There are many reasons for this. First, a chemical has one chemical structure but multiple properties, and although the chemical structure cannot be separated into the desirable and undesirable parts, properties can be. Thus, by assigning different required values for solvent properties, it is possible to get substitutes with more of the desirable behavior but less of the undesired behavior. For example, if one requires solvents with strong wetting ability but with less explosion potential, one can look for low surface tension but high flash point. The second reason for favoring property similarity is that for most properties there is correspondence between single chemicals and mixtures. For example, vapor pressure, viscosity, and surface tension are well-defined properties for single chemicals and mixtures. There are three basic requirements for successful solvent substitutes, as proposed in Section 2. Corresponding to these three basic requirements, one can roughly classify solvent properties into three categories: (1) properties related to effectiveness, (2) properties related to operability, and (3) properties related to toxicity and environmental concerns. These three categories of properties are discussed separately in Sections 3.2, 3.3, and 3.4. The use of these properties is discussed in Section 4.

3.2 PROPERTIES RELATED TO SOLVENT EFFECTIVENESS

In any practical application, the first and most important objective is to ensure that the solvent has the required effectiveness. As already discussed, the effectiveness of a solvent can be characterized by its properties. There are, however, two different classes of properties: (1) bulk solvent properties, such as those listed in Table 17.1, and (2) component-specific properties,

TABLE 17.1
Solvent properties of acetone, acrolein, and methyl ethyl ketone at 25°C

	Acetone	*Acrolein*	*Methyl ethyl ketone*
Molecular mass (kg/kmol)	58.1	56.1	72.1
Liquid density (kg/m^3)	787	834	800
Boiling temperature (K)	329	326	353
Vapor pressure (kPa)	30.8	36.5	12.3
Surface tension (kg/s^2)	0.023	0.023	0.024
Viscosity × 10^4 (kg/m-s)	3.07	3.32	3.96
Thermal conductivity (J/m-s-K)	0.161	0.158	0.145
Flash point (K)	255	247	267
Solubility parameter [(J/m^3)$^{1/2}$]	19,730	20,180	18,880

TABLE 17.2
The fictitious solutes used for solvent-solute interaction analysis

Solute	*Dipole moment (10^{-30} Coul m)*	*DIPPR chemical family*
n-Octane	0	*n*-Alkanes
Benzene	0	*n*-Alkylbenzene
Diethyl ether	3.84	Aliphatic ethers
Ethanol	5.64	*n*-Alcohols
Water	6.17	Polar inorganics
Acetone	9.61	Ketones

such as the dipole moments and the infinite dilution activity coefficients listed in Tables 17.2 and 17.3.

From Table 17.1, one can see that of the nine properties listed, acrolein has eight properties closer to acetone than does methyl ethyl ketone. The one exception, the liquid density, does not have much effect on solvent effectiveness. If the solvent properties listed in Table 17.1 are used to select substitutes for acetone, one may well choose acrolein rather than methyl ethyl ketone, but the fact is that both acetone and methyl ethyl ketone are ketones whereas acrolein is an aldehyde. It should, therefore, be expected that methyl ethyl ketone, rather than acrolein, should be the better choice. One cannot, therefore, ensure the effectiveness of a substitute solvent by considering bulk properties alone. Rather, it is necessary to include component-specific properties. It should also be noted that besides the properties listed in Table 17.1, other solvent properties, such as the dielectric constant, the dipole moment, and the solubility parameter, have been used to predict solvent effects. These are useful when both the solvents and the solutes are known and when both the solvents and solutes are single chemicals.

Solvents perform their most important functions through the molecular interactions between solute and solvent molecules. These interactions are

TABLE 17.3
Infinite dilution activity coefficients of six fictitious solutes in acetone, acrolein, and methyl ethyl ketone

Solute in infinite dilution	*Acetone*	*Acrolein*	*Methyl ethyl ketone*
n-Octane	8.17	1.48	5.26
Benzene	1.57	2.01	1.32
Diethyl ether	1.78	5.74	1.46
Ethanol	2.19	1.81	2.24
Water	5.95	1.83	7.38
Acetone	Not applicable	1.57	1.01

component-specific properties (i.e., they depend on the types of molecules making up the solvent and the solutes). There are three categories of molecular interactions: solvent-solvent, solvent-solute, and solute-solute. For solvent substitution problems, the most important are the solute-solvent and the solvent-solvent interactions. Because the composition and nature of many solutes commonly used with solvents are unknown, one cannot represent exactly the solute-solvent interactions. A feasible way of addressing the problem of not knowing solute information is to select "fictitious" solutes with typical compositions to represent the "real" solutes. One, for example, can choose a limited number of solutes, each of which represents a different chemical family. Different methods can be used to select fictitious solutes. Table 17.2 shows the six fictitious solutes selected by the authors to represent a wide spread of molecular polarities (dipole moments) and chemical families. The choice of solutes is based on the judgment of the authors and is not unique (i.e., there could be other sets of solutes that perform adequately). These six fictitious solutes were chosen because (1) they are relatively simple monofunctional compounds [e.g., the dominant molecular characteristics of ethanol are due to the presence of the —*OH* group (this is not the case for a long-chain alcohol, where the hydrocarbon "tail" can dominate molecular behavior)], (2) they each represent a distinct chemical family, and (3) they together cover a wide range of very different types of molecular interactions representing the interactions between solvents and many types of solutes. For example, solutes tend to have interactions like those of hydrocarbons, such as *n*-octane; aromatics, such as benzene; polar compounds, such as alcohols; strongly hydrogen-bonded molecules, such as water, etc; or a combination of these for the case of multifunctional compounds. The six solutes approximately capture and represent all of these types of interactions. The assumption is that matching the molecular interactions between these six fictitious solutes and the solvent approximates matching the molecular interactions between the "real" solutes and the solvent. These six fictitious solutes are used in our PARIS II program.

The ability of a solvent to dissolve solutes is primarily related to the molecular interactions between the solvent and the solutes. Again, solute-solute molecular interactions are of secondary importance. We characterize solute-solvent and solvent-solvent interactions at the macroscopic level from solute activity coefficients at infinite dilution in the solvent. We use the six fictitious solutes already mentioned as representatives of all solutes because, as already discussed, one does not generally know all of the solutes that the solvent is likely to be required to dissolve. The infinite dilution activity coefficients, which are also called *limiting activity coefficients*, are defined by

$$\gamma_i^\infty = \lim_{x_i \to 0} \gamma_i, \tag{1}$$

where γ_i^∞ is the activity coefficient of component i at infinite dilution and γ_i is the activity coefficient of component i. The infinite dilution activity coefficients characterize the interactions between a single molecule of solute and the molecules of solvent without the interference from solute-solute interactions.

There are various methods to estimate infinite dilution activity coefficients from experiment or prediction. For nonionic compounds, they can be estimated from different group contribution methods, of which the UNIFAC (Fredenslund et al., 1975) is most widely used. Most of the available group information used for calculating the solution activity coefficients are regressed at finite concentrations rather than at the infinite dilution, and this leads to some inaccuracy when infinite dilution activity coefficients are calculated by extrapolation, but in designing substitute solvents, each property for the current solvent and the potential substitutes is estimated using the same method and then compared. Thus, it is the property difference rather than the property values that is important. Therefore, the requirement for accuracy is not as critical. Table 17.3 shows the UNIFAC-predicted infinite dilution activity coefficients of the six fictitious solutes in acetone, acrolein, and methyl ethyl ketone. From Table 17.3 we can see that methyl ethyl ketone has infinite dilution activity coefficients closer to those in acetone than acrolein. This indicates that methyl ethyl ketone is closer to acetone than acrolein, as expected.

3.3 PROPERTIES AND BEHAVIOR RELATED TO EFFICIENT AND SAFE OPERATION

Finding substitute solvents that perform efficiently and safely involves four categories of properties and behaviors: dynamic properties, static properties, safety properties, and phase stability. The dynamic properties are viscosity, thermal conductivity, and diffusivity; the static properties are density, vapor pressure, boiling point, and surface tension; the safety properties are flash point and flammability limits. The dynamic properties simply characterize the three modes of transport—momentum, thermal energy, and mass. The static properties represent a reasonable collection of properties that are important in most applications. The safety properties represent the obvious issue of avoiding fires and explosions. Phase stability is important to ensure that a mixture substitute solvent does not form more than one phase.

For the dynamic properties, viscosity characterizes the flow resistance of fluids. For example, low viscosity is preferred in many solvent applications in order to use less power to pump liquid; however, high viscosity is required in some adhesives. High thermal conductivity is desirable when rapid heat transfer through the solvent is important, such as in heat exchangers. On the other hand, if the solvent is part of an insulator, then low thermal conductivity is desirable. Diffusivity is important in nearly all operations in chemistry. For example, in some chemical reactions the diffusivity sets the rate of reaction by determining how fast reactants can come together and how fast the products can move away from each other. Unfortunately, diffusivity is difficult to use in the design of substitute solvents. The reason is that most available liquid diffusivity estimation methods are only valid for dilute binary solutions, whereas in many solvent applications such as extraction, solutions have many components at finite concentrations.

For solvent static properties, the density is employed to avoid changes in capacity or transport power in the existing processes. For example, high-density fluids require more energy to be pumped, and a density difference is necessary for both stage-wise and continuous-contact separation operations. The vapor pressure is important in many applications. For example, a higher vapor pressure can reduce the drying time in printing or cleaning processes, but lower vapor pressure may be preferable to reduce the emission of volatile organic compounds. The boiling temperature is important because, except for a few applications, such as vapor degreasing, most solvents must be used below the boiling temperature of pure substances or the bubble point of mixtures. Achieving similar boiling temperature or bubble points enables substitute solvents to work in the same conditions as current solvents without boiling. Lastly, surface tension is important in many cases. For instance, surface tension is relevant for cleaning operations; it is important that substitute solvents will wet and penetrate porous surfaces. In extraction, the surface tension affects settling time and power consumption.

When a substitute solvent is designed for a single-phase solvent, it is critical that the substitute consists of only one liquid phase in the neighborhood of the expected operating temperature, pressure, and composition for the particular application. For substitutes in the form of liquid mixtures, the rigorous criterion for predicting single-phase stability is the minimization of the Gibbs free energy of mixing. Theoretically, a hyper tangent plane method can find the miscible composition region at the given condition of temperature and pressure, but using this method for mixtures with three or more components is difficult.

A solvent should be safe, effective, economical, and easy to use. Safety should always be the first priority in all operations. The most commonly used properties for the safe use of solvents are the flash point and flammability. The flash point is the temperature at which a solvent spontaneously combusts. It is also used to classify flammable liquids and combustible liquids. Usually, the flash point of the substitute solvents should be no lower than that of the current solvents. The flammability limit, however, is used differently. Below the lower flammability limit, the concentration of solvent vapor is too low to form a flammable mixture, and above the upper flammability limit, there is too little combustible oxygen to sustain a flame. For practical solvent applications, it is the lower flammability limit rather than the upper flammability limit that is the main concern. The reason is that usually it is better and safer to work in a condition below the lower flammability limit rather than in a condition above the upper flammability limit. In substitute solvent design, a more conservative but practical way is that, in the proposed substitute formulations, no component should have its lower flammability limit value below that of the component that has the smallest lower flammability limit of all components in the current formulation.

3.4 PROPERTIES FOR ENVIRONMENTAL CONSIDERATIONS

A main objective of solvent substitute design is to reduce the environmental impact of current solvent operations. A necessary condition to achieve this objective is, therefore, to be able to compare different solvent formulations based on potential environmental impacts. There are no standard or internationally agreed categories for classifying chemical potential environmental impacts. Recently, a database of basic environmental impact information has been developed for 1600+ chemicals with eight environmental impact categories (Young, 1999). The database is organized into the following categories:

Human toxicological: human toxicity potential by ingestion; human toxicity potential by dermal or inhalation exposure

Ecological toxicological: aquatic toxicity potential; terrestrial toxicity potential

Regional environmental: photochemical oxidation potential; acidification potential

Global environmental: stratospheric ozone depletion potential; global warming potential

The database has chemical environmental impact information in the form of LD_{50}s, LC_{50}s, OSHA Threshold Limit Values, etc., for each of the 1600+ chemicals. The details of this database are discussed further elsewhere (Young and Cabezas, 1999). For purposes of the present work, the database is used to obtain the normalized specific environmental impact scores $\psi_{i,j}^{s}$ used to calculate the environmental indexes and the air indexes discussed below. The scores are normalized by dividing by the average of all nonzero scores within each environmental impact category (e.g., if the human health score for a chemical i is represented by $1/LD_{50i}$ for rats, the normalized score is obtained by dividing $1/LD_{50i}$ by $1/n\sum_i 1/LD_{50i}$, where the sum is taken over all components having nonzero LD_{50}s. Thus, the normalized score for the average chemical within each environmental impact category in the database is 1.

For a pure chemical i, the environmental index ψ_i used in assessing the potential environmental impact of solvents is computed from

$$\psi_i = \sum_j \alpha_j \psi_{i,j}^{s}, \qquad (2)$$

where the summation is taken over the eight categories of environmental impacts mentioned above, α_j is a relative weighting factor for environmental impact category j, and $\psi_{i,j}^{s}$ is the normalized specific environmental impact score of chemical i in environmental impact category j. The weighting factor α_j can vary between 0 and 10, depending on the relative importance or unimportance of a particular impact category for a specific application. For instance, if the solvent is to be used where it normally comes into contact with worker's hands, then it may be prudent to overweight Human Toxicity Potential by Dermal or Inhalation Exposure relative to other effects. For a mixture m, the environmental index ψ_m is calculated from a mass fraction sum over all chemical components i as

$$\psi_m = \sum_i W_i \psi_i, \tag{3}$$

where the summation is taken over all components i, W_i is the mass fraction of component i in the mixture, and ψ_i is the environmental index of pure component i calculated from equation (2).

The air index is calculated by multiplying the environmental index of each chemical by its fugacity and dividing by the pressure. Thus, for a pure component i, the air index ψ_i^{air} is

$$\psi_i^{air} = \frac{P_i^{\nu} \psi_i}{P}, \tag{4}$$

where P_i^{ν} is the vapor pressure of component i, ψ_i is the environmental index of pure component i calculated from equation (2), and P is the operating pressure. For a mixture m, the air index ψ_m^{air} is calculated from

$$\psi_m^{air} = \frac{\sum_i x_i \gamma_i P_i^{\nu} \psi_i M_i}{P \sum_i x_i M_i}, \tag{5}$$

where x_i is the mole fraction of component i, γ_i is the activity coefficient of component i, M_i is the molecular mass of component i, and all the other symbols have their previously assigned interpretation.

4 PARIS II SOFTWARE

The PARIS II software provides a user-friendly interface to the sophisticated solvent replacement methods already described in this chapter. The user interface is written in Visual Basic and includes menus, graphical tools, on-line help, print utilities, etc. The software consists of several screens that correspond to the major steps in the solvent replacement algorithm. A typical sequence of steps followed in using the software is described in this section.

Step 1: Define Composition and Operating Conditions of Current Solvent

The process of finding a replacement for a solvent using PARIS II begins with entering information about the current solvent's composition and the operating conditions at which the solvent is used. The operating temperature and pressure are used to determine the list of chemicals in the database that exist in the liquid phase at the specified conditions. The user can choose up to nine chemicals as components of the current solvent and enter data on their composition. In this step, the user can also select if infinite dilution activity coefficients are to be used in the search for replacements.

Step 2: Assign Weights to Potential Environmental Impact Categories

Eight classes of environmental impacts are included in PARIS II to estimate the potential environmental effect of a solvent formulation. As already mentioned, the environmental impact classes are Human Toxicity by Ingestion, Human Toxicity by Dermal/Inhalation Exposure, Aquatic Toxicity, Terrestrial Toxicity, Photochemical Oxidation, Acidification, Global Warming, and Stratospheric Ozone Depletion. In this step, the user assigns a relative weight between 0 and 10 to each of the environmental impact classes to suit the particular application. These weights are used in subsequent steps of the program to rank and select chemicals for replacements.

Step 3: Static, Dynamic, Performance, and Environmental Properties of Current Solvent

Based on the information entered about the current solvent's composition and operating conditions, PARIS II calculates solvent properties that characterize the static, dynamic, performance, and environmental behavior of the solvent.

Step 4: Assign Property Tolerance Ranges for Designing the Replacement Solvent

The user assigns a precent tolerance and a desired value, an upper bound, and a lower bound for every static, dynamic, and performance property. For the environmental indices, however, the user can set a desired value and an upper or lower bound value but not the tolerance, which is internally set at 50%, based on the uncertainty associated with the environmental data. The boundary values (i.e., desired, upper, and lower bounds) for properties are used by PARIS II to find replacements that satisfy these constraints and to rank replacements by comparing the replacement's properties to the desired values specified.

Step 5: Search for Single Chemical Replacement

PARIS II first searches the included DIPPR (Design Institute for Physical Property Data, 1997) database for a single chemical replacement for the current solvent. If more than one single chemical replacement is found, then these replacements are ranked. The user can view the properties of the suggested replacements and compare these values to those of the current solvent. At this step, the user can exit the program, begin a new search, or modify the current search by changing the tolerance ranges or bounds to view more single chemical replacements.

Step 6: View Candidates for Mixture Replacements

If a single chemical replacement is not found, PARIS II displays a ranked list of candidates for a mixture replacement. The rank of a candidate is based on how close its properties are to the desired values. The properties of each candidate are displayed, and the user can compare a candidate's properties to the desired values and view which candidate property violated which boundary value specified in Step 4. The user can either go back to Step 4 and widen the tolerance range for the chemical properties that were violated and recalculate, or proceed to find a mixture replacement.

Step 7: Select Candidates for Mixture Replacement

In this step, the user selects the candidates for a mixture replacement. PARIS II is designed to try to find a replacement with the least number of components, so the user can add only one chemical at a time. The user can select any chemical from the ranked list of candidates as the first chemical to be used in the mixture and then successively add more components.

Step 8: Search for Mixture Replacement for Current Solvent

After the first two chemicals are selected, PARIS II tries to find a replacement that includes different proportions of these two chemicals. If a successful match is found (i.e., the mixture replacement's properties satisfy the tolerance ranges specified by the user), then the user can view the successful replacements. If there is more than one successful replacement, these replacements are ranked and up to five replacement formulations are shown to the user. The replacement's properties, along with those of the current solvent, are displayed. At this stage, the user may exit the program, begin a new search, or refine the search by changing the tolerance ranges specified in Step 4. If no successful replacement is found, then the ranked unsuccessful replacement mixtures are shown to the user. The user can view the properties of the unsuccessful replacements and check the property tolerance ranges that were violated.

Step 9: Select New Candidates for Mixture Replacement

If no successful mixture is found with the two candidates selected in Step 8, the user has three options: (1) replace one candidate, (2) replace both candidates, or (3) add another candidate to the mixture. PARIS II then searches for a mixture replacement with the new list of selected candidates, and Steps 7, 8, and 9 are repeated until a successful replacement is found.

5 DESIGNING SOLVENTS FOR THE ENVIRONMENT

The first and perhaps the most important set of criteria to be considered in designing solvents is their technological performance. The reason is that one chooses a solvent because it is needed to perform a given function (e.g., clean surfaces, dissolve certain compounds, etc.). Solvents that do nothing useful are simply not relevant (i.e., never used and never become an environmental concern). Within the field of useful solvents there are environmentally better and environmentally worse solvents. The focus of this chapter is, in a way, an exercise in designing useful and environmentally better solvents. The theory of this is discussed in detail in Sections 3 and 4. We now present some examples.

In this section we consider the design of replacement solvents under different environmental considerations. This is done by means of the illustrative exercise of designing replacements for methyl ethyl ketone under the following six different environmental scenarios: (1) neglecting all of the environmental effects of solvents, (2) considering only human toxicological effects, (3) considering only ecological toxicological effects, (4) considering only regional environmental effects, (5) considering only global environmental effects, and (6) considering all categories of environmental effects. Methyl ethyl ketone was chosen as an example because it is a widely used solvent that is listed under the Toxic Release Inventory and the Clean Air Act list of hazardous air pollutants because of its well-known adverse environmental effects. The normalized environmental impact scores of all the chemicals used in our example are listed in Table 17.4 in Appendix 1.

5.1 NEGLECTING ALL ENVIRONMENTAL EFFECTS

If one were to design a replacement solvent ignoring all the environmental impacts (i.e., considering all the technical and safety requirements but not the Air Index and the Environmental Index), the PARIS II algorithm would simply choose chemicals or chemical mixtures that have the optimal technological performance. The resulting solvent design represents the optimum that could be achieved in matching technological performance regardless of environmental consequences. As will be shown later, one need not always sacrifice technological performance to satisfy environmental requirements; although there is some trade-off among the various requirements.

Although the Air Index and the Environmental Index for methyl ethyl ketone and the replacement were not used to design the replacement, they were still calculated giving an equal weight of five to each of the eight environmental impact categories. The other properties were matched to those of methyl ethyl ketone within a 20% tolerance bound. For the case of methyl ethyl ketone, PARIS II designed a mixture of approximately 78% by mass methyl isopropyl ketone and 22% by mass acetone as a replacement. Whereas methyl ethyl ketone has an Air Index of 0.8 and an Environmental Index of 6, the replacement has an Air Index of 1 and an Environmental Index of 20. The difference in the air indexes is not particularly significant, but that for the environmental index is significant. From Table 17.4 we see that this difference is primarily driven by the fact that the scores for human toxicity by ingestion and for terrestrial toxicity for methyl ethyl ketone are 0.1, whereas those for methyl isopropyl ketone are 2. Therefore, one should be seriously concerned with the replacement solvent of 78% methyl isopropyl ketone and 22% acetone if there is a possibility of ingestion by humans or terrestrial animals; however, inspection of Tables 17.5 and 17.6 in Appendixes 2 and 3 reveals that the replacement mixture of approximately 78% by mass methyl isopropyl ketone and 22% by mass acetone meets the technical requirements of a replacement for methyl ethyl ketone as expressed by the various solvent properties.

5.2 HUMAN HEALTH ONLY

The PARIS II solvent design algorithm considers two categories of human health impacts: Human Toxicity Potential by Ingestion and Human Toxicity Potential by Dermal/Inhalation Exposure. The replacement solvent design that PARIS II developed for methyl ethyl ketone that meets technological requirements and human health requirements is, on a mass basis, approximately 72% ethyl acetate, 6% methanol, and 22% 2-pentanone. The Air Index for methyl ethyl ketone for this case, weighting Human Toxicity by Ingestion and Human Toxicity by Dermal/Inhalation exposure to 5 and all other categories to 0, is 0.08, and the Environmental Index is 0.7. The indexes for the replacement are, respectively, 0.04 and 0.4, which represents a significant reduction in human toxicological risk. This is driven by the fact that the normalized scores for all the chemicals in the replacement formulation are lower than those of methyl ethyl ketone, with the exception of methanol ingestion, but then there is very little methanol used in the replacement. From inspection of Tables 17.5 and 17.6 in Appendixes 2 and 3, it should be carefully noted that this replacement of approximately 72% ethyl acetate, 6% methanol, and 22% 2-pentanone also meets the technical requirements about as well as the replacement designed with no environmental considerations (i.e., the replacement properties are within a 20% bracket around the properties of methyl ethyl ketone).

5.3 ECOLOGICAL EFFECTS ONLY

The PARIS II solvent design algorithm considers two categories of ecological impacts: aquatic toxicity potential and terrestrial toxicity potential. The replacement solvent design that PARIS II developed for methyl ethyl ketone meeting technological requirements and ecological impact requirements is, on a mass basis, approximately 73% 2-pentanone and 27% acetone. The Air Index for methyl ethyl ketone for this case, weighting aquatic toxicity and terrestrial toxicity to 5 and all other categories to 0, is 0.08, and the Environmental Index is 0.7. The indexes for the replacement are, respectively, 0.04 and 0.4, which represents a significant reduction in ecological toxicological risk. This is driven by the fact that the normalized scores for all the chemicals in the replacement formulation are lower than those of methyl ethyl ketone, with the exception of the aquatic toxicity of 2-pentanone. From inspection of Tables 17.5 and 17.6 in Appendixes 2 and 3, it should be noted that this replacement of approximately 73% 2-pentanone and 27% acetone also meets the technical requirements about as well as the replacement designed with no environmental considerations (i.e., the replacement properties are within a 20 % bracket from the properties of methyl ethyl ketone).

5.4 REGIONAL ENVIRONMENTAL EFFECTS ONLY

The PARIS II solvent design algorithm considers two categories of regional environmental impacts: acidification or acid rain potential and photochemical oxidation or smog formation potential. The replacement solvent design that PARIS II developed for methyl ethyl ketone meeting technological requirements and regional environmental impact requirements is, on a mass basis, approximately 78 % methyl isopropyl ketone and 22 % acetone. The Air Index for methyl ethyl ketone for this case, weighting acidification and smog formation to 5 and all other categories to 0, is 0.6, the Environmental Index is 5, and the indexes for the replacement are 0.1 and 4, respectively, which represents a significant reduction in acidification and smog formation risk. This is driven by the fact that the normalized scores for all the chemicals in the replacement formulation are 0, with the exception of photochemical oxidation for acetone, which is 0.4. From inspection of Tables 17.5 and 17.6 in Appendixes 2 and 3, it should be noted that this replacement of approximately 78 % methyl isopropyl ketone and 22 % acetone again meets the technical requirements about as well as the replacement designed with no environmental considerations, (i.e., the replacement properties are within a 20 % bracket from the properties of methyl ethyl ketone).

5.5 GLOBAL ENVIRONMENTAL EFFECTS ONLY

The PARIS II solvent design algorithm considers two categories of global environmental impacts: enhancement of the greenhouse effect or global warming potential and stratospheric ozone depletion potential. The replacement solvent design that PARIS II developed for methyl ethyl ketone meeting technological requirements and global environmental requirements is, on a mass basis, approximately 78 % methyl isopropyl ketone and 22 % acetone. The Air Index for methyl ethyl ketone for this case, weighting global warming and stratospheric ozone depletion to 5 and all other categories to 0, is 0, the Environmental Index is 0, and the indexes for the replacement are also both 0. This is driven by the fact that none of these chemicals has known roles in either global warming or stratospheric ozone depletion. This essentially means that the environmental indexes play no role in the solvent design so that the resulting replacement is the same as was previously obtained when no environmental considerations were included.

5.6 ALL HUMAN HEALTH AND ENVIRONMENTAL CONSIDERATIONS

As already discussed, the PARIS II solvent design algorithm considers eight categories of potential toxicological and environmental effects: Human Toxicity by Ingestion, Human Toxicity by Dermal/Inhalation Exposure,

Aquatic Toxicity, Terrestrial Toxicity, Photochemical Oxidation, Acidification, Global Warming, and Stratospheric Ozone Depletion. The replacement solvent design that PARIS II developed for methyl ethyl ketone meeting technological requirements and all eight toxicological and environmental requirements all equally weighted to 5 is, on a mass basis, approximately 73% 2-pentanone and 27% acetone. The Air Index for methyl ethyl ketone for this case is 0.8, and the Environmental Index is 6. The indexes for the replacement are, respectively, 0.2 and 1, which represents a significant reduction in overall toxicological and environmental risk. This is driven by the fact that the normalized scores for all chemicals in the replacement formulation are lower than those of methyl ethyl ketone, with the exception of the aquatic toxicity of 2-pentanone. From inspection of Tables 17.5 and 17.6 in Appendixes 2 and 3, it should again be carefully noted that this replacement of approximately 73% 2-pentanone and 27% acetone meets the technical requirements about as well as the replacement designed with no environmental considerations at the beginning of this section (i.e., the replacement properties are within a 20% bracket from the properties of methyl ethyl ketone).

6 CONCLUSIONS

The most important conclusion that can be drawn from the work presented here is that, with the use of a sound theory and a well-designed computer algorithm, it is possible to design replacement solvents that can meet a wide range of environmental requirements while meeting strict standards of technical performance. This is important because it has been generally believed that when a solvent is replaced because of environmental concerns, there is a consequent and often detrimental compromise of technical performance. The work presented here would indicate that this may not be the case (i.e., the technical performance of the six replacement solvents designed under six different environmental constraints appears to be about the same). The conclusion is, of course, limited by the extent to which the properties included in the PARIS II software represent actual technical performance and by the extent to which the various toxicological and environmental scores and indexes represent actual potential environmental impact. Nonetheless, the results seem reasonable and encouraging. The second most important conclusion is that in replacement solvent design, it is necessary to include measures of solvent performance, such as the six activity coefficients at infinite dilution that are used in the PARIS II method. The six chemicals used, however, are not unique, and a different set of chemicals would probably do as well. Solubility parameters are adequate measures of solvent performance for pure chemicals, but they suffer from the limitations of Regular Solution Theory when applied to mixtures. The third most important conclusion is that one cannot, on the other hand, design a replacement solvent based solely on performance properties without considering operational properties, such as density, viscosity, etc. The reason is that the resulting solvent may well perform its function adequately, but it may or may not be practical because of reasons such as too high or too low a viscosity, too high or too low a vapor pressure, etc., none of which is represented by activity coefficients.

ACKNOWLEDGMENTS

The authors acknowledge the support and encouragement of Dr. Subhas K. Sikdar, the Director of the Sustainable Technology Division at the National Risk Management Research Laboratory. Renhong Zhao is a Research Fellow with the Oak Ridge Institute for Science and Education.

REFERENCES

Ahmad, B. S., and Barton, P. I. (1999). Solvent recovery targeting. *AIChE J.* 45:335.

Cabezas, H., Zhao, R., Bare, J., and Nishtala, S. (1996). A computer program for assisting the design and replacement of environmentally objectionable solvents. AIChE Annual Meeting, Chicago, IL, Paper, 46c.

Cabezas, H., Zhao, R. Bare, J. C., and Nishtala, S. R. (1999). Designing environmentally benign solvent substitutes. *Tools and Methods for Pollution Prevention* (S. K. Sikdar and U. Diwekar), p. 317. Kluwer Academic Publishers, Dordrecht, The Netherlands.

Duvedi, P., and Achenie, K. (1996). Designing environmentally safe refrigerants using mathematical programming. *Chem. Eng. Sci.* 51:3727.

Gani, R., and Brignole, E. A. (1983). Molecular design of solvents for liquid extraction based on UNIFAC. *Fluid Phase Equil.* 13:331.

Gani, R., Nielsen, B., and Fredenslund, A. (1991). Group contribution approach to computer-aided molecular design. *AIChE J.* 37:1318.

Hamad, A. A., and El-Halwagi, M. M. (1998). Simultaneous synthesis of mass separating agents and interception networks. *Trans. I. Chem. E.* 76:376.

Hansen, C. M. (1973). Solvent selection by computer. *Solvents Theory and Practice* (R. W. Tess), p. 48. Advances in Chemistry Series, Vol. 124, American Chemical Society, Washington, D.C.

Hermansen, R. D. (1993). Development of a solvent database software program. *Solvent Substitution for Pollution Prevention,* p. 159. Noyes Data Corporation, Park Ridge, NJ.

Joback, K. G., and Stephanopoulos, G. (1990). Designing molecules possessing desired physical property values. *Proceedings of the Third International Conference on Foundations of Computer-Aided Process Design*, Snowmass, Colorado, p. 631, Elsevier, Amsterdam.

Joback, K. G. (1994). Solvent substitution for pollution prevention. *Pollution Prevention Via Process and Product Modifications* (E. L. Gaden), p. 98. AIChE Symposium Series, Vol. 90, American Institute of Chemical Engineers, New York.

Kirschner, E. M. (1994). Environment, health concerns force shift in use of organic solvents. *Chem. Eng. News*, Issue 20, p. 13.

Klein, A., Wu, T., and Gani, R. (1992). Computer aided mixture design with specified property constraints. *Comp. Chem. Eng.* 16:S229.

Macchietto, S., Odele, O., and Omatsone, O. (1990). Design of optimal solvents for liquid-liquid extraction and gas absorption processes. *Trans. I. Chem. E.* 68:429.

Modi, K., and Stephanopoulos, G. (1996). SMART: A tool systematic solvents screening for batch-process development and revamping. AIChE Spring Meeting, New Orleans, LA, Paper 90c.

Pretel, J., Lopez, A., Bottini B., and Brignole, A. (1994). Computer-aided molecular design of solvents for separation processes. *AIChE J.* 40:1349.

Young, D. M. (1999). Unpublished database, U.S. Environmental Protection Agency, Office of Research and Development, National Risk Management Research Laboratory.

Young, D. M., and Cabezas, H. (1999). Designing sustainable processes with simulation: The waste reduction (WAR) algorithm. *Comp. Chem. Eng.* 23:1477.

Zhao, R., Cabezas, H., Govind, R., and Fang, Y. (1996). Computer aided solvent substitution for pollution prevention. AIChE Spring Meeting, New Orleans, LA, Paper 40c.

Zhao, R., and Cabezas, H. (1998). The design of technically effective and environmentally benign solvent substitutes. *Green Chemistry & Engineering Conference Proceedings*, p. 215.

APPENDIX 1

TABLE 17.4
Potential environmental impact (PEI) scores for six chemicals in units of PEI per kilogram of chemical

	PEI/kg					
	Methyl ethyl ketone	***Methyl isopropyl ketone***	***Acetone***	***Ethyl acetate***	***Methanol***	***2-Pentanone***
Human Toxicity: Ingestion	1.32E-1	2.45	6.24E-2	6.44E-2	6.43E-2	9.71E-2
Human Toxicity: Dermal and Inhalation	5.06E-4	4.23E-4	1.24E-4	2.13E-4	1.15E-3	4.26E-4
Aquatic Toxicity	6.36E-4	2.37E-3	2.52E-4	8.91E-3	6.97E-5	1.65E-3
Terrestrial Toxicity	1.32E-1	2.45	6.24E-2	6.44E-2	6.43E-2	9.71E-2
Photochemical Oxidation	9.72E-1	0	3.66E-1	4.48E-1	2.53E-1	0
Acidification	0	0	0	0	0	0
Global warming	0	0	0	0	0	0
Ozone depletion	0	0	0	0	0	0

APPENDIX 2

TABLE 17.5

Operational, safety and environmental properties of methyl ethyl ketone and substitute solvents designed under (1) no environmental considerations, (2) human toxicological considerations weighted 5 and all other considerations weighted 0, (3) ecological toxicological considerations weighted 5 and all other considerations weighted 0, (4) regional environmental considerations weighted 5 and all other considerations weighted 0, (5) global environmental considerations weighted 5 and all other considerations weighted 0, and (6) all aforementioned considerations weighted equally to 5

		Environmental considerations					
	Methyl ethyl ketone	*None (1)*	*Human toxicity (2)*	*Ecological toxicity (3)*	*Regional (4)*	*Global (5)*	*All (6)*
Molecular mass (kg/kmol)	72.1	77.7	79.3	76.3	77.7	77.7	76.3
Liquid density (kg/m^3)	800	801	866	799	801	801	799
Boiling temperature (K)	353	350	345	351	350	350	351
Vapor pressure (kPa)	12.3	14.3	14.8	14.0	14.3	14.3	14.0
Surface tension (kg/s^2)	0.0240	0.0243	0.0232	0.0236	0.0243	0.0243	0.0236
Viscosity $\times 10^4$ (kg/m-s)	3.96	3.92	4.28	4.13	3.92	3.92	4.13
Thermal conductivity (J/m-s-K)	0.145	0.146	0.147	0.147	0.146	0.146	0.147
Flash point (K)	267	268	273	271	268	268	271
Air index (impact/kg) substitute/methyl ethyl ketone	N/A	N/A	0.04/0.08	0.04/0.08	0.1/0.6	0/0	0.2/0.8
Env. index (impact/kg) substitute/methyl ethyl ketone	N/A	N/A	0.4/0.7	0.4/0.7	0.4/5	0/0	1/6

APPENDIX 3

TABLE 17.6
Solvent function properties; i.e., activity coefficients at infinite dilution for six representative solutes in methyl ethyl ketone and substitute solvents designed under (1) no environmental considerations, (2) human toxicological considerations weighted 5 and all other considerations weighted 0, (3) ecological toxicological considerations weighted 5 and all other considerations weighted 0, (4) regional environmental considerations weighted 5 and all other considerations weighted 0, (5) global environmental considerations weighted 5 and all other considerations weighted 0, and (6) all aforementioned considerations weighted equally to 5

		Environmental considerations					
Infinite dilution activity coefficient	*Methyl ethyl ketone*	*None (1)*	*Human toxicity (2)*	*Ecological toxicity (3)*	*Regional (4)*	*Global (5)*	*All (6)*
Ethanol	2.24	2.27	2.12	2.26	2.27	2.27	2.26
Diethyl ether	1.46	1.37	1.22	1.39	1.37	1.37	1.39
Acetone	1.01	N/A	1.00	N/A	N/A	N/A	N/A
Water	7.38	7.94	6.95	7.79	7.94	7.94	7.79
Octane	5.26	4.60	4.81	4.75	4.60	4.60	4.75
Benzene	1.32	1.24	1.29	1.26	1.24	1.24	1.26

Index

Page references followed by f or t refer to figures or tables, respectively.

F

G

H

P

T

X

Z